T0138935

Mechanical Properties of Nanocrystalline Materials

edited by
James C. M. Li

Mechanical
Properties of
Nanocrystalline
Materials

PAN STANFORD PUBLISHING

Published by

Pan Stanford Publishing Pte. Ltd.
Penthouse Level, Suntec Tower 3
8 Temasek Boulevard
Singapore 038988

Email: editorial@panstanford.com
Web: www.panstanford.com

Mechanical Properties of Nanocrystalline Materials

ISBN 978-981-4241-97-7 (Hardcover)
ISBN 978-981-4241-75-5 (eBook)

Printed in Singapore

Contents

Foreword

About 30 years ago, the development of nanocrystalline materials was motivated by the idea that it should be possible to generate a new class of crystalline materials by increasing the fraction of atoms located in the cores of defects so that the fraction in the defect cores becomes comparable or even larger than the fraction of atoms situated in the lattice of the crystals surrounding the defects. It turned out that one way to prepare materials with this kind of structure was to reduce the size of the crystallites of polycrystals to a few nanometers so that the diameters of the crystallites become comparable to the thickness of the grain boundaries between them. These materials were called nanocrystalline materials. As the atomic arrangements in grain boundaries (and in the cores of other defects as well) differ from the ones in crystallites, nanocrystalline materials represent a separate class of materials in the sense that their atomic structures differ from the structures of single-crystalline or coarse-grained materials as well as of glasses with the same chemical composition. However, the reduction of the crystal size to a few nanometers not only results in new atomic structures, it also opens the way to other attractive features such as new electronic structures, it permits the alloying of components that were considered previously to be immiscible, or it results in materials with new dynamic features and hence new thermodynamic and new diffusional properties.. In fact, in recent years, nanocrystalline materials with mobile charge carriers (such as metals) containing nanometer-sized pores (called nanoporous materials) were studied. They were discovered to permit the tuning of the electronic structure and hence the tuning of all electron-structure-dependent properties such as the electric, optical, and magnetic properties. This tuning of the electronic structure was performed by filling the nanopores with an electrolyte and by applying an electric potential between the electrolyte in the pores and the material surrounding them.

In addition to nanocrystalline materials, a new generation of noncrystalline, nanometer-structured materials got recently on its way. They were called nanoglasses. Nanoglasses are noncrystalline materials characterized by a microstructure that consists of nanometer-sized glassy regions which are separated by interfaces between them (called glass/glass interfaces). At present, these nanoglasses are produced by consolidating glassy spheres with a size of a few nanometers. By analogy to nanocrystalline materials, nanoglasses may exhibit a comparable large variety of parameters that may be utilized to control their microstructure. For example, all of the consolidated glassy spheres may have the same size and the same chemical composition. However, their size and/or chemical composition may also be different. Indeed, one more option is to have only a fraction of the consolidated spheres to be glassy, and the others may be nanometer-sized crystallites or pores, leading to nanoporous nanoglasses, etc.

As it is primarily the number (per unit volume), the topological arrangement, the atomic structure, etc., of the intercrystalline or the glass/glass boundaries that control the new features of nanomaterials, the properties of nanomaterials may be changed by varying one or several of these parameters, for example, the diameter and/or shape of the nanometer-sized crystal or glassy regions, their chemical composition (e.g., all crystals or glassy regions may have the same composition or their chemical compositions may differ), etc. Moreover, the intercrystalline interfaces and/or the nanometer-sized crystallites may contain defects (e.g., dislocations), or solute atoms may have segregated to the interfaces or to the crystallites/glassy regions.

In other words, it seems attractive for basic research as well as for the technological applications of nanomaterials to vary the microstructure and hence the properties of nanocrystalline or nanoglassy materials in a controlled way by tuning one or several of these intercrystalline parameters In fact, this approach would permit to generate nanocrystalline/nanoglassy materials, the structure and properties of which could be varied within an even wider range than the structure and properties of, for example, coarse-grained polycrystals. On the other hand, the selection of the microstructural parameters that result in nanocrystalline/nanoglassy materials with interesting new properties requires the understanding of the relationship between the microstructure and properties of nanocrystalline/nanoglassy materials. The need for this understanding was convincingly demonstrated by the discovery of the Giant Magneto Resistance (GMR) effect. The GMR effect exists in nanostructured, multilayer metallic films or nanocrystalline metallic materials of specific chemical compositions. The first step in the discovery of the GMR effect was the understanding of the relationship between the spin scattering in thin metallic films and the microstructure of these films. On the basis of this knowledge, thin, multilayer films with specific chemical compositions and specific film morphologies were prepared. In these selectively prepared multilayer films, a specific property (the magnetoresistance) was studied, which led to the discovery of the GMR effect. In principle, a similar approach is likely to apply to all future attempts to search for interesting new effects in nanocrystalline or nanoglassy materials. However, at present, we seem still far away from being able to do this. At present, we do not have a general physical understanding of the relationship between the properties of nanomaterials and their microstructure. In fact, if one recalls how long it took us to develop this kind of understanding for the materials we are using today, it may be obvious that it will be a long way to develop the scientific basis for the controlled utilization of the novel properties of nanocrystalline or nanoglassy materials by controlling their microstructure.

Moreover, one has to keep in mind that nanostructured materials are far away from thermodynamic equilibrium. As a consequence, their properties may vary even if their microstructural parameters (e.g., the crystallite size, the chemical composition, etc.) seem to be unchanged. Indeed, comparable effects have been reported in the past, for example, for metallic glasses. Depending on the cooling rate applied to quench the melt, the atomic mobility in the chemically identical glass may differ by an order of magnitude or even more due to relaxation processes in the glassy structure.

If one considers the situation summarized so far, it may be clear that the compendium of review articles presented in this book—summarizing our present understanding of the mechanical properties of nanocrystalline materials—is an important first step toward the goal of understanding the relationship between the properties of nanomaterials and their microstructure. In fact, this compendium not only deepens and broadens our knowledge in the specific area of the mechanical properties of nanocrystalline materials, it also evidences that the properties of nanomaterials cannot be understood by simply extrapolating the properties of coarse-grained polycrystals to the nanometer scale. It is the new physical effects that go along with the reduction of the size of the crystallites/glassy regions to atomic dimension. In fact, it is these new physical effects that result in the novel features of nanomaterials, and hence it is the understanding of these new physical effects that provides the basis for the understanding of the new properties of these materials. Unless we have understood these new physical effects, we have no reliable way to utilize and optimize their properties. The GMR effect, the novel ferromagnetic properties reported for several nanomaterials produced in the form of crystallized metallic glasses, the high ductility of some nanocrystalline materials that are brittle in the coarse-grained polycrystalline form, the electronic tuning of magnetic and chemical properties of nanoporous materials, and the observation of the formation of solid solutions of conventionally immiscible components if prepared in the form of nanomaterials are quite likely to represent nothing more but the tip of the iceberg. However, unless we develop a broadly based physical understanding of the properties of nanomaterials, we cannot optimize nanomaterials in a comparable way as we can do today, for example, for coarse-grained materials. Moreover, without developing a broadly based physical understanding of nanomaterials, we are likely to overlook interesting new properties such as the GMR effect.

In that sense this compendium is a first and remarkable attempt to create the basis that will allow us to utilize eventually the full potential of nanomaterials.

Herbert Gleiter
Karlsruhe Institute of Technology (KIT)
Institute of Nanotechnology
Hermann-von-Helmholtz-Platz
76344 Eggenstein
Germany

Preface

The purpose of the present book project was to obtain an advanced undergraduate/graduate text on the relatively new subject of the mechanical properties of nanocrystalline materials—and to do so at reasonable cost of the book for the intended audience.

That the strength properties of nanocrystalline materials are generally superior is made clear in a number of the included articles, and the general result is evidenced by the demonstration of theoretical limiting strength levels being approached in many cases. Production of nanocrystalline materials, as described in the current book, is an important topic that was initiated in pioneering work done by Herbert Gleiter. Beyond producing the material, much analysis is provided to better understand the crystal grain and grain boundary bases that are involved in determining the property behaviors over a range in temperatures and applied loading rates. As usual in evaluating the strength of materials, all of the methods of materials science are joined with the principles of continuum mechanics and applied to the subject in the present collection of articles; and, in particular, there is the ubiquitous presence of dislocations threading through the various descriptions.

Issues of dislocation density, dislocation pile-ups, rate-controlling mechanisms, and ideally perfect or porous grain boundary structures, including their composite representations and mechanistic contributions to shearing processes, are described for the development of constitutive deformation relationships. What an interesting subject! And what practical design possibilities are on the horizon for the newly to-be-fabricated nanocrystalline materials being developed for load-bearing and other structural applications. We join with the current authors in providing with enthusiasm a current summary of our knowledge on the topic.

R.W. Armstrong
University of Maryland

Hans Conrad
North Carolina State University

James C. M. Li
University of Rochester

Chapter 1

MECHANISMS GOVERNING THE PLASTIC DEFORMATION OF NANOCRYSTALLINE MATERIALS, INCLUDING GRAIN-SIZE SOFTENING

Hans Conrad and Jay Narayan

Materials Science and Engineering Department,
North Carolina State University,
Raleigh, NC 27696-7907, USA
E-mail: hans_conrad@ncsu.edu

The physical mechanisms which govern the grain-size (GS) dependence of the flow stress in metals and metal compounds are considered in this chapter. Included are (a) the Hall–Petch (H–P) relation, (b) the corresponding plastic deformation kinetics, and (c) GS softening. Special attention is given to the GS uniformity, the effect of GS on the dislocation density and structure, and grain boundary (GB) shear as a mechanism for significant plastic deformation.

1.1 INTRODUCTION

Microscopy observations on the dislocation structure reveal three regimes in the effect of GS on the plastic deformation flow stress σ of metals at low homologous temperatures; see, for example, Figs. 1.1 and 1.2. Regime I ($d \geq \sim 0.5$ μm) is characterized by the presence of well-defined dislocation cells, regime II ($d \approx 10$–500 nm) by a more uniform distribution of dislocations (i.e., by the absence of well-defined cells), and regime III ($d < \sim 10$ nm) by the absence of dislocations.

Mechanical Properties of Nanocrystalline Materials
Edited by James C. M. Li
Copyright © 2011 Pan Stanford Publishing Pte. Ltd.
www.panstanford.com

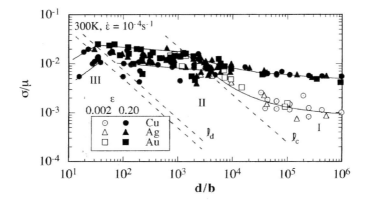

Figure 1.1 Modulus-normalized tensile flow stress σ/μ vs. Burgers vector-normalized GS d/b for the plastic deformation of Cu, Ag, and Au at 300 K, showing three GS regimes: I, II, and III. The line l_c refers to the dislocation cell size and the lines l_d to the dislocation elastic interaction spacing for edge and screw dislocations; ε is the plastic strain. Reproduced with permission from Ref. 1.

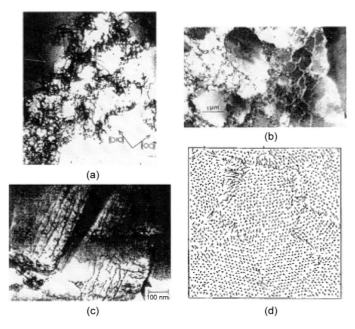

Figure 1.2 Dislocation structures in polycrystalline Cu deformed at 300 K: (a) TEM, $\varepsilon = 10\%$ with $d = 20$ μm, (b) TEM, $\varepsilon = 10\%$ with $d = 25$ μm, (c) TEM, $\varepsilon = 3\%$ with $d = 0.5$ μm, and (d) computer simulation, $\varepsilon = 3.2\%$ with $d = 5$ nm. Reproduced with permission from Ref. 2. *Abbreviation*: TEM, transmission electron microscopy.

At high stress, or GS in the nanocrystalline range (d = 10–100 nm), the microstructure may include partial dislocations, stacking faults, and twins; see, for example, Figs. 6.13 and 6.14 in Chapter 6, by Lu and Lu. To be noted in Fig. 1.1 is that GS hardening occurs in regimes I and II and GS softening in regime III.

The physical mechanisms which govern the effect of d on σ in the three GS regimes are considered in the present chapter. Included are the effects of d on the following parameters: (a) the flow stress σ, (b) the dislocation density ρ, (c) the apparent activation volume $v = KT\partial\ln\dot\varepsilon/\partial\sigma$, and the strain-rate sensitivity parameter $m = \partial\log\sigma/\partial\log\dot\varepsilon$. The variation of these parameters with GS in Cu is shown in Fig. 1.3. To be noted is that σ, m, and ρ increase with decrease in GS, while v decreases. Also, a kink in the curves for v and ρ occurs near the transition from GS regime I to II. As mentioned earlier, the dislocations in regime I form a cellular structure (see Fig. 1.2a,b), whose spacing l_c is given by Staker and Holt [13] and Raj and Pharr [21].

$$l_c = Ab/(\sigma/\mu) \tag{1.1}$$

where A = 10–50, (A = 21 for Cu), b the Burgers vector, and μ the shear modulus.

1.2 GRAIN-SIZE HARDENING: REGIMES I AND II

1.2.1 Hall–Petch Equation

The effect of GS d on the flow stress σ is usually expressed in terms of the H–P equation [22, 23]

$$\sigma = \sigma_i + K_{H-P}\, d^{-1/2} \tag{1.2}$$

where σ_i is the lattice friction stress and K_{H-P} the H–P constant. It is usually considered that a single H–P equation applies over the entire GS range from nanometers (d = 10–100 nm) to micrometers. However, Conrad [24] reported for the face-centered-cubic (FCC) metals Cu, Ag, Au, and Ni that there exist two separate H–P relations over this GS range, one for $d > \sim200$ nm and one for $d < \sim200$ nm, the latter having a larger σ_i and a smaller K_{H-P} than the former; see Fig. 1.4. This behavior has also been reported for Ni by Torre *et al.* [25] and by Asaro and Suresh [26]. It was determined by Conrad and Yang [27] that the occurrence of two GS regimes in the H–P plots can result from the nature of the variation of ρ with d and in turn the relationship between σ and ρ. This will now be reviewed.

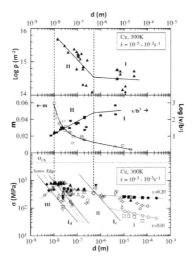

Figure 1.3 Variation of the tensile flow stress σ, the strain-rate sensitivity parameter $m = \partial\log\sigma/\partial\log\dot{\varepsilon}$, the apparent activation volume $v = kT\partial\ln\dot{\varepsilon}/\partial\sigma$, and the dislocation density ρ with GS d for Cu deformed at 300 K. Data for σ from [2], that for m from [3–8], that for v from [3–8], and that for ρ from [9–20].

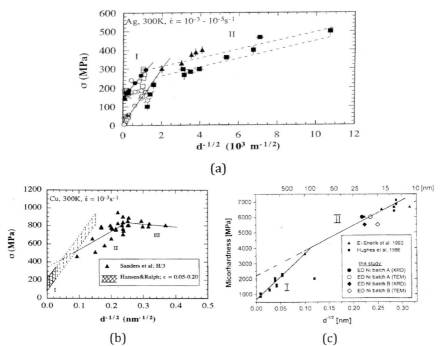

Figure 1.4 H–P plots showing two H–P regions (I and II) for (a) Ag, (b) Cu, and (c) Ni. Reproduced with permission from Ref. 24.

Three mechanisms have been proposed for the effect of d on ρ (Fig. 1.5): (a) GB ledges provide sources for dislocations [28], (b) grain boundaries reduce the average free-slip distance of mobile dislocations [29], and (c) additional "geometric" dislocations are required to accommodate the inhomogeneity of strain which occurs in the transfer of slip from one grain to its neighbor [30]. All three mechanisms give that ρ increases with d^{-1} according to the relation

$$\rho = \beta/bd \qquad (1.3)$$

where $\beta \approx 0.05$ for mechanism (a), ~ 1 for (b), and 0.25 for (c). In keeping with Eq. 1.3 a plot of ρ versus d^{-1} is presented for Cu in Fig. 1.6. The data lies on three lines designated A, B, and C, whose slopes decrease in this order. The GS range for each line, the values of the intercept ρ_i, and the value of the constant β calculated from the slope of each line are given in Table 1.1. To be noted is that β for line A ($d \geq 16$ μm) is approximately that for the freeslip length mechanism, β for line B ($d = 0.2$–16 μm) is approximately that for the geometric dislocation mechanism, and β for line C ($d \leq 0.2$ μm) is approximately that for the GB ledge mechanism. Also to be noted is that ρ_i for line C is about an order of magnitude larger than that for lines A and B.

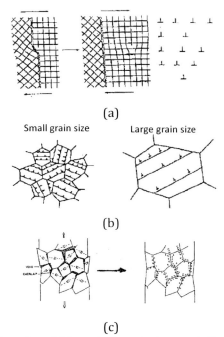

(a)

Small grain size   Large grain size

(b)

(c)

Figure 1.5 Mechanisms by which grain boundaries can influence the dislocation density: (a) generation of dislocation at GB ledges [28], (b) reduction of the free-slip distance [29], and (c) geometric dislocations [30].

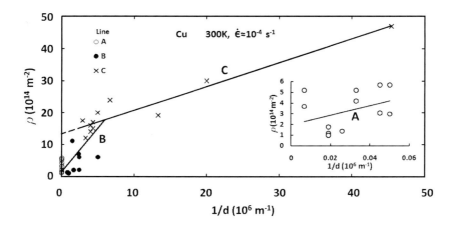

Figure 1.6 Dislocation density ρ vs. reciprocal of the GS for the plastic deformation of Cu at 300 K. Reproduced with permission from Ref. 27.

Table 1.1 The GS range and values of ρ_i and β determined from Fig. 1.6. Also included are the calculated and experimental values of σ_i and K_{H-P} for Cu.

Line	GS Range (μm)	ρ_i (10^{14}m^{-2})	β	σ_i(MPa) Calc. Expt.[a]		K_{H-P}(MPa-m$^{1/2}$) Calc, Expt.[a]	
A	>16	2.0	1.1	177	26–223	0.67	0.16–0.14
B	0.2–16	1.0	0.10	108	26–223	0.21	0.16–0.14
C	<0.2	18.0	0.013	457	320	0.07	0.07
II[b]	<0.2						

[a] Experimental values of d_{II}, σ_I, and K_{H-P} taken from Ref. 24.
[b] Line II is the line representing region II in the H–P plot in Fig. 1.4.

It is well established that the relationship between the tensile flow stress σ and dislocation density ρ in FCC metals deformed at low homologous temperatures is given by

$$\sigma = \alpha \mu b \rho^{1/2} \tag{1.4}$$

where $\alpha \approx 1$.

A plot of σ/μ versus $\rho^{1/2}$ for the GS range from 10 nm to 150 μm in Cu is presented in Fig. 1.7. The data falls on a single line through the origin with a slope $\alpha = 1.03$. Taking ρ_i to represent the dislocations which result from multiplication within the grains (e.g., by double cross-slip) and $\rho_{gb} = (\beta/bd)$ to result from the action of grain boundaries, and assuming that ρ_i and ρ_{gb} independently contribute to the flow stress, one obtains

$$\sigma = \alpha\,\mu\,b\,\rho_i^{1/2} + \alpha\,\mu\,b\,\rho_{gb}^{1/2} \tag{1.5}$$

$$\text{and} \quad \sigma = \alpha\,\mu\,b\,\rho_i^{1/2} + \alpha\,\mu\,b\,(\beta/b)^{1/2}\,d^{-1/2} \tag{1.6}$$

Equation 1.6 is equivalent to the H–P equation with $\sigma_i = \alpha\,\mu\,b\,\rho_i^{1/2}$ and $K_{H-P} = \alpha\,\mu\,b\,(\beta/b)^{1/2}$. The values of σ_i and K_{H-P} calculated employing Eq. 1.6 with $\alpha = 1.03$ are compared in Table 1.1 with those determined from conventional H–P plots [24]. There is reasonable agreement between the calculated values and those from the H–P plots. Especially noteworthy is the agreement between the calculated and experimental magnitudes of (a) σ_i and K_{H-P} for $d < 200$ nm and (b) in the transition GS d_{B-C} between lines B and C in the plot of ρ versus $d^{-1/2}$ (Fig. 1.6) and the transition GS d_{I-II} between regions I and II in the H–P plot (Fig. 1.4b).

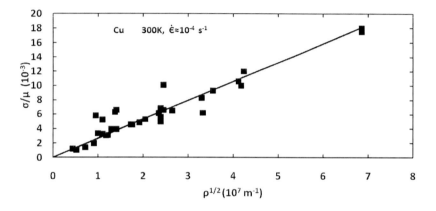

Figure 1.7 Modulus-normalized tensile flow stress (σ/μ) vs. square root of the dislocation density ($\rho^{1/2}$) for Cu deformed at 300 K. Reproduced with permission from Ref. 27.

These results show that for a fixed set of plastic deformation test conditions, the dislocation density increases with decrease in GS, thereby giving a GS dependence of the flow stress, which is equivalent to the H–P equation, that is, the so-called dislocation density model for the H–P relation [2, 29]. Moreover, the GS can have an influence on the mechanism by which most of the dislocations are generated, and on their form and arrangement, so that the H–P parameters for nanocrystalline ($d = 10-100$ nm) metals can differ from those with a micron GS. In the case of FCC metals, σ_i is larger and K_{H-P} smaller for a nanometer GS compared with micrometers. This can be attributed to the difference in the mechanism by which grain boundaries affect the dislocation density and structure [27].

1.2.2 Rate-Controlling Mechanism

Three dislocation mechanisms have been proposed to be strain-rate controlling during the plastic deformation of polycrystalline FCC metals at low homologous temperatures (Fig. 1.8): (a) intersection of dislocations [2, 4, 7, 31], (b) GB shear promoted by the stress concentration due to the pileup of dislocations at the boundary [2, 6, 32], and (c) cross slip in the adjacent grain due to the stress concentration there resulting from the pileup of dislocations [33–35].

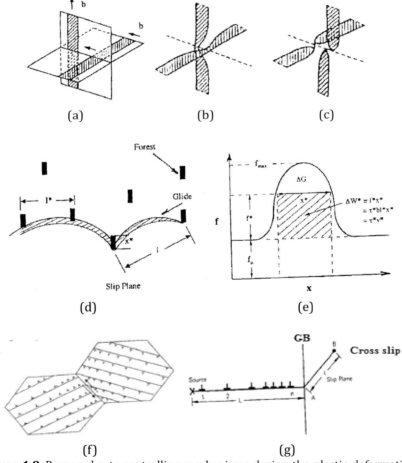

Figure 1.8 Proposed rate-controlling mechanisms during the plastic deformation of FCC metals at low homologous temperatures: (a–e) intersection of dislocations, including the force–distance curve, (f) GB shear promoted by the pileup of dislocations, and (g) generation of dislocations in the adjacent grain by cross-slip promoted by the pileup of dislocations.

There exists considerable evidence that for modest strains and strain rates, the intersection of dislocations is rate controlling in FCC metals with GS in the micron range; that is, in regime I [4, 7, 31, 36, 37]. Of special significance is that the magnitude of v is $\geq -100b^3$ and is proportional to σ^{-1}; see Fig. 1.9. Further, in the intersection mechanism, dislocations provide both the short-range thermal (σ^*) and the long-range athermal (σ_μ) components of the flow stress $\sigma = (\sigma^* + \sigma_\mu)$. A schematic of the force (f) versus distance (x) curve for the intersection process is included in Fig. 1.8. The resolved shear strain-rate for this mechanism is given by

$$\dot\gamma = \rho_m b^2 v_D \exp\left(\frac{-\Delta G(\tau)}{kT}\right) \tag{1.7}$$

where ρ_m is the mobile dislocation density, b the Burgers vector, v_D the Debye frequency, and ΔG Gibbs free energy, which is a decreasing function of the applied resolved shear stress $\tau = \sigma/M = (\sigma^* + \sigma_\mu)/M$. Since in general $\partial\ln\rho_m/\partial\tau$ is small compared with $\partial\Delta G/\partial\tau$, the "true" activation volume for the intersection process is [38]

$$v^* = -\partial\,\Delta G/\partial\tau = l^* b x^* = \frac{3}{2}kT\frac{\partial\ln\dot\gamma}{\partial\tau} = \frac{3}{2}MkT\frac{\partial\ln\dot\varepsilon}{\partial\sigma} = \frac{3}{2}Mv \tag{1.8}$$

where l^* is the spacing of the dislocation forest "trees" in contact with the mobile dislocations and $\dot\varepsilon$ the uniaxial strain-rate. Further, taking $l^* = l_f$ the average dislocation forest spacing and $l_f = \rho_f^{-1/2} = (\Phi\rho)^{-1/2}$, we obtain

$$l^* = (\Phi\rho)^{-1/2} \tag{1.9}$$

where Φ is the fraction of the total dislocation density ρ, which comprises the forest density. Considering the bowing of the mobile dislocations between the forest dislocations, Friedel [39] calculated that

$$l^* = (2\Gamma l_f^2/\tau^* b)^{1/3} \tag{1.10}$$

where Γ is the dislocation line tension and l_f the average spacing between the dislocations penetrating the slip plane. Taking $\Gamma = 1/2\mu b^{1/2}$, $l_f = \rho_f^{1/2} = (a\rho)^{1/2}$, and $\tau^* = c\tau = c\sigma/M = c\mu b\rho^{1/2}/M$, and inserting into Eq. 1.10 one obtains

$$l^* = (M/a)^{1/3}\rho^{-1/2} = (\Phi_F\,\rho)^{-1/2} \tag{1.11}$$

where a and ρ are proportionality constants, which can vary with dislocation density and/or structure and $\Phi_F = (ac/M)^{2/3}$ is the fraction of the total density, which comprises ρ_f. Inserting Eq. 1.9 or 1.11 for l^* into Eq. 1.8 one obtains for the activation distance

$$x^*/b = \frac{3Mv}{2l^* b^2} = \left(\frac{3M}{2b^2}\Phi^{1/2}\right)\rho^{1/2}v \tag{1.12}$$

and

$$f^*/\mu b^2 = \frac{\tau^* b^*}{\mu b^2} = c/M\Phi^{1/2} \qquad (1.13)$$

for the effective, thermal component of the force for the intersection process.

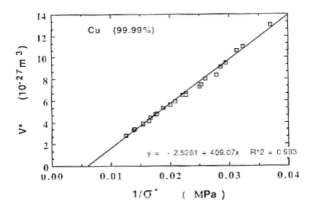

Figure 1.9 Apparent activation volume v vs. $1/\sigma^*$ for the plastic deformation of Cu ($d = 18\ \mu m$) at 78–300 K and $\dot{\varepsilon} = 10^{-4}\ s^{-1}$. Reproduced with permission from Ref. 4.

Taking $\Phi = \Phi_F$ and $c = 1/3$ from Conrad and Cao [4] and the reasonable value $a = 3/4$ (four nonparallel slip planes), one obtains $x^*/b = 3.5$ and $f^*/\mu b^2 = 0.23$ for Cu with $d = 18\ \mu m$ [4]. These values along with those in Table 1.2 provide strong support that the intersection of dislocations is rate controlling in GS regime I.

Table 1.2 Comparison of experimentally derived parameters for the plastic deformation kinetics of polycrystalline Cu ($d = 18\ \mu m$, regime I) at low homologous temperatures (T-$\leq 0.3T_M$) with theoretical predictions for the intersection of dislocations. Reproduced with permission from Ref. 2.

Parameter	Predicted	Experiment
v^*/b^3	10^2–10^3 (decreases with ε, σ, and T^{-1})	10^2–10^3 (decreases with ε, σ, and T^{-1})
x^*/b	3–7	3.1–4.2
$\Delta F^*/\mu b^3$	~0.2	0.23
$\Delta S^*/eV\ K^{-1}$	~(2–4) × 10^{-4}	3 × 10^{-4}
$\dot{\gamma}_0/s$	10^7	7 × 10^6
ρ/m^2	10^{13}–10^{15}	10^{13}–10^{15}

The fit to a single straight line *through the origin* of the plot of σ/μ versus $\rho^{1/2}$ in Fig. 1.7 suggests that the same mechanism is rate controlling in GS regime II as in I. Moreover, the work ΔW expended by the total applied stress during the thermally activated process is also in accord with this hypothesis. ΔW is given by

$$\Delta W = \tau b l^* x^* = \sigma v^*/M = 3/2 \sigma v = 3/2 \alpha \mu b \rho^{1/2} v \qquad (1.14)$$

Rearranging the strain-rate sensitivity parameter equation $m = \partial \ln \sigma / \partial \ln \dot{\varepsilon}$ and multiplying the numerator and denominator by kT, one obtains

$$\frac{kT}{m} = \sigma v \qquad (1.15)$$

Equations 1.4, 1.14, and 1.15 then give

$$m(v/b^3)\rho^{1/2} = kT/\mu b^4 \qquad (1.16)$$

A plot of the product $m(v/b^3)\rho^{1/2}$ versus log d is presented in Fig. 1.10.

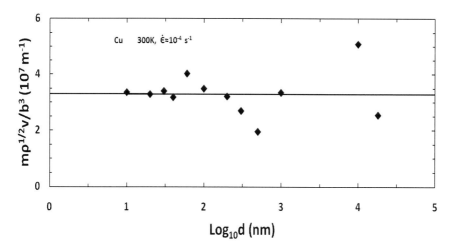

Figure 1.10 The product $m\rho^{1/2}v/b^3$ vs. log d for Cu. Reproduced with permission from Ref. 40.

To be noted is that the product is relatively independent of GS, including both regimes I and II. Also, the average value of the product $(3.3 \times 10^7 \text{ m}^{-1})$ is in reasonable accord with the predicted value $(kT/\mu b^4 = 2.3 \times 10^7 \text{ m}^{-1})$, considering that the pertinent data was obtained from a number of sources. The results in Figs. 1.7 and 1.10 thus indicate that the same mechanism is rate controlling in both regimes I and II, namely, the intersection of dislocations.

For further analysis, a plot of $(x^*/b)/\Phi^{1/2}$ (given by Eq. 1.12) versus log d is presented in Fig. 1.11. The data lies along two lines with a sharp change in slope at d = 300–500 nm, which is the transition between GS regimes I and II. The behavior in Fig. 1.11 indicates that the parameter Φ varies with dislocation density and/or structure. An estimate of the variation of Φ with ρ (and in turn d) can be obtained from the results in Fig. 1.11 by assuming that for constant $\dot{\varepsilon}$ and T, x^*/b is relatively independent of GS and moreover assuming that $\Phi = \Phi_F$ at the transition GS d_{I-II} = 500 nm. At this GS the corresponding dislocation structure is consistent with that considered by Friedel [39]. The magnitude of Φ_F employed at d = 500 nm was obtained by taking a = 3/4 (four nonparallel slip planes), c = 1/3 [4], and M = 3 in Eq. 1.10. For d = 500 nm, this gives Φ_F = 0.19, x^*/b = 0.64, and $f^*/\mu b^3$ = 0.25, which are all reasonable values. A log–log plot of Φ_F (obtained employing Eqs. 1.10 and 1.12) versus d (and the corresponding ρ) is presented in Fig. 1.12. It is seen that Φ_F decreases with increase in d (decrease in ρ) in regime II. A sharp drop occurs between d = 500 and 1,000 nm, that is, at the transition GS d_{I-II} and thereafter only a slight change with further increase in GS in regime I. These results show that the dislocation structure can have a significant effect on the magnitude of Φ_F and in turn on the influence of GS on the plastic deformation kinetics parameters v and m. These considerations thus give that the rate-controlling mechanism during the plastic deformation at low homologous temperatures of FCC metals is the intersection of dislocations in both GS regimes I and II. Pertinent parameters are the dislocation density ρ and its arrangement (structure) and their influence on Φ.

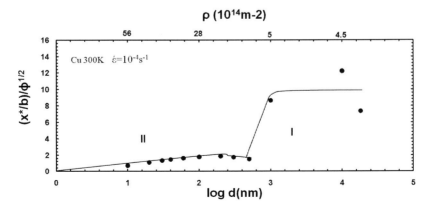

Figure 1.11 $(x^*/b)/\Phi^{1/2}$ vs. log d for Cu. Reproduced with permission from Ref. 40.

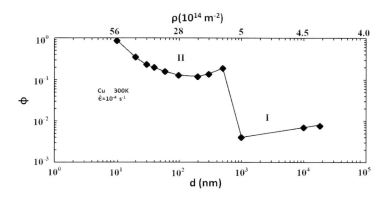

Figure 1.12 Φ_F vs. log d for Cu. Reproduced with permission from Ref. 40.

Within the context of the intersection mechanism, the so-called "anomalous" increase in v with a decrease in temperature for Cu with $d \approx 500$ nm [5, 6] can result from an increase in the dislocation forest contact spacing l^*, that is, from a decrease in either Φ_F or ρ (or both) with a decrease in temperature. GB shear promoted by the pileup of dislocations at the boundary (Fig. 1.8f) can, however, also give an anomalous increase in v with a decrease in T [2, 24]. The constitutive equation for this mechanism is [24]

$$\dot{\gamma} = \gamma_{o,c} \exp - \Delta F_{o,c} - v_c^* \ [ac]$$

$$\dot{\gamma} = \gamma_{o,c} \exp - \left\{ \frac{\Delta F_{o,c} - v_c^* \ [\alpha_c \ \pi \tau_e^2 \ (d/\mu b) - \tau_c^\mu]}{kT} \right\} \tag{1.17}$$

where $\dot{\gamma}_{o,c} = 1.3 \times 10^5$ s^{-1}, $\Delta F_{o,c}$ is Helmholtz free energy, v_c^* the true activation volume, α_c a constant (which includes the distance ahead of the pileup, the orientation of the promoted shear, and the type of dislocation [edge or screw] in the pileup), $\tau_e = \tau - \tau_i$ (where τ_i is the lattice friction stress), and τ_c^μ the long-range back stress at the specific location at which the thermally activated event occurs due to the stress concentration from the pileup. The subscript c refers to the parameter corresponding to the thermally activated event. The fit of the plastic deformation kinetics data by Embury and Lahaie [5] for Cu ($d = 0.5$ μm) to Eq. 1.17 is shown in Fig. 1.13. The apparent activation volume v for the GB shear mechanism is given by [24]

$$\frac{1}{v} = \frac{1}{v_i^{II}} + \frac{M^2 \mu b}{\alpha_c \ 2\pi K_{H-P}^{II} \ d \ (\sigma - \sigma_i^{II})} \tag{1.18}$$

where the subscript i refers to the friction stress and the superscript II refers to GS regime II. Employing Eqs. 1.17 and 1.18 the magnitudes of ΔH_c^*, v_c^*/b^3, and $\dot{\gamma}_{o,c}$ determined from the plastic deformation kinetics of Cu and Ni at low homologous temperatures are given in Table 1.3. Included are predicted values, which are in reasonable accord with those obtained experimentally. The agreement thus indicates that GB sliding promoted by the pileup of dislocations could be rate controlling in GS regime II. Moreover, the relative independence of the H–P parameters in regime II on the stacking fault energy [24] is in accord with this mechanism. However, microscopic evidence of classical dislocation pileups and of significant GB shear is lacking for an unequivocal acceptance of the GB shear mechanism. A lack of visible pileups could, however, occur if the existing dislocation structure gives an effective stress concentration that is equivalent to a classical pileup.

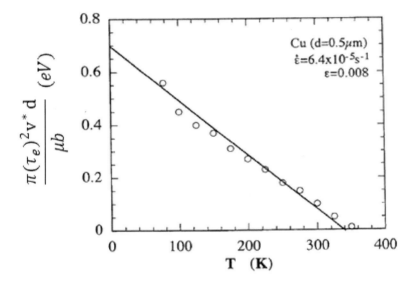

Figure 1.13 $\pi(\tau_e)^2 v^* d/\mu b$ vs. temperature for vapor-deposited Cu foil ($d = 0.5$ μm) deformed at 77–350 K. $\tau_e = \tau_c = \tau - \tau_i$. Reproduced with permission from Ref. 2.

Table 1.3 Comparison of the experimental deformation kinetics parameters with those predicted for GB shear promoted by the pileup of dislocations. Reproduced with permission from Ref. 24.

Metal	ΔH_c^* (eV)		v_c^*/b^3		$\dot{\gamma}_{o,c}$ (s^{-1})	
	Exptl	**Pred.**	**Exptl**	**Pred.**	**Exptl**	**Pred.**
Cu	0.70	0.64–1.08[a]	1.1	1–10	1.3×10^5	1.1×10^5
Ni	0.96	0.70–1.20[b]	9.4	1–10	3.7×10^4	4.4×10^6

[a] GB diffusion in Cu.

[b] GB diffusion in Ni.

Regarding GB shear, Gleiter *et al.* [41] and Hirth [42] have pointed out that GB shear can occur by the glide of dislocations within the boundary, for which direct evidence exists [41, 43, 44].

Instead of GB shear the rate-controlling mechanism could be the generation of lattice dislocations from sources contained within the boundary. There exists considerable direct evidence that lattice dislocations emanate from GBs; see, for example, Fig. 1.14. A model for the emission of perfect and partial dislocations from grain boundaries has recently been developed by Asaro *et al.* [26, 46]. They propose that the resolved shear stress for the emission of a perfect dislocation is given by

$$\tau/\mu = b/d \tag{1.19}$$

and that for a partial dislocation

$$\tau/\mu = (\alpha_p - 1)/\alpha_p \Gamma_e + \frac{1}{3}\frac{b}{d} \tag{1.20}$$

where $\alpha_p = d/\delta_{eq}$ (δ_{eq} is the equilibrium spacing between the partial dislocations) and $\Gamma_e = \gamma_{SF}/\mu b$ is the reduced stacking fault energy γ_{SF}. These authors found qualitative agreement between the experimental data on FCC metals (especially for GS near the transition between regimes II and III) and the predictions of their model.

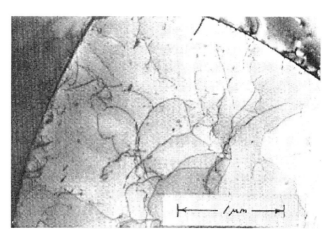

Figure 1.14 TEM micrograph of α–Ti deformed 1.3% at 500 K, showing emission of dislocations from the grain boundaries. Reproduced with permission from Ref. 45.

The mechanism of dislocation multiplication in the adjacent grain due to the pileup of dislocations at the boundary is considered in the chapter by Armstrong. This mechanism has been developed over the years by Armstrong *et al.* [33–35] and leads directly to the H–P equation. Since the temperature dependence of the parameter K_{H-P} was in accord with that for cross-slip, it was proposed that the mechanism governing the behavior was the multiplication of dislocations ahead of the pileup by double cross-slip. Although there exists considerable support for this model, a major shortcoming is the lack of microscopic evidence for the existence of classical dislocation pileups in unalloyed metals. Moreover, it still needs to be established that the temperature and strain-rate dependence of K_{H-P} is universally the same as that for cross-slip. Also, the mechanism governing σ_i needs clarification. Further, the finding [24] that the H–P parameter K_{H-P} in regime II (Fig. 1.4) is relatively independent of γ_{SF} is not expected for dislocation multiplication by cross-slip.

In summary some agreement exists between the available experimental data and all of the models that have been proposed for the GS dependence of the flow stress in regimes I and II. This suggests that the governing mechanism may vary with the GS and test conditions or that several mechanisms operate concurrently in series or parallel. Needed for resolution of the governing mechanism(s) is data which includes the dislocation density and structure, along with the effects of strain, strain-rate, and temperature on the flow stress as a function of GS with a narrow, uniform distribution.

1.3 GRAIN-SIZE SOFTENING: REGIME III

1.3.1 Production of Nanocrystalline Materials of Controlled Structure and Properties

Nanocrystalline materials have been produced by a variety of chemical, physical, and mechanical routes [47–53]. The chemical routes are based upon nucleation and growth in a supersaturated solution medium. In the chemical processing of nanostructured materials, nucleation is kinetically controlled before Ostwald ripening can take over, thereby leading to a uniform size distribution. Using surface-active agents it is possible to induce self-organization of nanoparticles in the first layer and in certain cases in the second layer. However, in general, chemical processing does not lend itself to three-dimensional (3-D) self-assembly processing. In the related technique of electrodeposition, the nucleation rate and size are controlled by reducing the electric current pulse duration and repetition rate to produce nanostructured materials. However, this control works only down to a certain size range, below which impurities are needed to control the nucleation. Thus, impurity contamination leads to undesirable effects in nanostructured materials produced by chemical or electrodeposition methods.

Mechanical deformation-based processing produces a highly defective state, which usually crystallizes to give nanostructured materials [54–56].

Since the nucleation and dynamic crystallization cannot be easily controlled, mechanical processing often leads to a large size distribution of grains, whose defect content and boundaries are not well defined. However, this technique can be easily implemented on a laboratory scale, and in certain cases a size distribution can lead to desirable mechanical properties, such as enhanced toughness and ductility [56]. A related technique based upon nucleation and growth of nanograins involves recrystallization of amorphous materials. In this case, control of nucleation and growth is often difficult, and the process also usually leads to a nonuniform size distribution of grains [57]. In certain cases, an explosive recrystallization can lead to very large grains of several microns with nanosize grains in neighboring regions. Figure 1.15 shows an example of explosive recrystallization in amorphous silicon, which was triggered by focusing an electron beam inside a transmission electron microscope at a certain nucleation center.

Figure 1.15 Recrystallization in amorphous silicon, showing large micron-size grains mixed with nanocrystalline grains. Reproduced with permission from Ref. 57.

Physical methods based upon pulsed laser deposition (PLD) can be used to create layered or nanodot nanostructures and thereby provide control over size distribution and defect content [51]. Using this method, nanostructured materials of uniform size in the range of 5–100 nm can be created for controlled structure–property correlations. Figure 1.16 shows a schematic of a PLD system, where a high-powered excimer laser ablates material in a controlled way with the precision of a single atomic layer. The typical laser parameters are pulse energy density 3–5 J cm^{-2}, pulse duration 25 ns, and laser wavelength with photon energy 4–6 eV. The large laser power in the range of 120–600 MW leads to nonequilibrium stoichiometric evaporation, where the average energy of the evaporated species ranges from 100 to 1000 kT.

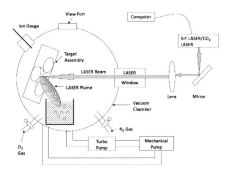

Figure 1.16 Schematic of the PLD system to synthesize controlled nanostructured materials of uniform GS.

To create nanostructured materials of uniform GS to investigate the correlation between hardness and GS, we utilized the 3-D island growth mode of thin films, known as the Volmer–Weber growth mode. As illustrated in Fig. 1.17, there are three distinct modes of film growth: (a) two-dimensional (2-D) layer by layer growth (Frank–Vander Merwe growth), (b) 3-D island growth (Volmer–Weber growth), and (c) a mixture of 2-D and 3-D growth (Stranski–Krastnov growth).

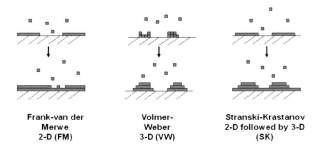

Figure 1.17 Thin-film growth modes to create layered and self-assembled nanostructured materials.

These growth modes depend upon the surface free energy of the substrate (γ_s), surface free energy of the film (γ_f), and the interfacial free energy (γ_i). The 2-D layer-by-layer growth occurs when $\gamma_s > \gamma_f + \gamma_i$, where γ_s is high and γ_f and γ_i are low. This 2-D growth mode can be used to create layered nanostructures. The 3-D growth mode occurs when $\gamma_s < \gamma_f + \gamma_i$, where γ_s is low and γ_f and γ_i are high. Islands or nanodots of uniform size can be created by controlling the laser deposition and substrate variables. The laser deposition variables include pulse energy density, pulse duration, laser wavelength, and repetition. The substrate variables include substrate temperature and surface free energy, which can be varied by introducing appropriate surfactant interlayers. To produce nanostructured materials consisting of islands, we lay down the first layer of islands by controlling the kinetics of clustering, where nanoclusters are self-organized as a result of a kinetically driven free-energy minimum. The total energy of the system has four components: volume free energy, surface free energy, interface free energy, and interaction free energy. The total free energy has two components: nanodot size-dependent and nanodot size-independent terms. The total energy of the system shows a local minimum in free energy, which can be used to obtain self-organized nanodots. The idea is to create self-organization driven kinetically by the local minimum, while avoiding the global thermodynamic minimum which leads to Ostwald ripening. After the formation of the first layer of nanodots, an interfacial layer is introduced to start the renucleation process for the second set of nanodots. The characteristics of this interfacial layer are critical for renucleation and self-organization of the second layer. As examples, in the case of Cu we have used a monolayer of W; in the case of WC we used a monolayer of NiAl. Figure 1.18a shows WC with 11 nm grains of uniform size with a monolayer of interfacial NiAl. In the selected area diffraction, we observed only WC lines due to the small amount of NiAl interfacial layering. It is remarkable to see the uniformity of GS, where interfacial layering and alloying can be controlled so precisely. Figure 1.18b shows a smaller GS of 6 nm with a high degree of uniformity. In this case, the electron diffraction pattern also showed only WC lines. Figure 1.19 shows the character of the grain boundaries in nanocrystalline Zn produced by PLD.

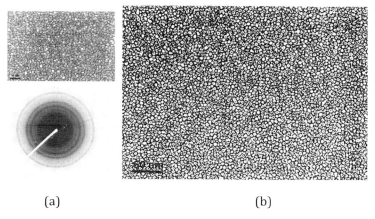

(a) (b)

Figure 1.18 TEM micrographs showing nanostructured WC with uniform GS: (a) average GS 11 nm and (b) average GS 6 nm.

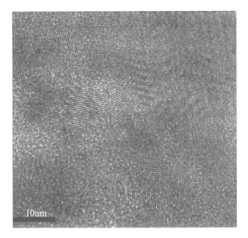

Figure 1.19 HRTEM micrograph of the structures of grain boundaries in nanocrystalline Zn produced by PLD. *Abbreviation*: HRTEM, high-resolution transmission electron microscopy.

Samples of uniform GS with over 0.5 μm thickness were used for hardness measurements using an instrument from Nanoindentor, Inc. The results of hardness with GS (H vs. $d^{-1/2}$) in WC are quite interesting, as shown in Fig. 1.20. The hardness increases linearly with decreasing GS down to a certain GS, and below this critical size the hardness decreases. We envisioned that intragrain deformation dominates above the critical size, and below this size intergrain deformation dominates, for example, GB shear [58]. This phenomenon leads to softening, and in turn hardness decreases with a decrease in GS. With decreasing GS the boundary area increases, which leads to softening. The boundary deformation and shear can be a function of the structure of the GB. If the boundaries can act as an effective source for twinning dislocations, as proposed in recent papers [59, 60], then twinning shear can dominate the deformation characteristics.

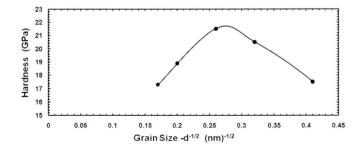

Figure 1.20 Hardness of WC as a function of GS (H vs. $d^{-1/2}$).

There has been a fair amount of controversy regarding the softening characteristics below a critical GS. Most of the controversial data is based upon samples with nonuniform GS distribution, where only an average GS is considered. In these samples, defect content within the grains is varied and not characterized. Under a nonuniform size distribution, the deformation characteristics are expected to be quite different. The larger grains will deform preferentially as the smaller grains stopped deforming below their critical size. In this mode, the hardness will continue to increase with the decreasing GS, as has been observed experimentally.

1.3.2 Plastic Deformation Mechanism

The effect of GS produced by PLD on the hardness of Cu is shown in Fig. 1.21. GS hardening occurs for $d > {\sim}10$ nm and GS softening for smaller values of d. Similar behavior occurred for nanocrystalline Zn [61, 62] and the metal compounds WC [63] and TiN [64] produced by PLD. The results on these materials along with those on nanocrystalline metals and compounds prepared by other methods are presented in Figs. 1.21 and 1.22. For each material, with a decrease in GS there occurs a transition from GS hardening to GS softening at a critical GS d_{II-III} on the order of 10 nm. This value of d is of the order of the elastic interaction between lattice dislocations given by [66]

$$l_{d,s} = \mu b/2\pi\sigma \qquad \text{(screw)} \tag{1.19}$$

$$l_{d,c} = \mu b/2\pi(1 - v)\sigma \ \text{(edge)} \tag{1.20}$$

This then explains the absence of lattice dislocations in materials with GS in regime III, since when l_d becomes greater than d_{II-III} the interaction stress becomes larger than the applied stress, thereby preventing the occurrence of dislocations in the grain interior. As seen from Fig. 1.23 the volume fraction of atoms located at the grain boundaries when $d = 10$ nm is of the order of 10%.

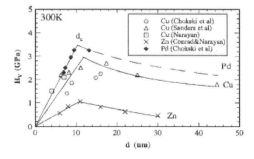

Figure 1.21 Hardness vs. GS for Zn, Cu, and Pd, showing a transition from GS hardening to GS softening at $d \approx 10$ nm. Reproduced with permission from Ref. 65.

Two mechanisms have been proposed for the GS softening (Fig. 1.24): (a) Coble creep [67, 68] and (b) GB shear accommodated by atomic adjustments [58, 69, 70]. The constitutive equation which applies to both mechanisms is

$$\dot{\varepsilon} = \frac{A^g}{d^p} \sin h \left(\frac{\sigma_e \Omega}{kT} \right) \exp \left(\frac{-\Delta F}{kT} \right) \tag{1.21}$$

where A_g and p are constants defining the specific mechanism, $\sigma_e = \sigma - \sigma_o$ the effective stress, σ_o a back stress or threshold stress, Ω the atomic volume, and ΔF Helmholtz free energy, which is either the GB diffusion energy Q_{gb} or that for vacancy migration Q_m^V.

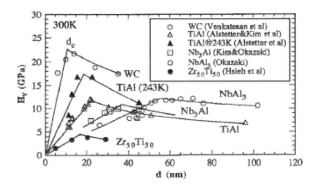

Figure 1.22 Hardness vs. GS for a number of metallic compounds showing a transition from GS hardening to GS softening at d = 10–55 nm. Reproduced with permission from Ref. 65.

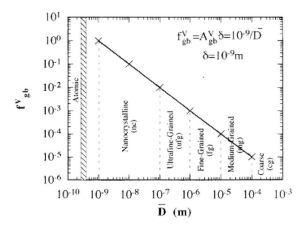

Figure 1.23 Volume fraction of atoms located at the GB vs. GS. A_{gb}^V = GB area per unit volume; δ = GB width; D = mean linear intercept GS.

Regime III

Grain boundary Diffusion Creep: Coble Creep

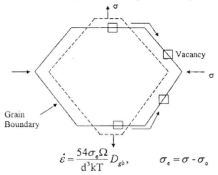

$$\dot{\varepsilon} = \frac{54\sigma_e\Omega}{d^3kT}D_{gb}, \qquad \sigma_e = \sigma - \sigma_o$$

Grain Boundary Shear: Hahn-Padmanabhan (1997)
Swygenhoven-Caro (1997-1998), Conrad-Narayan (2000)

$$Conrad - Narayan : \dot{\gamma} = \frac{2\delta v_D}{d}\sinh(\frac{\tau_e{}^*v^*}{kT})\exp-(\frac{\Delta F^*}{kT})$$

$$\delta \approx 3b, \ \tau_e{}^* = \tau - \tau_0, \ v^* \approx 1b^3, \ \Delta F^* = \approx Q_b$$

(a)

Figure 1.24 Schematics of two mechanisms proposed for the plastic deformation of materials with GS in regime III. Reproduced with permission from Ref. 2.

A major difference between the two mechanisms is in the GS exponent p, which is 1 for GB shear and 3 for Coble creep. Also, σ is the hydrostatic stress in the Coble mechanism and the shear stress in the GB shear mechanism. The experimental values of the parameters ΔF, p, and $v^* = \Omega$ determined employing Eq. 1.21 are listed in Table 1.4. The magnitudes of $v_b^*/a^3 \approx 1$ and $\Delta F \approx Q_b$ or Q_m are in accord with either mechanism. However, a value of $\rho = 1$ is only in accord with the GB shear mechanism. Experimental evidence that $\rho = 1$ rather than 3 is presented in Fig. 1.25 for Cu and in Fig. 1.26 for Zn. The limited available data thus gives that the mechanism governing GS softening is GB shear accommodated by atomic rearrangements. Additional support for this mechanism is provided by the computer simulations of Sygenhoven *et al.* [70] and the analyses by Asaro *et al.* [26, 71, 72].

Table 1.4 Values of the parameters pertaining to GS softening determined by linear regression analysis of hardness. All tests at 300 K except the one on TiAl at 243 K. Reproduced with permission from Ref. 65.

Material	$a^{(1)}$ (nm)	$\mu^{(2)}$ (GP$_a$)	$H_{vic}^{(3)}$ (GP$_a$)	$d_c^{(4)}$ (nm)	v_b^* (10^{-30} m³)	v_b^*/a^3	ΔF^* (kJ/mole)	$R^{(5)}$	Q_m' (kJ/mole)	Q_b (kJ/mole)	$d_c/\ell_{d,a}$	$d_c/\ell_{d,c}$
Cu[1-3]	0.256	42.1	2.65	10.5	18.8	1.1	75.1	0.996	75.1	104	1.2	1.8
Pd[1]	0.275	40.4	3.45	10.0	3.70	0.2	87.5	0.998	87.5	133	1.4	2.2
Zn[4]	0.267	49.3	1.12	11.4	13.4	0.7	72.6	0.994	54.4	60.5	0.5	0.7
TiAl[6-8]	0.128	66	11.8	22	3.08	1.5	93.7	0.983	—	—	4.8	7.3
TiAl(243K)[7]	0.128	66	18.0	17	0.42	0.20	71.3	0.983	—	—	3.8	5.6
Nb₃Al[9]	0.138	—	10.5	35	1.84	0.70	90.3	0.987	—	68 to 91	—	—
NbAl₃[10]	0.138	—	12.2	55	1.57	0.60	87.1	0.990	—	—	—	—
Zr₅₀/T₅₀[11]	0.174	—	4.0	18	6.43	1.22	93.2	0.993	—	—	—	—
WC[12]	0.136	265	23.2	9	2.25	0.93	97.2	0.993	—	300	1.4	2.0

Notes: (1) Atomic diameter for pure metals (largest ionic diameter in compounds); (2) shear modulus; (3) maximum hardness; (4) critical grain size; and (5) correlation coefficient.

The effects of $\dot{\varepsilon}$ and T on the GS dependence of σ in the nanocrystalline range are illustrated in Fig. 1.27, where the critical GS d_c occurs at the intersection of the constitutive equation for GB shear (Eq. 1.21) with the equation for the elastic interaction of dislocations (Eqs. 1.19 and 1.20). If σ given by the dislocation kinetics equation intersects the elastic interaction line below σ_c, then deformation with a further decrease in GS must occur either elastically according to Eqs. 1.19 and 1.20 until the stress increases to σ_c or plastically by the generation of stacking faults, partial dislocations, and twins, requiring a stress higher than that given by the equation for the dislocation kinetics.

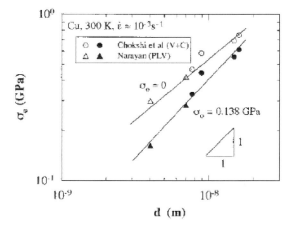

Figure 1.25 Log $(\sigma_c = \sigma - \sigma_0)$ vs. log d for Cu, taking $\sigma_0 = 0$ and $\sigma_0 = 0.138$ GPa. Reproduced with permission from Ref. 2.

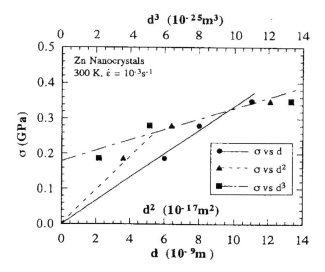

Figure 1.26 σ vs. d^n with $n = 1, 2$, and 3 for Zn with GS in regime III. Reproduced with permission from Ref. 62.

Hence, the transition GS $d_{II–III}$ is not a constant but will depend on strain-rate $\dot{\varepsilon}$ and temperature T, increasing with a decrease in $\dot{\varepsilon}$ or an increase in T. An example of the increase in d_c with a decrease in $\dot{\varepsilon}$ is given in Fig. 1.28.

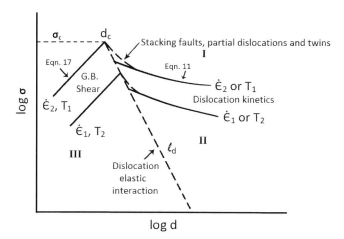

Figure 1.27 Schematic of the influence of strain-rate $\dot{\varepsilon}$ or temperature T on the transition from dislocation kinetics in GS regime II to GB shear in regime III. $\dot{\varepsilon}_2 > \dot{\varepsilon}_1$ and $T_2 > T_1$.

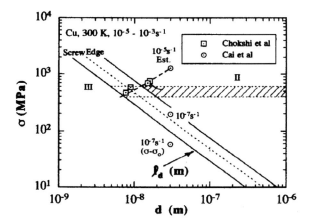

Figure 1.28 Log σ vs. log d for Cu, showing that d_{II-III} depends on the strain-rate. Reproduced with permission from Ref. 65.

1.4 SUMMARY COMMENTS

It is recognized that the above analyses are based on limited data from a number of sources with some scatter. It is hoped that our chapter will stimulate further work to evaluate the concepts and ideas that are presented. Especially desirable are definitive studies of the plastic deformation kinetics on specimens with a uniform GS distribution over a wide GS size range and corresponding advanced microscopy observations and measurements on the GB structure and lattice defects (e.g., complete and partial dislocations, stacking faults, and twins). Especially important in the nanocrystalline GS range is that the materials be free of so-called artifacts (e.g., porosity, impurities and oxides) and that the GS distribution be very narrow. The PLD process provides a means for meeting these requirements to a high degree.

1.5 ACKNOWLEDGMENTS

The authors wish to acknowledge the assistance of Christopher Ledford in the preparation of the manuscript.

References

1. H. Conrad and K. Jung (2005), *Mater. Sci. Eng. A*, **406**, 78.
2. H. Conrad (2004), *Metall. Mater. Trans. A*, **35A**, 2681.

3. L. Lu, M. Dao, T. Zhu, and J. Li (2009), *Scripta Mater.*, **60**, 1062.

4. H. Conrad and Wei-di Cao (1996), in *The Johannes Weertman Symposium*, eds., R. J. Arsenault *et al.*, TMS, Warrendale, PA, p. 321.

5. J. D. Embury and D. J. Lahaie (1993), in *Mechanical Properties and Deformation Behavior of Materials Having Ultrafine Microstructure*, eds., M. Natasi *et. al.*, Kluwer-Academic, Dordrech, p. 287.

6. H. Conrad and Di Yang (2002), *J. Electronic Mater.*, **31**, 304.

7. H. Conrad (1965), in *High-Strength Materials*, ed., V. F. Zackay, John Wiley, NY, p. 436.

8. R. P. Carreker and W. R. Hibbard (1935), *Acta Mater.*, **1**, 656.

9. P. Gordon (1955), trans. *AIME*, **203**, 1043.

10. L. M. Clareborough, M. E. Hargreaves, and M. H. Lorretto (1958), *Acta Metall.*, **6**, 725.

11. J. E. Bailey (1963), *Phil. Mag.*, **8**, 223.

12. M. Doner, H. Chang, and H. Conrad (1974), *J. Mech. Phys. Solids*, **22**, 555.

13. M. R. Staker and D. L. Holt (1972), *Acta Metall.*, **20**, 569.

14. N. Hanson and B. Ralph (1982), *Acta Metall.*, **30**, 411.

15. M. Hommel and O. Kraft (2001), *Acta Mater.*, **49**, 3935.

16. T. Uger, S. Oth, P. Sanders, A. Borbely, and J. R. Weertman (1998), *Acta Mater.*, **46**, 3693.

17. Y. Zheng, N. Tao, and K. Lu (2008), *Acta Mater.*, **56**, 2429.

18. L. Balough, T. Unger, Y. Zhao, Y. Zhu, Z. Morito, C. Yu, and T. Langdon (2008), *Acta Mater.*, **56**, 809.

19. S. Brandstetter, P. M. Derlet, S. Van Petegem, and H. Van Swygenhoven (2008), *Acta Mater.*, **56**, 165.

20. Y. S. Li, Y. Zhang, N. Tao, and K. Lu (2009), *Acta Mater.*, **57**, 761.

21. S. V. Raj and G. M. Pharr (1986), *Mater. Sci. Eng.*, **81**, 217.

22. E. O. Hall (1951), *Proc. Phys. Soc. Lond.*, **B64**, 747.

23. N. J. Petch (1953), *J. Iron Steel Inst.*, **17**, 25.

24. H. Conrad (2007), *Nanotechnology*, **18**, 325701.

25. F. Dalla Torre, H. Swygenhoven, and M. Victoria (2002), *Acta Mater.*, **50**, 3857.

26. R. J. Asaro and S. Suresh (2005), *Acta Mater.*, **53**, 3369.

27. H. Conrad and Di Yang (2009), *J. Mechanical Behaviour Mater.*, **19**(6), 365.

28. J. C. M. Li (1963), trans. *TMS AIME*, **227**, 239.

29. H. Conrad (1963), in *Electron Microscopy and Strength of Crystals*, eds., G. Thomas and J. Washburn, Interscience, NY, p. 299; (1970) in *Ultrafine-Grain Metals*, eds., J. Burke, M. L. Reed, and V. Weiss, Syracuse University Press, Syracuse, NY, p. 213.

30. M. F. Ashby (1970), *Phil. Mag.*, **21**, 399.

31. H. Conrad (2003), *Mater. Sci. Eng. A*, **341**, 216.

32. H. Conrad and K. Jung (2005), *Scripta Mater.*, **53**, 581.

33. R. W. Armstrong (1970), *Adv. Mater. Res.*, **4**, 101.
34. R. W. Armstrong (1972), in *Defect Structures in Solids*, eds., K. I. Vasu, K. Ramman, D. Sastry, and Y. Prasad, Indian Institute of Science, Bangalore, p. 306.
35. R. W. Armstrong and P. Rodrigues (2006), *Phil. Mag.*, **86**, 5787.
36. H. Conrad (1964), *J. Met.*, **16**, 582.
37. H. Conrad and K. Jung (2005), *Mater. Sci. Eng. A*, **391**, 272.
38. H. Conrad, B. deMeester, C. Yin, and M. Doner (1975), in *Rate Processes in Plastic Deformation of Materials*, eds., A.K. Mukerjee and J.C.M. Li, ASM, Materials Park, OH, p. 175.
39. J. Friedel (1967), *Dislocations*, Pergamon Press, NY, p. 225.
40. H. Conrad and Di Yang (2010), *J. Mater. Sci.*, 45(22), 6166.
41. W. Gleiter, E. Hornbogen, and G. B *ä* ro (1968), *Acta Met.*, **16**, 1053.
42. J. P. Hirth (1972), *Metall. Trans.*, **3**, 3047.
43. Y. Ishide and M. H. Brown (1967), *Acta Met.*, **15**, 857.
44. G. Buzzichelli and A. Mascanzoni (1971), *Phil Mag.*, **24**, 497.
45. H. Conrad (1981), *Prog. Mater. Sci.*, **26**, 123.
46. R. J. Asaro, P. Krysl, and B. Ked (2003), *Phil. Mag. Lett.*, **83**, 733.
47. J. Narayan, Y. Chen, and R. M. Moon (1981), *Phys. Rev. Lett.*, **46**, 1491; US Patent # 4,376,755(1984).
48. *Nanomaterials: Synthesis, Properties and Applications* (1998), eds., A. Edelstein and R. Cammarata, Institute of Physics, London.
49. C. J. Brinker, Y. Lu, A. Sellinger, and H. Fan (1999), *Adv. Mater.*, **11**, 579.
50. G. M. Whitesides and B. Grzybowski (2002), *Science*, **295**, 2418.
51. J. Narayan and A. Tiwari (2004), *J. Nanosci. Nanotech.*, **4**, 726; US Patent # 7,105,118 (2006).
52. *Nanostructured Materials: Processing, Properties and Applications* (2007), ed., C. Koch, William Andrews Publishing, Norwich, NY.
53. J. Narayan (2009), *Int. J. Nanotech.*, 6, 493.
54. X. Zhang, H. Wang, M. Kassem, J. Narayan, and C. C. Koch (2002), Scripta *Mater.*, **46**, 661.
55. X. K. Zhu, X. Zhang, H. Wang, A. V. Sergueeva, A. K. Mukherjee, R. O. Scattergood, J. Narayan, and C. C. Koch (2003), Scripta *Mater.*, **49**, 429.
56. C. C. Koch (2009), *J. Phys. Conference Series*, **144**, 012081.
57. J. Narayan, S. J. Pennycook, D. Fathy, and O. W. Holland (1984), *J. Vac. Sci. Tech.*, **2**, 1495.
58. H. Conrad and J. Narayan (2000), Scripta *Mater.*, **42**, 1025.
59. Y. T. Zhu, J. Narayan, J. P. Hirth, S. Mahajan, X. L. Wu, and X. Z. Liao (2009), Acta *Mater.*, **57**, 3763.
60. J. Narayan and Y. T. Zhu (2008), *Appl. Phys. Lett.*, **92**, 151908.
61. H. Conrad and J. Narayan (2002), *Acta Mater.*, **50**, 5067.
62. H. Conrad and J. Narayan (2002), *Appl. Phys. Lett.*, **81**, 2241.

63. H. Conrad and J. Narayan (2003), in *Electron Microscopy: Its Role in Materials Science*, eds., J. R. Weetman, M. Fine, K. Faber, W. King, and P. Liaw, TMS, Warrendale, PA, p. 141.

64. H. Conrad, J. Narayan, and K. Jung (2005). *Int. J. Refract. Mat. Hard Mater.*, **23**, 301.

65. H. Conrad and K. Jung (2006), *J. Electronic Mater.*, **35**, 857.

66. A. H. Cottrell (1953), *Dislocations and Plastic Flow in Crystals*, Oxford University Press, London.

67. R. L. Coble (1963), *J. Appl. Phys.*, **14**, 1679.

68. B. Cai, Q. Kang, L. Lu, and K. Lu (1999), *Scripta Mater.*, **41**, 755.

69. H. Hahn and K. Padnamabham (1997), *Phil. Mag.*, **B76**, 559.

70. H. Van Swygenhoven, M. Spaczer, and A. Caro (1999), *Acta Mater.*, **47**, 3117.

71. B. Zhu, R. J. Asaro, P. Krysl, and R. Bailey (2005), *Acta Mater.*, **53**, 4825.

72. B. Zhu, R. J. Asaro, P. Krysl, K. Zhang, and J. R. Weertman (2006), *Acta Mater.*, **54**, 3307.

Chapter 2

ENHANCED MECHANICAL PROPERTIES OF NANOSTRUCTURED METALS PRODUCED BY SPD TECHNIQUES

Ruslan Z. Valiev[a] and Terence G. Langdon[b,c]
[a]*Institute of Physics of Advanced Materials, Ufa State Aviation Technical University, 12 K. Marx St., Ufa 450000 Russia*

E-mail: rzvaliev@mail.rb.ru

[b]*Departments of Aerospace and Mechanical Engineering and Materials Science, University of Southern California, Los Angeles, CA 90089-1453, USA*
[c]*Materials Research Group, School of Engineering Sciences, University of Southampton, Southampton SO17 1BJ, UK*

E-mail: langdon@usc.edu

Despite rosy prospects, the use of nanostructured metals and alloys as advanced structural materials has remained controversial until recently. Only in recent years has a breakthrough been outlined in this area, associated both with development of new routes for the fabrication of bulk nanostructured materials and with investigation of the fundamental mechanisms that lead to the new properties of these materials.

This chapter discusses new concepts and principles of using severe plastic deformation (SPD) to fabricate bulk nanostructured metals with advanced properties. Special emphasis is laid on the relationship between microstructural features and properties, as well as the first applications of SPD-produced nanomaterials.

2.1 INTRODUCTION

It is now over 20 years since Herbert Gleiter presented the first concepts for developing nanocrystalline materials (i.e., ultrafine-grained [UFG] materials with a grain size under 100 nm) with the potential for producing special properties [1]. Since that time, the field of nanostructured materials has developed rapidly, owing to the considerable interest in this topic and the scientific and technological importance.

Gleiter's original idea was that, owing to the very small grain size, nanocrystalline materials will contain an extremely large fraction of grain boundaries having a special atomic structure. As a result, nanomaterials should have unusual properties [2]. As regards the mechanical properties, one may expect very high strength, toughness, fatigue life, and wear resistance. Nanostructuring seemed likely to lead to a revolutionary use of nanomaterials for many functional and structural applications.

But in practice these interesting prospects were put in jeopardy. It was shown in numerous investigations [3–5] that, although nanocrystalline materials demonstrated very high strength or hardness, they were of very low ductility or even brittle, thereby producing insuperable problems for use in advanced structural applications. When addressing the reasons for the low ductility, many researchers point out drawbacks in their synthesis on the basis of the compaction of nanopowders obtained using various methods [4, 5]. Nanomaterials produced by compacting usually have residual porosity, contaminations, and, as a rule, small geometric dimensions. All this may lead to a decline in their ductility. Another possible reason is of a fundamental nature: the plastic deformation mechanism associated with the generation and movement of dislocations may not be effective in ultrafine grains (as described later). In this connection, recent findings of extraordinarily high strength and ductility in several bulk UFG metals are especially interesting [6–9].

Various nanomaterials possess microstructural features which are closely linked to the processing methods and regimes. Therefore, it is necessary to initially address the fundamental principles involved in the processing techniques that are currently used to produce these materials and the structural features that are inherent features of these processed materials.

2.2 THE PRINCIPLES OF GRAIN REFINEMENT BY SPD PROCESSING

In recent years there has been a growing interest in a new approach to the fabrication of bulk nanostructured metals and alloys, where this new approach forms an alternative to the conventional compaction of nanopowders. This new approach is based on microstructure refinement in bulk billets through the application of SPD; that is, heavy straining under high imposed pressure

[10]. SPD-produced nanomaterials are fully dense, and their large geometric dimensions make it possible to attain excellent properties thorough mechanical testing. The fabrication of bulk nanostructured materials by SPD is now becoming one of the most actively developing areas in the field of nanomaterials [11, 12].

Since the pioneering work on tailoring of UFG structures by SPD processing [13, 14], two SPD techniques have attracted close attention and have lately experienced further development. These techniques are high-pressure torsion (HPT) [10, 15] and equal-channel angular pressing (ECAP) [10, 16]. The fundamental principles of these two techniques are illustrated schematically in Fig. 2.1.

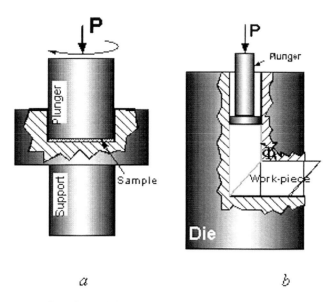

a   *b*

Figure 2.1 Principles of SPD: (a) HPT—a sample is held between anvils and strained in torsion under an applied pressure (*P*). (b) Equal-channel angular pressing—a work-piece is repeatedly pressed through a special die.

Samples processed under HPT are disc shaped (Fig. 2.1a). In this process, the sample, with a diameter ranging from 10 mm to 20 mm and thickness of 0.2–0.5 mm, is put between anvils and compressed under an applied pressure (*P*) of several gigapascals (GPa). The lower anvil turns, and friction forces result in shear straining of the sample. As a result of a high imposed pressure, the deforming sample does not break even at high strains [10].

Essential structural refinement is observed after deformation through one-half or one complete (360°) turn. But to produce a homogeneous nanostructure, with a typical grain size of about 100 nm or less, deformation by several turns is necessary (Fig. 2.2). The important role of applied pressure in the formation of a more homogeneous nanostructured state during HPT is also shown in recent work on nickel [17].

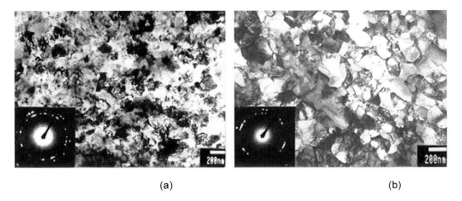

(a)   (b)

Figure 2.2 TEM images of UFG Cu. (a) Cu processed by HPT at room temperature (P = 6 GPa, 5 turns) and (b) Cu processed by ECAP (12 passes). *Abbreviation*: TEM, transmission electron microscopy.

Processing by ECAP (Fig. 2.1b) is now an established procedure for use with metals and it is an attractive technique for several reasons. First, it is relatively easy to set up and use an ECAP die. Second, exceptionally high strains may be imposed either through the repetitive pressing of the same sample or by developing special multipass dies [18], rotary dies [19], or side-extrusion facilities [20]. Third, the ECAP samples can be easily scaled up to produce relatively large bulk materials [21], where these large samples have a potential for use in a range of applications from the biomedical to aerospace industries [22]. Fourth, although ECAP is generally used with samples in the form of bars or rods, the process may be applied also to plate samples: for example, a recent report described the application of ECAP to an Al plate [23]. Fifth, ECAP can be incorporated into conventional rolling mills for use in continuous processing [24–27] or into the ECAP-conform process for the production of long-sized rods and wires [28]. In view of these many advantages, it is appropriate to examine the principles of processing through the use of ECAP.

Figure 2.3 shows a schematic illustration of the ECAP procedure [29]. A die is constructed containing a channel that is bent through an abrupt angle, where this angle is 90° in Fig. 2.3. A sample is machined to fit in the channel, and the sample is then pressed through the die using a plunger. It is apparent from the illustration that the sample has the same cross-sectional dimensions before and after pressing, thereby permitting repetitive pressings of the same sample. It is important to note also that the retention of the same cross-sectional area differentiates processing by ECAP in a very significant way from more conventional industrial processes, such as rolling, extrusion, or drawing, where the sample dimensions are reduced on each consecutive pass. Three orthogonal planes are illustrated in Fig. 2.3, where X is the transverse plane perpendicular to the flow direction, Y is the flow plane parallel to the side face at the point of exit from the die, and Z is the longitudinal plane parallel to the top surface at the point of exit from the die.

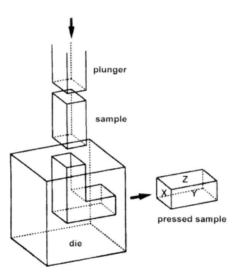

Figure 2.3 The principle of ECAP, including a definition of the three orthogonal planes X, Y, and Z. Reproduced with permission from Ref. 29.

The strain imposed on the sample in each passage through the die is dependent primarily upon the angle, Φ, between the two parts of the channel (90° in Fig. 2.3) and also to a minor extent upon the angle of curvature, Ψ, representing the outer arc of curvature where the two channels intersect (0° in Fig. 2.3). To achieve optimum results, ECAP is generally conducted using a die having a channel angle of $\Phi = 90°$ [30], and with this configuration it can be shown from first principles that for these angles the imposed strain on each pass is approximately equal to 1, with only a small, and almost insignificant, dependence upon the arc of curvature, Ψ [31].

When samples are pressed repetitively, different slip systems may be introduced by rotating the samples about the X-axis between consecutive passes through the die. In practice, four separate processing routes have been identified for use in ECAP: route A, in which the sample is pressed repetitively without rotation; route B_A, in which the sample is rotated through 90° in alternate directions between each pass; route B_C, in which the sample is rotated by 90° in the same sense between each pass; and route C, where the sample is rotated by 180° between passes [32]. These different processing routes are illustrated schematically in Fig. 2.4, and the distinction between these routes is important because the various routes introduce different shearing patterns into the samples [33], leading to variations both in the macroscopic distortions of the individual grains in polycrystalline matrices [34] and in the capability to develop a reasonably homogeneous and equiaxed UFG microstructure [35–37].

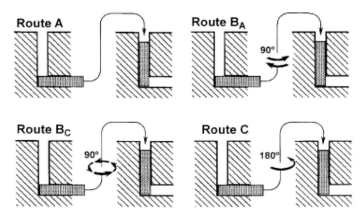

Figure 2.4 The four processing routes in ECAP.

When a metal is processed by ECAP, a high density of dislocations is introduced into the material on each separate pass through the die. There have been numerous investigations of the microstructural development in the processing of polycrystalline materials by ECAP, but by contrast, there has been only a limited number of studies describing the application of ECAP to single crystals.

A significant advantage in using single crystals for fundamental studies of microstructural development in ECAP is that the crystals may be placed in different orientations prior to pressing through the die. Furthermore, the results are not adversely affected by the presence of grain boundaries. It is convenient to consider an Al single crystal oriented at 20° to the theoretical shear plane in a clockwise sense. This crystal was pressed using a die having an angle of $\Phi = 90°$ between the two parts of the channel, and it gave an angle of $\Psi = 30°$ at the outer arc of curvature [38].

Using TEM, Fig. 2.5 shows the microstructure visible on the Y or flow plane in the center of the single crystal after a single pass, where the X and Z directions lie parallel to the lower and side edges of the photomicrograph, and the selected-area electron diffraction (SAED) pattern was taken with a beam diameter of 12.3 μm [38].

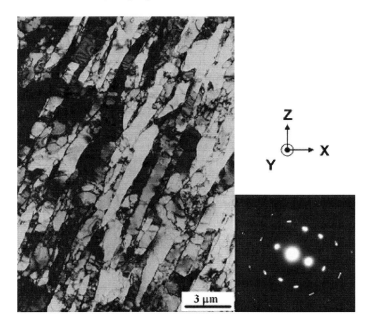

Figure 2.5 An Al single crystal in the 20° orientation after one pass of ECAP, showing the formation of subgrain bands on the Y plane and an SAED pattern: the X and Z directions lie parallel to the lower edge and the side edge of the photomicrograph, respectively. Reproduced with permission from Ref. 38.

It is apparent from the SAED pattern that the boundaries visible in the photomicrograph have low angles of misorientation, and therefore they are subgrain boundaries.

Regarding microstructural development in polycrystalline samples, it is worth noting that early investigations were conducted to record the microstructures produced in polycrystalline pure Cu [39] and Al [35, 36] after processing by ECAP. For the experiments on high-purity Al [36], the initial grain size was ~1 mm and the ECAP die contained a channel angle of $\Phi = 90°$ and an arc of curvature at the intersection of the channels of $\Psi = 20°$.

Figure 2.6 shows the microstructures visible on the X, Y, and Z planes after a single pass through the die, together with the SAED patterns recorded using a beam diameter of 12.3 μm [36].

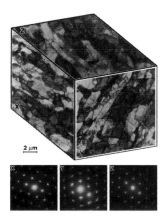

Figure 2.6 Microstructures on the *X*, *Y*, and *Z* planes in high-purity polycrystalline Al after one pass with the associated SAED patterns. Reproduced with permission from Ref. 36.

It is apparent that the microstructures again consist of arrays of elongated subgrains, where the boundaries primarily have low angles of misorientation and the subgrain arrays are aligned either horizontally on the *X* plane, at an angle of approximately 45° to the *X* axis when viewed on the *Y* plane and perpendicular to the flow direction when viewed on the *Z* plane.

Close inspection of the photomicrographs in Fig. 2.7 shows that, even after a total of four passes, the subgrains tend to be elongated when using routes A and C, but the microstructures on the three orthogonal planes are more reasonably homogeneous and equiaxed when using route B_C. Furthermore, the SAED patterns show there are predominantly high-angle boundaries after processing by route B_C but a higher fraction of low-angle boundaries when using routes A and C.

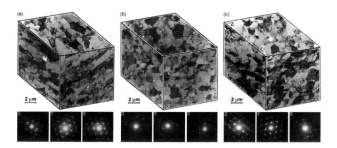

Figure 2.7 Microstructures on the *X*, *Y*, and *Z* planes in high-purity polycrystalline Al after four passes, together with the associated SAED patterns, for (a) route A, (b) route B_C, and (c) route C. Reproduced with permission from Ref. 36.

It is important to note also that the average size of the equiaxed grains visible on each plane after four passes when processing by route B_C, as illustrated in Fig. 2.7b, was measured as ~1.2–1.3 µm, which is consistent both with the width of the subgrain arrays recorded after a single pass through the die and with other detailed measurements of the equiaxed grain size in polycrystalline pure Al after ECAP [40]. These results provide a clear demonstration, therefore, that the ultimate size of the equiaxed grains attained in ECAP corresponds to the width of the elongated subgrain arrays developed in the first pass through the die.

A model for grain refinement by SPD processing was initially developed to explain the results obtained on Armco iron processed using HPT [41]. In this model, a very high dislocation density is introduced in the early stages of torsional straining, and this leads to the formation of an intragranular structure consisting of cells with thick cell walls and low angles of misorientation. As the strain increases, the thickness of the cell walls decreases by recovery through dislocation annihilation, leading, ultimately, to an excess of dislocations of only one sign on each boundary and the formation of an array of ultrafine grains separated by high-angle nonequilibrium boundaries. This type of model is attractive because, when applied to processing by ECAP, it provides an explanation for the well-established increase in the average boundary misorientation angle with increasing numbers of passes through the die [40, 42–47]. Nevertheless, the model is insufficient because it predicts a gradually increasing refinement in the microstructure with increasing strain, and this is not consistent with the photomicrographs shown in Figs. 2.6 and 2.7. In addition, it incorporates no relationship between the formation of subgrains and the occurrence of slip within the grains. This limitation was addressed recently by proposing an alternative model based on an interrelationship between the formation of subgrain boundaries and slip [48], and the approach was later further developed to provide a more detailed mechanism for the occurrence of grain refinement in ECAP [49].

Figure 2.8 is a schematic illustration of the model, where each illustration represents the appearance of the microstructure on the Y plane, the three rows correspond to the conditions after one, two, and four passes of ECAP, the three columns represent the appearance when using processing routes A, B_C, and C, and the total angular range for the various slip systems, η, is included beneath each illustration [49]. The color coding in Fig. 2.8 follows the sequence used with red, mauve, green, and blue representing the slip lines introduced in the first, second, third, and fourth pass, respectively. It is apparent by inspection that the shearing system introduced on each pass in Fig. 2.8 is consistent with the shearing orientations documented theoretically [32, 33] so that, for example, shearing is confined exclusively to planes lying at

45° to the *X*-axis when using route C. Furthermore, the width of the subgrain bands is set equal to *d* for each separate passage through the die.

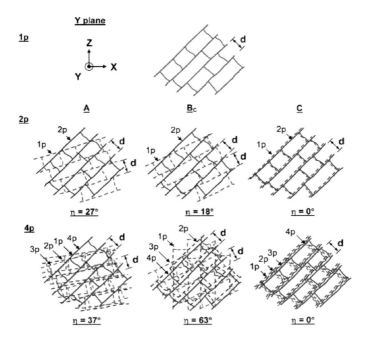

Figure 2.8 A model for grain refinement in ECAP, where subgrain bands are formed, with a width of *d*, with their longer sides lying parallel to the primary slip plane. The results are illustrated on the *Y* plane for one, two, and four passes using routes A, B_c, and C, with the colors red, mauve, green, and blue corresponding to the first, second, third, and fourth pass through the die, respectively. Reproduced with permission from Ref. 49.

The presence of several intersecting slip systems when using route B_c gives a high density of dislocations, and it is reasonable to assume these dislocations rearrange and annihilate consistent with the low-energy dislocation structures (LEDS) theory [50, 51] and in a manner similar to the evolution proposed for Armco iron processed by HPT [41]. This leads, for route B_c, to the formation of a reasonably equiaxed array of grains on each orthogonal section, as illustrated schematically in Fig. 2.9, which shows equiaxed grains on the *Y* plane after a total of four passes. By contrast, the development of an equiaxed microstructure is less advanced when using routes A and C, and there remain elongated grains or subgrains after processing through four passes, because of the low value for the angular range, *η*, for these two routes.

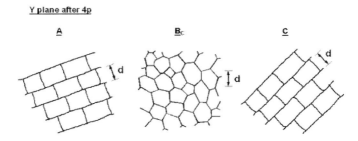

Figure 2.9 The appearance of the microstructures on the Y plane after four passes using routes A, B_C, and C. Reproduced with permission from Ref. 49.

It should be noted that the occurrence of more elongated microstructures when processing using routes A and C is consistent with experimental observations [52]. For route C, the shearing plane lies consistently at 45° with the X direction when viewed on the Y plane, as documented in Fig. 2.8, and the schematic illustration in Fig. 2.9 is consistent with the appearance of the Y plane in Fig. 2.7c. For route A, the shearing planes lie at angles of 8°, 11°, 18°, and 45° to the X direction for the slip traces associated with one, two, three, and four passes [53], respectively, thereby suggesting that the elongated grains or subgrains lie at an average angle of the order of ~20° with the X-axis after four passes: this is depicted schematically for route A in Fig. 2.9, and it is consistent with the microstructure visible on the Y plane in Fig. 2.7a.

The consistency between the schematic illustrations in Fig. 2.9 and the microstructures visible on the Y plane in Fig. 2.7 provide strong support for a mechanism of grain refinement in ECAP of the form depicted in Figs. 2.8 and 2.9.

Though the basic principles of grain refinement by SPD processing were primarily developed for pure metals, they can be applied to alloys as well, including commercial ones [10, 15]. Moreover, the basic rules, determining the requirements to SPD processing regimes and routes to form a UFG structure in metals and alloys, have been established recently by Valiev [54].

2.3 MECHANICAL PROPERTIES ACHIEVED USING SPD PROCESSING

The small grain sizes and high defect densities inherent in UFG materials processed by SPD lead to much higher strengths than in their coarse-grained counterparts. Moreover, according to the constitutive relationship for superplasticity, it is reasonable to expect the appearance in UFG metals of low-temperature and/or high-rate superplasticity [55–58]. The realization of these capabilities will become important for the future development of high-strength and wear-resistant materials, advanced superplastic alloys, and metals having high-fatigue life. The potential for achieving all of these possibilities has raised a keen interest among scientists and engineers in studying the mechanical and functional properties of these UFG materials.

In this connection, the fabrication of bulk samples and billets using SPD, in particular ECAP, was a crucial first step in initiating investigations into the properties of UFG, because the use of ECAP processing permitted, and subsequently fully supported, a series of systematic studies using various nanostructured metallic materials, including commercial alloys [10, 59–63]. The results of these recent studies are discussed in this section.

2.4 STRENGTH AND DUCTILITY AT AMBIENT TEMPERATURE

During the last decade it has been widely demonstrated that a major grain refinement, down to the nanometer range, may lead to very high hardness in various metals and alloys, but nevertheless these materials invariably exhibit low ductility under tensile testing [64, 65]. A similar tendency is well known for metals subjected to heavy straining by other processes such as rolling, extrusion, and drawing. Strength and ductility are the key mechanical properties of any material, but these properties typically have opposing characteristics. Thus, materials may be strong or ductile, but they are rarely both.

The reason for this dichotomy is of a fundamental nature. As discussed in more detail in an earlier report [66], the plastic deformation mechanisms associated with the generation and movement of dislocations may not be effective in ultrafine grains or in strongly refined microstructures. This is generally equally true for SPD-processed materials. Thus, most of these materials have a relatively low ductility, but they usually demonstrate significantly higher strength than their coarse-grained counterparts. Despite this limitation, it is important to note the experiments show that SPD processing leads to a reduction in the ductility, which is generally less than in more conventional deformation processing techniques, such as rolling, drawing, and extrusion [67].

Recent findings of extraordinary high strength and good ductility in several bulk UFG metals produced by SPD are also of special interest [6, 8, 68–70]. It is important to consider in detail the three different approaches that were used in these investigations.

In the first study, high-purity (99.996%) Cu was processed at room temperature using ECAP with a 90° clockwise rotation around the billet axis between consecutive passes in route B$_c$ [6]. The strength and ductility were measured by uniaxial tensile tests, and the resulting engineering stress–strain curves are shown in Fig. 2.10 for the Cu samples tested at room temperature in the initial coarse-grained condition and in three processed states [6]. It is apparent that the initial coarse-grained Cu, with a grain size of about 30 μm, has a typical low yield stress with significant strain hardening and a large elongation to failure. At the same time, cold-rolling (CR) the Cu to a thickness reduction of 60% significantly increases the strength, as shown by curve 2 in Fig. 2.10, but dramatically decreases the elongation to failure. This result is consistent with the classical mechanical behavior of metals that are deformed plastically. The same tendency is true also for Cu subjected to two passes of ECAP. However, further straining of Cu to 16 passes of ECAP, as shown by curve 4 in Fig. 2.10, simultaneously increases both the strength and the ductility. Furthermore, the increase in ductility is much more significant than the relatively minor increase in strength.

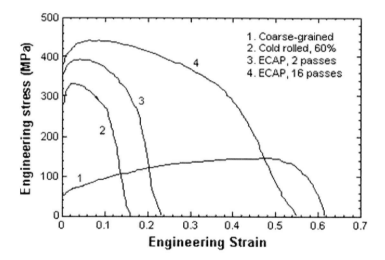

Figure 2.10 Tensile engineering stress–strain curves for Cu tested at 22°C with a strain-rate of 10^{-3} s^{-1}: the processing conditions for each curve are indicated. Reproduced with permission from Ref. 6.

Thus, the data shown in Fig. 2.10 for Cu processed by ECAP clearly demonstrate an enhancement of strength as well as ductility with accumulated deformation due to an increase in the number of passes from 2 to 16. This is a very remarkable result that, at the time of the investigation in 2002, had never been observed in metals processed by plastic deformation.

Accordingly, the effect was termed the "paradox of strength and ductility in SPD-processed metals," and the principles of this paradox are illustrated in Fig. 2.11, where it is apparent that conventional metals lie within the lower shaded quadrant [6]. As seen in Fig. 2.11 for Cu and Al, CR (the reduction in thickness is marked by each data point) increases the yield strength but decreases the elongation to failure or ductility [71, 72]. The extraordinary combination of high strength and high ductility shown in Fig. 2.11 for the nanostructured Cu and Ti after processing by SPD clearly sets them apart from the other coarse-grained metals.

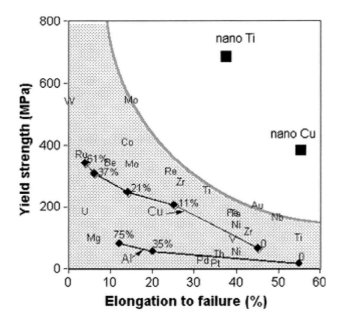

Figure 2.11 The paradox of strength and ductility in metals subjected to SPD ("VAZL paradox"): the extraordinary combination of high strength and high ductility in nanostructured Cu and Ti processed by SPD (two upper points) clearly sets them apart from conventional coarse-grained metals (lower points relating to metals of 99.5%–99.9% purity). Reproduced with permission from Ref. 6.

In recent years, similar tendencies have been reported in a number of metals, including Al [73, 74], Cu [75], Ni [76], and Ti [6, 70], after processing through various types of SPD such as ECAP, HPT, or accumulative roll bonding. Concerning the origin of this phenomenon, it has been suggested that it is associated with an increase in the fraction of high-angle grain boundaries with increasing straining and with a consequent change in the dominant deformation mechanisms due to the increasing tendency for the occurrence of grain boundary sliding and grain rotation [6, 66]. An increase in the fraction of high-angle grain boundaries is an example of grain boundary engineering, which can be referred to as tailoring interfaces with a specific structure to control the materials' properties through variation of SPD processing regimes and routes [77].

Another approach to the problem of ductility enhancement was suggested through the introduction of a bimodal distribution of grain sizes [8]. In this study, nanostructured Cu was produced through a combination of ECAP and subsequent rolling at the low temperature of liquid nitrogen prior to heating to a temperature close to ~450 K. This procedure gave a bimodal structure of micrometer-sized grains, with a volume fraction of around 25%, embedded in a matrix of nanocrystalline grains. The material produced in this way exhibited an extraordinarily high ductility but also retained very high strength. The reason for this behavior is that, while the nanocrystalline grains provide strength, the embedded larger grains stabilize the tensile deformation of the material. Other evidence for the importance of grain size distribution comes from investigations on Zn [78], Cu [79], and an Al alloy [80]. Furthermore, the investigation of Cu [79] showed that bimodal structures may increase the ductility not only during tensile testing but also during cyclic deformation. This observation is important in improving the fatigue properties of materials.

A third approach has been suggested for enhancing strength and ductility on the basis of the formation of second-phase particles in the nanostructured metallic matrix [65], where it is anticipated these particles will modify the shear band (SB) propagation during straining and thereby lead to an increase in the ductility.

The principle of achieving high strength and high ductility through the introduction of intermediate metastable phases was successfully realized recently in a commercial Al–Zn–Mg–Cu–Zr alloy [81] and an Al–10.8% Ag alloy subjected to ECAP and subsequent ageing [68]. The principle of this approach is illustrated in Fig. 2.12 for the Al–Ag alloy, where the Vickers microhardness is plotted against the ageing time at 373 K for samples in a solution-treated (ST) condition and after CR and ECAP [68]. For the ST condition, the hardness is initially low but increases with ageing time to a peak value after 100 hours (3.6×10^5 s). For the CR condition, the hardness is higher, but there is only a minor increase with ageing. The hardness is even higher after ECAP and further increases with ageing to a peak value after 100 hours. The relatively lower values of hardness recorded after CR by comparison with ECAP are due to the lower equivalent strain imposed on the sample: these strains were ~1.4 in CR and ~8 in ECAP so that the microstructure after CR consisted of subgrains or cell boundaries having low angles of misorientation. It was shown, using scanning TEM, that the peak hardness achieved after ECAP and ageing for 100 hours is due to precipitation within the grains of spherical particles with diameters of ~10 nm and elongated precipitates with lengths of ~20 nm. The spherical particles were identified as η-zones consisting of arrays of solute atoms lying parallel to the (001) planes, and the elongated precipitates were identified as the plate-like γ' particles. It was shown also that additional ageing up to 300 hours led to a growth in the γ' particles and a very significant reduction in the density of the fine η-zones, thereby giving a consequent loss in hardening at the longest ageing time recorded in Fig. 2.12.

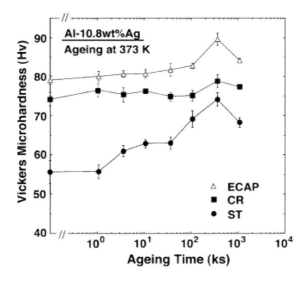

Figure 2.12 Variation of the Vickers microhardness with ageing time for the Al–10.8% Ag alloy after ST, CR, and ECAP. Reproduced with permission from Ref. 68.

The introduction of ageing after ECAP has an important influence on the stress–strain behavior at room temperature, as demonstrated in Fig. 2.13, where the tensile stress–strain curves are shown after ECAP and after ST, CR, and ECAP with additional ageing for 100 hours at 373 K [68]. Thus, ST and ageing give reasonable tensile strength, an extensive region of uniform strain and good ductility, whereas CR and ageing give increased strength but very limited uniform strain and a marked reduction in the total ductility. For the ECAP condition, the strength is high in the absence of ageing but there is a negligible region of uniform strain and no significant strain hardening. By contrast, the sample processed by ECAP and aged for 100 hours shows similar high strength, a region of strain hardening, and good ductility. In practice, the uniform strain of ~0.14 achieved in this specimen is similar to the uniform strain of ~0.17 in the sample after ST and ageing, and the elongation to failure of ~0.40 is comparable to, and even slightly exceeds, the elongation of ~0.37 recorded in the ST and aged condition. These results demonstrate, therefore, the potential for producing high strength and good ductility in precipitation-hardened alloys. Furthermore, although the results documented in Figs. 2.12 and 2.13 relate to a model Al–Ag alloy, it is reasonable to anticipate it should be possible to achieve similar results in commercial engineering alloys, where the ageing treatments are generally well documented.

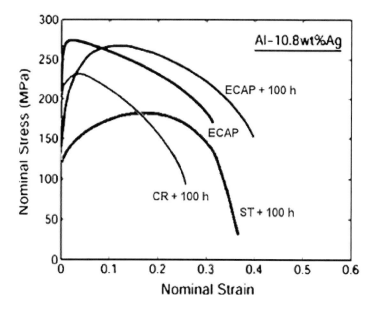

Figure 2.13 Tensile plots of stress vs. strain at room temperature for the Al–10.8% Ag alloy after ST or CR with ageing at 373 K for 100 hours or ECAP without subsequent ageing and ECAP with ageing at 373 K for 100 hours. Reproduced with permission from Ref. [68].

It is worth noting also that in UFG metals processed by ECAP, both strength and ductility can be improved by performing mechanical tests at lower temperatures. As an example, Fig. 2.14 displays the tensile engineering stress–strain curves of UFG Ti with a grain size of 260 nm tested at room temperature and 77 K [82]. At room temperature, the Ti has some ductility and a small uniform elongation, as shown by curve A obtained at a strain-rate of 1×10^{-3} s^{-1}. However, at 77 K the strength of the material is drastically elevated to ~1.4 GPa. There is also a simultaneous increase in the elongation to failure, and this increases with the strain-rate up to a maximum close to ~20%, as shown in Fig. 2.14, where curves B–D are for strain rates of 1×10^{-3} s^{-1}, 1×10^{-2} s^{-1}, and 1×10^{-1} s^{-1}, respectively. These results for strength and ductility are better than, or at least comparable to, those of Ti alloys with a large percentage of alloying elements. Here, pronounced necking is delayed even for this very strong metal, resulting in a large area under the stress–strain curve and a generally tough behavior of the material. For comparison, curve E shows the initial 18% of strain for a conventional coarse-grained Ti sample tested at 77 K.

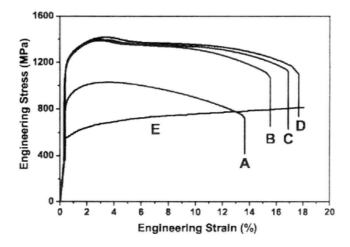

Figure 2.14 Engineering stress–strain curves for nanostructured Ti where curve A is for testing at room temperature at a strain-rate of 1×10^{-3} s^{-1}and curves B–D for the same Ti tested at 77 K for strain rates of 1×10^{-3} s^{-1}, 1×10^{-2} s^{-1}, and 1×10^{-1} s^{-1}, respectively; for comparison, curve E shows the behavior of coarse-grained Ti over the initial 18% of strain when testing at 77 K. Reproduced with permission from Ref. 81.

It is well known that UFG Cu with high ductility was found to have a higher strain-rate sensitivity, m, where m is defined as $\{d \ln \sigma / d \ln \dot{\varepsilon}\}$ where σ is the applied stress and $\dot{\varepsilon}$ is the strain-rate [6]. The value of m was equal to ~0.14 for ECAP-processed Cu taken through 16 passes compared with a value of ~0.06 for ECAP-processed Cu taken through only 2 passes. A high value for the strain-rate sensitivity indicates viscous flow and renders the material more resistant to necking and therefore more ductile. Increased values for the strain-rate sensitivities were also revealed in a number of other studies [74, 83]. At the same time, there are also some reports illustrating low values of *m* after ECAP. It is possible these apparent differences are due to microstructural features in the samples since the microstructures produced by SPD processing may differ significantly depending upon the processing conditions.

In conclusion, recent results show that grain refinement by ECAP can lead to a unique combination of strength and ductility in metallic materials. Such superior mechanical properties are highly desirable in the development of advanced structural materials for the next generation [66]. However, the achievement of these properties is associated with the tailoring of specific microstructures, which, in turn, are determined by the precise processing regimes and the nature of any further treatments. In general, any microstructural refinement by SPD will lead usually to a hardening of the material, but the formation of specific UFG structures appears to be a necessary condition for obtaining advanced properties and the development of new and viable structural materials.

2.5 FATIGUE BEHAVIOR

Fatigue is associated with the processes of damage accumulation and the resulting fracture of materials under cyclic loading at stress levels below the tensile strength. The total fatigue life has been conventionally divided into two regions corresponding to the times required for crack nucleation and crack propagation [84, 85]. The resistance to crack initiation naturally requires strength, while the tolerance to crack advance requires ductility. The most promising feature of SPD materials, which suggests the possibility of obtaining significantly enhanced fatigue properties, is associated with a combination of high strength and good ductility in the nanostructured state [66, 68]. The low-cycle fatigue (LCF) and high-cycle fatigue (HCF) regimes are conventionally distinguished in accordance with the applied stain amplitude. Testing in the HCF regime corresponds to probing the resistance of a material to crack initiation, whereas testing in the LCF regime corresponds to assessing the defect tolerance of a material.

Experimental results concerning the cyclic behavior are currently available for several UFG SPD materials in the form of both pure metals and metal alloys, and the data show that the ultimate tensile strength and the fatigue limit follow the standard Hall–Petch relationship [86]. Hence, it is reasonably anticipated that UFG materials produced by SPD should generally demonstrate a great potential for enhancement of the HCF life.

It has been established experimentally for a number of materials that ultrafine structures usually exhibit higher fatigue resistance than materials with a conventional grain size under stress-controlled loading. For example, the fatigue life and endurance limit for constant stress amplitude cycling of Ni and an Al–Mg alloy increases when the microstructure changes from microcrystalline to UFG and nanocrystalline [87]. In the case of Cu, which is probably the most closely studied material, an improvement in the fatigue lifetime was observed in UFG Cu when compared with Cu with a conventional grain size [88–92].

A very pronounced improvement of the fatigue limit was observed in a planar slip material, ECAP-processed Ti produced at an elevated temperature of 400°C, subjected to further strengthening by CR to an area reduction of 75%, and further annealing for structural stabilization at 300°C for 1 to 2 hours [93]. The fatigue limit of 500 MPa in pure SPD-processed Ti is close to that of conventional Ti alloys, as shown in Fig. 2.15 [93]. However, the results of strain-controlled tests expressed on the basis of the Coffin–Manson plot show nearly the same or even a somewhat shorter lifetime for UFG Cu than for its counterpart with a conventional grain size [88, 91, 92, 94]. The effect of lifetime shortening is more pronounced at the higher plastic strain amplitudes. The explanation of this behavior lies in the often-reported low thermal and mechanical stability and a strong tendency to recover the highly deformed UFG structure.

Cyclic softening, grain growth, and strain localization appear to be the main mechanisms responsible for the lower fatigue resistance of UFG structures under the same plastic strain amplitude [88]. Cyclic softening caused by total strain-controlled cyclic loading was observed in early

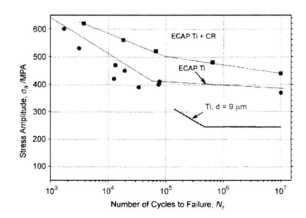

Figure 2.15 Characteristic S–N plots and fatigue limits for Ti processed by ECAP and Ti processed by ECAP and CR: data is also shown for conventional Ti with a grain size of 9 μm. Reproduced with permission from Ref. 92.

studies of UFG Cu [94]. It was concluded that the softening is due to a general decrease in the defect density and perhaps changes in the boundary misorientations. At low strain amplitudes, the softening is less pronounced. The cyclic hardening/softening curve was found to be nearly flat during most of the fatigue life, and no softening was observed under a plastic strain amplitude, ε_{ap}, below 10^{-3} [95]. The softening behavior depends, in practice, upon both the loading and the microstructural parameters. A microstructure characterized by nearly equiaxed subgrains (termed Type A) exhibits a nearly stable cyclic behavior when cycled at $\varepsilon_{ap} = 10^{-3}$, whereas a microstructure consisting of mainly elongated subgrains or lamellar subgrains (termed type B) undergoes marked softening under the same loading conditions [88, 91].

Grain growth and changes of microstructure due to fatigue are frequently observed in UFG Cu, and there is evidence also for a significant increase of cell size [94]. Areas with recrystallized grains develop in UFG structures of type B, whereas recrystallization and grain growth are not evident in structures of Type A [96]. There is evidence for a pronounced recrystallization in UFG Cu of high purity after constant plastic strain-controlled loading with a plastic strain amplitude of the order of 10^{-4} [97]. In recent work it was assumed that no characteristic dislocation structure can develop in ultrafine grains [92], whereas a specific dislocation structure develops in coarse grains. At higher strain amplitudes, a well-defined cell and subgrain structure is observed in UFG Cu.

Cyclic strain localization, resulting in fatigue crack initiation, is an important feature of the fatigue process in UFG materials. The development of macroscopic SBs is considered the major form of fatigue damage in wavy-slip UFG materials processed by ECAP. Thus, SBs oriented at 45° to the loading axis were observed on the surface of cyclically loaded Cu specimens [94, 95, 98–100], where these bands are macroscopically parallel to the shear plane of the last extrusion pass. These SBs have been shown to be "persistent" in the sense that they appear on the surface after repolishing [91]. This persistent nature suggests that they are sites of easier and more intensive cyclic deformation. Observations after LCF also reveal extrusion and intrusions, which are similar to the features known from fatigue studies on Cu single crystals [94]. For UFG materials, the average dimensions of the extrusions, in terms of both their lengths along the slip lines and their elevations above the surrounding surface, are larger than the very small grain size. Accordingly, the details for the formation of these SBs are not entirely clear. It has been proposed that the mechanism responsible for the formation of SBs is an interaction between cyclically induced grain coarsening and grain rotation [101]. In this mechanism, local grain coarsening takes place first and shear localization appears later. It was also suggested that recrystallization occurs preferentially in some regions of microstructural inhomogeneity in the form of banded structures [98]. The occurrence of grain growth triggered by cyclic deformation is believed to be a very common feature of cyclically deformed UFG metals. On the other hand, no detectable grain coarsening was observed in the vicinity of SBs, which suggests there is no relationship between SBs and grain coarsening [100].

There has been some discussion of the effect of ECAP on fatigue and the procedure for the optimization of fatigue performance [86]. It was noted that pressing through an increasing number of passes in ECAP leads to an increasing monotonic strength and fatigue limit [102–104]. Furthermore, the fatigue properties of UFG metals can be improved by gaining some ductility and by reducing the constraints for dislocation motion as, for example, by decreasing the tendency for shear banding and strain localization, which is common in many hardened metals. Thus, it may be advantageous in improving the fatigue properties to employ materials with partially recovered structures. The positive effect of the heat treatment on LCF was revealed in early fatigue studies of ECAP-processed materials [105]. It was shown, using the acoustic emission technique and microscopic surface observations, that susceptibility to shear banding in ECAP-processed Cu decreases dramatically after short-term annealing of only 10 minutes at the relatively low temperature of 250°C [106], and the LCF life can be improved by a factor of ~5–10 after heat treatment that gives no significant grain growth [107, 108]. While processing by ECAP leads to a considerable reduction in the tensile and cyclic ductility, the same materials subjected to post-ECAP annealing can potentially obtain a higher ductility than their conventional coarse-grained counterparts, with shifts in the Coffin–Manson line toward higher fatigue lives [100]. Since metals processed by SPD retain some ductility after fabrication, their tensile and high cyclic strengths can be further improved after processing through the use of conventional CR with or without intermediate annealing at a moderate temperature. This has been shown for several Al–Mg alloys [109] and commercial purity Ti [108, 110].

The effect of precipitation in SPD-processed nanocrystalline metals is complex. On the one hand, it was noted earlier that precipitates can dramatically increase the thermal stability of SPD-processed metals, while, on the other hand, the grain boundaries may recover during ageing, thereby reducing their susceptibility to strain localization and premature cracking. As an example, it was shown that optimal ageing of an ECAP-processed UFG Cu–Cr–Zr alloy gives a structure having high strength with a grain size of ~200 nm, which remains fine after subsequent annealing at temperatures as high as 500°C [103]. It was shown that ECAP of the solid ST Al-2024 alloy through one pass, followed by low-temperature ageing, can impressively enhance both the strength and the ductility. Thus, samples aged at 100°C for 20 hours exhibited an ultimate tensile strength of $\sigma_{UTS} \approx 715$ MPa and a total elongation to failure of $\delta = 16\%$ [111]. It was shown also that the yield stress and tensile strength of the Al-6061 alloy benefits from multiple ECAP up to four passes by comparison with an ST sample pressed by ECAP through only a single pass at 125°C with no subsequent heat treatment [112]. Although the effect of the single pass of ECAP on the 6061 Al alloy is impressive, it is not clear whether it would be possible to achieve the same strength and ductility in Al alloys after conventional treatments.

Hence it appears there are several principal competing approaches for the enhancement of fatigue properties via ECAP processing [86, 113]. The first is the achievement of a compromise between strength and ductility in a minimum number of passes of ECAP, where only a single pass is used, whenever possible, since this is a cost-effective procedure employing relatively small imposed strains. The second is the achievement of the maximum possible strength, leading to an HCF life. The third is the achievement of both high strength and high ductility through multipass ECAP, leading to enhanced LCF and HCF lives.

It is reasonable to conclude that there is a good body of results available to date on the effects of ECAP on the subsequent fatigue properties, but nevertheless there remain significant opportunities for developing optimum processing schemes for attaining the desired fatigue properties of SPD-produced materials.

2.6 CONCLUDING REMARKS

The results summarized in this brief review demonstrate the considerable potential that is made available through the processing of UFG materials having submicrometer or nanometer grain sizes. These bulk solids are produced using the imposition of SPD and the materials can be manipulated to achieve excellent strength and good ductility at ambient temperature. There is also an opportunity to attain exceptional superplastic properties when testing at elevated temperatures. Early experiments provided a direct confirmation of the potential for achieving high strain-rate superplasticity in commercial Al alloys when testing in tension after ECAP [114], and it was shown also that a high superplastic-forming capability is available for producing complex parts in the as-processed alloys [115]. The results to date are very encouraging because of the need to make use of this technology for rapid-forming operations in industrial applications. An additional development is the recent demonstration that exceptional superplastic flow may be achieved in Mg alloys after processing by ECAP with elongations to failure up to more than 3,000% [116]. These and other results confirm that processing by SPD leads to enhanced properties that cannot be achieved by more conventional processing. They also demonstrate that development of new and superior properties of metals and alloys through their nanostructuring by SPD techniques becomes one of the most topical trends in modern materials science.

References

1. H. Gleiter (1981), in eds., N. Hansen, A. Horswell, T. Leffers, and H. Lilholt, *Proceedings of the 2nd Riøo International Symposium on Metallurgy and Materials Science*, Risø National Laboratory, Roskilde, Denmark, p. 15.

2. H. Gleiter (1989), *Prog. Mater. Sci.*, **33**, 223.

3. J. R. Weertman (1993), *Mater. Sci. Eng.*, **A 166**, 161.

4. C. C. Koch (2003), *Scripta Mater.*, **49**, 657.

5. D. G. Morris (1998), *Mechanical Behaviour of Nanostructured Materials*, Trans Tech, Uetikon-Zürich.

6. R. Z. Valiev, I. V. Alexandrov, Y. T. Zhu, and T. C. Lowe (2002), *J. Mater. Res.*, **17**, 5.

7. R. Valiev (2002), *Nature*, **419**, 887.

8. Y. Wang, M. Chen, F. Zhou, and E. Ma (2002), *Nature*, **419**, 912.

9. Y. M. Wang and E. Ma (2004), *Acta Mater.*, **52**, 1699.

10. R. Z. Valiev, R. K. Islamgaliev, and I. V. Alexandrov (2000), *Prog.Mater. Sci.*, **45**, 103.

11. Z. Horita, ed. (2006), *Nanomaterials by Severe Plastic Deformation*, Trans Tech Publications, Switzerland.

12. M. J. Zehetbauer and Y. T. Zhu, eds., (2009), *Bulk Nanostructured Materials*, WILEY-VCH Verlag GmbH & Co. KGaA, Weinheim.

13. R. Z. Valiev, A. V. Korznikov, and R. R. Mulyukov (1993), *Mater. Sci. Eng.*, **A 186**, 141.

14. R. Z. Valiev, N. A. Krasilnikov, and N. K. Tsenev (1991), *Mater. Sci. Eng.*, **A 137**, 35.

15. A. P. Zhilyaev and T. G. Langdon (2008), *Prog. Mater. Sci.*, **53**, 893.

16. R. Z. Valiev and T. G. Langdon (2006), *Prog. Mater. Sci.*, **51**, 881.

17. A. P. Zhilyaev, S. Lee, G. V. Nurislamova, R. Z. Valiev, and T. G. Langdon (2001), *Scripta Mater.*, **44**, 2753.

18. K. Nakashima, Z. Horita, M. Nemoto, and T. G. Langdon (2000), *Mater. Sci. Eng.*, **A 281**, 82.

19. Y. Nishida, H. Arima, J.-C. Kim, and T. Ando (2001), *Scripta Mater.*, **45**, 261.

20. A. Azushima and K. Aoki (2002), *Mater. Sci. Eng.*, **A 337**, 45.

21. Z. Horita, T. Fujinami, and T. G. Langdon (2001), *Mater. Sci. Eng.*, **A 318**, 34.

22. Y. T. Zhu, T. C. Lowe, and T. G. Langdon (2004), *Scripta Mater.*, **51**, 825.

23. M. Kamachi, M. Furukawa, Z. Horita, and T. G. Langdon (2003), *Mater. Sci. Eng.*, **A 361**, 258.

24. Y. Saito, H. Utsunomiya, H. Suzuki, and T. Sakai (2000), *Scripta Mater.*, **42**, 1139.

25. J.-C. Lee, H.-K. Seok, and J.-Y. Suh (2002), *Acta Mater.*, **50**. 4005.

26. J.-H. Han, H.-K. Seok, Y.-H. Chung, M.-C. Shin, and J.-C. Lee (2002), *Mater. Sci. Eng.*, **A 323**, 342.

27. J.-C. Lee, H.-K. Seok, J.-Y. Suh, J.-H. Han, and Y.-H. Chung (2002), *Metall. Mater. Trans.*, **33A**, 665.

28. G. J. Raab, R. Z. Valiev, T. C. Lowe, and Y. T. Zhu (2004), *Mater. Sci. Eng.*, **A 382**, 30.

29. P. B. Berbon, M. Furukawa, Z. Horita, M. Nemoo, and T. G. Langdon (1999), *Metall. Mater. Trans.*, **30A**, 1989.

30. K. Nakashima, Z. Horita, M. Nemoto, and T. G. Langdon (1998), *Acta Mater.*, **46**, 1589.

31. Y. Iwahashi, J. Wang, Z. Horita, M. Nemoto, and T. G. Langdon (1996), *Scripta Mater.*, **35**, 143.

32. M. Furukawa, Y. Iwahashi, Z. Horita, M. Nemoto, and T. G. Langdon (1998), *Mater. Sci. Eng.*, **A 257**, 328.

33. M. Furukawa, Z. Horita, M. Nemoto, and T. G. Langdon (2002), *Mater. Sci. Eng.*, **A 324**, 82.

34. Y. Iwahashi, M. Furukawa, Z. Horita, M. Nemoto, and T. G. Langdon (1998), *Metall. Mater. Trans.*, **29A,** 2245.

35. Y. Iwahashi, Z. Horita, M. Nemoto, and T. G. Langdon (1997), *Acta Mater.*, **45**, 4733.

36. Y. Iwahashi, Z. Horita, M. Nemoto, and T. G. Langdon (1998), *Acta Mater.*, **46,** 3317.

37. K. Oh-ishi, Z. Horita, M. Furukawa, M. Nemoto, and T. G. Langdon (1998), *Metall. Mater. Trans.*, **29A**, 2011.

38. Y. Fukuda, K. Oh-ishi, M. Furukawa, Z. Horita, and T. G. Langdon (2006), *Mater. Sci. Eng.*, **A 420**, 79.

39. S. Komura, Z. Horita, M. Nemoto, and T. G. Langdon (1999), *J. Mater. Res.*, **14**, 4044.

40. S. D. Terhune, D. L. Swisher, K. Oh-ishi, Z. Horita, T. G. Langdon, and T. R. McNelley (2002), *Metall. Mater. Trans.*, **33A**, 2173.

41. R. Z. Valiev, Y. V. Ivanisenko, E. F. Rauch, and B. Baudelet (1996), *Acta Mater.*, **44**, 4705.

42. C. P. Chang, P. L. Sun, and P. W. Kao (2000), *Acta Mater.*, **48**, 3377.

43. J.-Y. Chang, J.-S. Yoon, and G.-H. Kim (2001), *Scripta Mater.*, **45**, 347.

44. Z. C. Wang and P. B. Prangnell (2002), *Mater. Sci. Eng.*, **A 328**, 87.

45. A. Goloborodko, O. Sitdikov, T. Sakai, R. Kaibyshev, and H. Miura (2003), *Mater. Trans.*, **44**, 766.

46. M. J. Starink, N. Gao, M. Furukawa, Z. Horita, C. Xu, and T. G. Langdon (2004), *Rev. Adv. Mater. Sci.*, **7**, 1.

47. R. Kaibyshev, K. Shipilova, F. Musin, and Y. Motohashi (2005), *Mater. Sci. Eng.*, **A 396,** 341.

48. C. Xu, M. Furukawa, Z. Horita, and T. G. Langdon (2005), *Mater. Sci. Eng.*, **A 398,** 66.

49. T. G. Langdon (2007), *Mater. Sci. Eng.*, **A462**, 3.

50. D. Kuhlmann-Wilsdorf (1989), *Mater. Sci. Eng.*, **A 113**, 1.

51. D. Kuhlmann-Wilsdorf (1997), *Scripta Mater.*, **36**, 173.

52. A. Gholinia, P. B. Prangnell, and M. V. Marushev (2000), *Acta Mater.*, **48**, 1115.

53. M. Furukawa, Z. Horita, and T. G. Langdon (2002), *Mater. Sci. Eng.*, **A 332,** 97.

54. R.Z. Valiev (2009), *Int. J.Mat. Res.*, **100**, 6.

55. R. Z. Valiev (2009), *Int. J. Mat. Res.*, **100,** 6.

56. S. X. McFadden, R. S. Mishra, R. Z. Valiev, A. P. Zhilyaev, and A. K. Mukherjee (1999), *Nature*, **398**, 684.

57. T. G. Langdon, M. Furukawa, M. Nemoto, and Z. Horita (2000), *JOM*, **52**(4), 30.

58. M. Kawasaki and T. G. Langdon (2007), *J. Mater. Sci.*, **42**, 1782.

59. T. C. Lowe and R. Z. Valiev (2000), *Investigations and Applications of Severe Plastic Deformation*, Kluwer, Dordrecht, The Netherlands.

60. Y. T. Zhu, T. G. Langdon, R. Z. Valiev, S. L. Semiatin, D. H. Shin, and T. C. Lowe, eds. (2004), *Ultrafine-Grained Materials III*, The Minerals, Metals and Materials Society, Warrendale, PA.

61. M. J. Zehetbauer and R. Z. Valiev, eds. (2004), *Nanomaterials by Severe Plastic Deformation*, Wiley-VCH Verlag, Weinheim, Germany.

62. Y. T. Zhu and V. Varyukhin, eds. (2006), *Nanostructured Materials by High-Pressure Severe Plastic Deformation*, Springer, Dordrecht, The Netherlands.

63. Y. T. Zhu, T. G. Langdon, Z. Horita, M. J. Zehetbauer, S. L. Semiatin, T. C. Lowe, eds. (2006), *Ultrafine Grained Materials IV*, The Minerals, Metals and Materials Society, Warrendale, PA.

64. D. G. Morris (1998), *Mechanical Behaviour of Nanostructured Materials*, Trans Tech, Uetikon-Zürich, Switzerland.

65. C. C. Koch (2003), *Scripta Mater.*, **49**, 657.

66. R. Z. Valiev (2004), *Nature Mater.*, **3**, 511.

67. Z. Horita, T. Fujinami, M. Nemoto, and T. G. Langdon (2000), *Metall. Mater. Trans.*, **31A**, 691.

68. Z. Horita, K. Ohashi, T. Fujita, K. Kaneko, and T. G. Langdon (2005), *Adv. Mater.,* **17**, 1599.

69. Y. Zhao, T. Topping, J. F. Bingert, J. J. Thornton, A. M. Dangelewicz, Y. Li, Y. Zhu, Y. Zhou, and E. J. Lavernia (2008), *Adv.Mater.,* **20**, 3028.

70. R. Z. Valiev, A. V. Sergueeva, and A. K. Mukherjee (2003), *Scripta. Mater.,* **49**, 669.

71. E. R. Parker (1967), *Materials Data Book for Engineers and Scientists*, McGraw-Hill, New York, NY.

72. E. A. Brandes, G. B. Brook (1992), *Smithells Metals Reference Book*, 7th ed., Butterworth-Heinemann, Oxford, UK, Ch. 22.

73. H. W. Höppel, J. May, P. Eisenlohr, and M. Z. Göken (2005), *Metallkd* **96**, 566.

74. J. May, H. W. Höppel, and M. Göken (2005), *Scripta. Mater.,* **53**, 189.

75. F. Dalla Torre, R. Lapovok, J. Sandlin, P. F. Thomson, C. H.J. Davies, and E. V. Pereloma (2004), *Acta. Mater.,* **52**, 4819.

76. N. Krasilnikov, W. Lojkowski, Z. Pakiela, R. Valiev (2005), *Mater Sci Eng.,* **A397**, 330.

77. R. Z. Valiev (2008), *Mater. Sci. Forum,* **584–586**, 22.

78. X. Zhang, H. Wang, R. O. Scattergood, J. Narayan, C. C. Koch, A. V. Sergueeva, and A. K. Mukherjee (2002), *Acta Mater.,* **50**, 4823.

79. H. Mughrabi, H. W. Höppel, M. Kautz, and R. Z. Valiev (2003), *Z. Metallkd* **94**, 1079.

80. Y. S. Park, K. H. Chung, N. J. Kim, and E. J. Lavernia (2004), *Mater. Sci. Eng.,* **A374**, 211.

81. R. K. Islamgaliev, N. F. Yunusova, I. N. Sabirov, A. V. Sergueeva, and R. Z. Valiev (2001), *Mater. Sci. Eng.,* **A319–A321**, 877.

82. Y. Wang, E. Ma, R. Z.Valiev, and Y. Zhu (2004), *Adv. Mater.,* **16**, 328.

83. Li YJ, X. H. Zeng, and W. Blum (2004), *Acta Mater.,* **52**, 5009.

84. S. Suresh (1991), *Fatigue of Materials*, Cambridge University Press, Cambridge, UK.

85. J. Polák (1991), *Cyclic Plasticity and Low Cycle Fatigue Life of Metals*, Elsevier, Amsterdam, The Netherlands.

86. A. Yu. Vinogradov and S. R. Agnew (2004), in *Dekker Encyclopedia of Nanoscience and Nanotechnology*, Marcel Dekker, NY, p. 2269.

87. T. Hanlon, Y. N. Kwon, and S. Suresh (2003), *Scripta Mater* **49**, 675.

88. S. R. Agnew, A. Yu. Vinogradov, S. Hashimoto, and J. R. Weertman (1999), *J. Electron. Mater.,* **28**, 1038.

89. H. Mughrabi (2000), in eds., T. C. Lowe, and R. Z. Valiev, *Investigations and Applications of Severe Plastic Deformation*, Kluwer Academic Publishers, Dordrecht, The Netherlands, p. 241.

90. H. W. Höppel, M. Brunnbauer, H. Mughrabi, R. Z. Valiev, and A. P. Zhilyaev (2001), in *Materials Week 2000*, Available from: http//www.materialsweek.org/proceedings, Munich, Germany.

91. A. Vinogadov and S. Hashimoto (2001), *Mater. Trans.* **42**, 74.

92. H. Mughrabi, H. W. Höppel, and M. Kautz (2004), *Scripta Mater.,* **51,** 807.

93. A. Yu Vinogradov, V. V. Stolyarov, S. Hashimoto, and R. Z. Valiev (2001), *Mater. Sci. Eng.,* **A318,** 63.

94. S. R. Agnew and J. R. Weertman (1998), *Mater. Sci. Eng.,* **A244**, 145.

95. A. Vinogradov and S. Hashimoto (2004), in eds., M. J. Zehetbauer and R. Z. Valiev, *Nanomaterials by Severe Plastic Deformation*, Wiley-VCH Verlag, Weinheim, Germany, p. 663.

96. S. Hashimoto, Y. Kaneko, K. Kitagawa, A. Vinogradov, and R. Z. Valiev (1999), *Mater. Sci. Forum.,* **312–314**, 593.

97. S. D. Wu, Z. G. Wang, C. B. Jiang, G. Y. Li (2002), *Phil. Mag. Lett.,* **82**, 559.

98. S. D. Wu, Z. G. Wang, C. B. Jiang, G. Y. Li, I. V. Alexandrov, and R. Z. Valiev (2004), *Mater. Sci. Eng.,* **A387–A389**, 560.

99. A. Vinogradov, V. Patlan, S. Hashimoto, and K. Kitagawa (2002), *Phil. Mag.,* **A82**, 317.

100. S. D. Wu, Z. G. Wang, C. B. Jiang, G. Y. Li, I. V. Alexandrov, and R. Z. Valiev (2003), *Scripta Mater.,* **48,** 1605.

101. H. W. Höppel, C. Xu, M. Kautz, N. Barta-Schreiber, T. G. Langdon, and H. Mughrabi (2004), in eds., M. J. Zehetbauer and R. Z. Valiev, *Nanomaterials by Severe Plastic Deformation*, Wiley-VCH Verlag, Weinheim, Germany, p. 677.

102. V. V. Stolyarov, I. V. Alexandrov, Yu. R. Kolobov, M. Zhu, Y. Zhu, and T. Lowe (1999), in eds., X. R. Wu and Z. G. Wang, *Fatigue'99–Proceedings of the 7th International Fatigue Congress*, Vol. 3, Higher Education Press, Beijing, P. R. China, p. 1345.

103. A. Vinogradov, Y. Suzuk, V. I. Kopylov, V. Patlan, and K. Kitagawa (2002), *Acta Metall.,* **50,** 1636.

104. A. Vinogradov, V. Kopylov, and S. Hashimoto (2003), *Mater. Sci. Eng.,* **A355,** 277.

105. A. Vinogradov , Y. Kaneko, K. Kitagawa, S. Hashimoto, V. Stolyarov, R. Valiev (1997), *Scripta Mater.,* **36**, 1345.

106. A. Vinogradov (1998), *Scripta Mater.,* **38**, 797.

107. A. Vinogradov, V. Patlan, K. Kitagawa (1998), *Mater. Sci. Forum.,* **312–314**, 607.

108. H. W. Höppel, Z. M. Zhou, H. Mughrabi, and R. Z. Valiev (2002), *Phil. Mag.,* **A82**, 1781.

109. A. Vinogradov, V. Patlan, K. Kitagawa, and M. Kawazoe (1999), *Nanostruct. Mater.,* **11**, 925.

110. Z. G. Wang, S. D. Wu, C. B. Jiang, S. M. Liu, and I. V. Alexandrov (2002), in ed., A. F. Blom, *Fatigue 2002–Proceedings of the 8th International Fatigue Congress*, EMAS, Warrington, UK, p. 1541.

111. W. J. Kim, C. S. Chung, D. S. Ma, S. I. Hong, and H. K. Kim (2003), *Scripta Mater.,* **49,** 333.

112. C. S. Chung, J. K. Kim, H. K. Kim, and W. J. Kim (2002), *Mater. Sci. Eng.,* **A337**, 39.

113. M. Kautz, H. W. Höppel, C. Xu, M. Murashkin, T. G. Langdon, R. Z. Valiev, and H. Mughrabi (2006), *Intl. J. Fatigue.,* **28,** 1001.

114. R. Z. Valiev, D. A. Salimonenko, N. K. Tsenev, P. B. Berbon, and T. G. Langdon (1997), *Scripta Mater.,* **37**, 1945.

115. Z. Horita, M. Furukawa, M. Nemoto, A. J. Barnes, and T. G. Langdon (2000), *Acta. Mater.,* **48**, 3633.

116. R. B. Figueiredo and T. G. Langdon (2008), *Adv. Eng. Mater.,* **10**, 37.

Chapter 3

STRENGTH AND STRAIN-RATE SENSITIVITY OF NANOPOLYCRYSTALS

Ronald W. Armstrong

Center for Energetic Concepts Development, Department of Mechanical Engineering
University of Maryland, College Park, MD 20742, USA
E-mail: rona@umd.edu

The strength and strain-rate sensitivity properties of nanopolycrystals are attributed, at effective low temperatures, to a combination of Taylor-type, dislocation-density-based deformation within the grain volumes and Hall–Petch (H–P)-characterized obstacle resistance of the grain boundaries, consistent with results on conventional-grain-size materials but now scaled downward even to single-dislocation loop behavior. Because of the generally substantial grain-boundary resistances associated with the yield point behaviors of body-centered cubic (BCC) metals, only their grain volume deformation is thermally activated. For face-centered cubic (FCC) metals, there is thermal activation both of controlling dislocation intersections within the grain volumes and, especially, of needed cross-slip at higher local stresses in the grain-boundary regions, thus providing an increasingly larger strain-rate sensitivity at smaller grain sizes. Certain hexagonal close-packed (HCP) metals follow a BCC-like behavior, but for others, the thermally activated grain-boundary resistances and enhanced nanopolycrystal strain-rate sensitivities are FCC-like and controlled by the local operation of prism or pyramidal slip systems. Under effective high temperature, creep-like conditions, an inverse H–P grain size dependence occurs and, then, the strain-rate sensitivity is predicted to be smaller at smaller grain size.

3.1 INTRODUCTION

The current importance attached to investigating the deformation properties of nanopolycrystal materials has substantiated the assertion that "there

Mechanical Properties of Nanocrystalline Materials
Edited by James C. M. Li
Copyright © 2011 Pan Stanford Publishing Pte. Ltd.
www.panstanford.com

are exciting reasons for investigating the strength properties of ultrafine grain metals both from an experimental and theoretical point of view."[1] Emphasis was given to extension of the experimental H–P observation that higher strengths were to be obtained for ultrafine-grain-size metals at effective lower temperatures and opposite (creep-based) lower strengths were to be obtained at effective higher temperatures. Thus, it was proposed that advantage could be taken of the lower material strength during high-temperature fabrication, and then additional advantage would be gained with a stronger material in the design application. From a theoretical viewpoint, emphasis was given to finer-grain-size material properties producing (1) easier distinctions between continuum and discrete dislocation model descriptions of strength properties and (2) greater differences between alternative dislocation mechanisms proposed for quantifying the properties.

Figure 3.1 provides an updated description[2] on a log/log basis of the strength properties achieved over a range from conventional to ultrafine grain sizes for iron and steel materials.

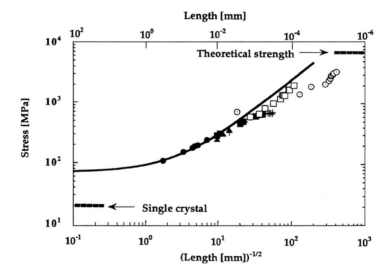

Figure 3.1 The yield stress of iron and steel materials as a function of the inverse square root of their effective grain size. Reproduced with permission from Ref. 2.

These material results are of special importance to begin with because Hall[3] and Petch[4] initially established an inverse square root of grain size dependence for the lower yield stress.

In the present chapter, the H–P-based dislocation pileup model description proposed to explain the Fig. 3.1 results, and other similar results obtained for HCP and FCC metals, is combined with a thermal activation–strain-rate analysis (TASRA) model description for interpreting the temperature and strain-rate dependencies of nanopolycrystal material deformations. Experimental results are described, in addition to those for iron and steel shown in Fig. 3.1, for copper, magnesium, molybdenum, niobium, zinc, cadmium, aluminum, and nickel materials.

3.2 THE HALL–PETCH RELATION

The Hall–Petch (H–P) relation was initially measured over a conventional range of grain sizes, say, between a largest grain diameter of ~1.0 mm and a smallest diameter of ~10 microns, expressed in direct linear plots of lower yield stress versus the inverse square root of the average grain diameter as

$$\sigma_y = \sigma_{0y} + k_y \ell^{-1/2} \tag{3.1}$$

In Eq. 3.1, σ_y is the lower yield stress, ℓ is the average grain diameter, and σ_{0y} and k_y are experimental constants.[3, 4] The log σ/log [length]$^{-1/2}$ representation of Fig. 3.1 is taken to be an appropriate method of plotting strength/grain size results, say, when a range in grain diameter of more than 100 times is covered. In this case, as can be seen in the figure, σ_y is essentially determined at larger grain diameters by the value of the constant friction stress, σ_{0y}, and at very small grain diameters, a straight line H–P dependence of slope 1.0 is obtained.

In Fig. 3.1, the relatively low "single crystal" reference stress of ~20 MPa was taken as two times the critical resolved shear stress reported for essentially pure iron crystals by Atshuler and Christian.[5] The upper-limiting "theoretical strength" was taken somewhat arbitrarily as $E/30$, in which E is Young's modulus for steel. The compiled experimental measurements cover conventional and nanopolycrystal results.[1, 6–10] The drawn H–P curve was taken from Ref. 6. Attention is directed particularly to both the open-square measurements made by Embury and Fisher,[7] for which the effective length on the abscissa scale was determined from the interlamellar spacings of eutectoid steel materials, and to the open-circle results of Jang and Koch,[10] for which the length was determined from the average grain diameters of ball-milled iron material. Both nanopolycrystal materials show at the finest grain sizes a somewhat reduced H–P k_y value. The same result will be shown here to apply for other materials also as will be connected even with an inverse H–P dependence observed in certain cases at grain diameters of length in the vicinity of ~10 nm and smaller.

3.3 THE THERMAL ACTIVATION—STRAIN-RATE ANALYSIS

The viscoplasticity behavior of polycrystal materials has a long-standing history leading to the "invention" of crystal dislocations and subsequent modeling of their generation and motion, as recently reviewed.[11] For our purpose, important references are to articles by Orowan,[12] Schoeck,[13] Li,[14] and Armstrong.[15] At conventional internal dislocation shear strain rates, say, of $\sim 10^{-4}$ to $\sim 10^{+4}$ s^{-1}, less than those involved in shock loading, there is control of the externally measured strain-rate by the average dislocation velocity, v, in the expression[12]

$$d\gamma/dt = \rho \mathbf{b} v \tag{3.2}$$

In Eq. 3.2, $d\gamma/dt$ is the resolved shear strain-rate determined by an average mobile dislocation density, ρ, with Burgers vector, $\mathbf{b}$, moving at an average velocity, v. The dislocation velocity, in turn, is taken to be thermally activated through overcoming a Gibbs free energy barrier, G_0, with assistance of a thermal component of stress, τ^*, in the expression[13]

$$d\gamma/dt = (\rho \mathbf{b} v_0)\exp\{-(G_0 - \smallint \mathbf{b}A^*d\tau^*)/k_B T\} \tag{3.3}$$

In Eq. 3.3, v_0 is the reference upper-limiting dislocation velocity, A^* is the dislocation activation area, k_B is Boltzmann's constant, and T is absolute temperature. The activation area is a critical parameter relating to the strain-rate sensitivity of thermally activated plasticity, as characterized also by the activation volume, $V^* = \mathbf{b}A^*$, in the relation

$$V^* = k_B T[\partial \mathbf{ln}(d\gamma/dt)/\partial \tau^*]_T \tag{3.4}$$

Li[14] has described thermodynamic aspects of the dislocation velocity dependence on shear stress and hydrostatic pressure for dynamic plasticity and high-temperature creep conditions, respectively. Of interest here, as will be described, is the connection with an alternative power–law (P–L) relationship involving the velocity stress exponent, m^*, in

$$v = B(\tau^*)^{m^*} \tag{3.5}$$

with B being constant, which relation is a suitable condensation of v in Eq. 3.3 when the shear-based activation area, A^*, follows an inverse proportionality to τ^*. Figure 3.2 shows a compilation of such inversely proportional measurements[16-22] in terms of $V^* = \mathbf{b}A^*$ as reported in the TASRA description given by Armstrong.[15] In agreement with Eq. 3.5, Prekel and Conrad[21] determined a linear log/log dependence of v on τ^* over more than three orders of magnitude of v. Zerilli and Armstrong[23] employed the inverse dependence of V^* on τ^* in development of the so-called Z–A constitutive equations employed for material dynamics calculations.[24] It's important to note[22] that it's τ^*, the thermal shear stress component, and not τ, the total shear stress, in Eq. 3.5. Also, as will be made use of in the present report, the open-square measurements for magnesium in Fig. 3.2 apply for basal (0001)[11–20] slip,[18] whereas the closed-square measurements are for prism (10–10)[11–20] slip.[19]

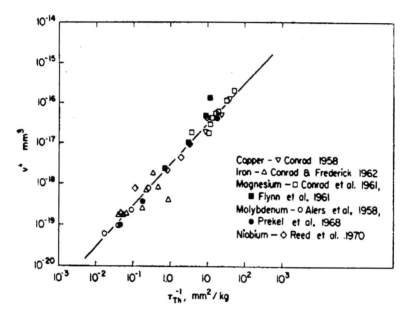

Figure 3.2 Thermal activation volume, V^*, measurements compiled for a number of metals as a function of τ_{Th}^{-1} ($= \tau^{*-1}$ in the text), as reported by Armstrong[15]; 1 kg/mm^2 = 9.81 MPa.

3.4 THE HALL–PETCH-BASED DISLOCATION PILEUP MODEL

Further application of the dislocation pileup analysis to explain H–P dependence of the yield stress of BCC, FCC, and HCP metals with or without a yield point and of the flow stress at constant values of strain, σ_ε, led to the model relationship[6]

$$\sigma_\varepsilon = m\tau_{0\varepsilon} + mk_{s\varepsilon}\ell^{-1/2} \tag{3.6}$$

In Eq. 3.6, m is an orientation factor of the type originally described by Taylor for grain volume deformations within a polycrystal,[25] $\tau_{0\varepsilon}$ is the critical resolved shear stress for multislip within the grains, for example, as described by Kocks,[26] and $k_{s\varepsilon}$ is the microstructural shear stress intensity characterizing the average grain-boundary resistance to plasticity spreading between the grains.[8] In relation to Eq. 3.1, now $m\tau_{0\varepsilon} = \sigma_{0\varepsilon}$ and $mk_{s\varepsilon} = k_{\varepsilon}$. The determination of $\sigma_{0\varepsilon}$ and k_{ε} values over the full stress–strain behavior of mild steel materials was described by Armstrong.[27]

The potential temperature and strain-rate dependencies of σ_{ε} that are implicitly contained in the modeled H–P relation are in $\sigma_{0\varepsilon}$ and k_{ε}. For $\sigma_{0\varepsilon}$,

$$\sigma_{0\varepsilon} = m[\tau_{o\varepsilon G} + \tau_{0\varepsilon}{}^{*}] \tag{3.7}$$

in which $\tau_{0\varepsilon G}$ and $\tau_{0\varepsilon}{}^{*}$ are the athermal and thermal components of the grain volume–resolved shear stress, respectively, in the manner described by Seeger.[28] The result obtained for k_{ε} from the dislocation pileup model is[8]

$$k_{\varepsilon} = m[\pi m_{s} G \mathbf{b} \tau_{c}/2\alpha]^{1/2} \tag{3.8}$$

in which m_{s} is a Sachs orientation factor, G is the shear modulus, τ_{c} is the local resolved shear stress needed in the grain-boundary region, and α is the average orientation factor between edge and screw characterizations of the dislocation stress. In Eq. 3.8, τ_{c} has its own athermal and thermal stress components, also in the same manner as given for Eq. 3.7. Equations 3.7 and 3.8 build onto the original Taylor idea[25] of relating single-crystal and polycrystal flow stresses, not, however, in terms of a single factor m making the connection, but through consideration of the coupled stress requirements within the grain volumes and at the grain boundaries.[29]

3.4.1 Application to HCP Polycrystal Deformations

Figure 3.3 shows for magnesium the match both of the temperature dependence of σ_{0} for an initial proof stress at yield and that of τ_{0} for the basal slip system,[18] as well as of k_{ε}^{2} with that of τ_{c} for the prism slip system.[19, 30] The H–P constants were determined from the measurements reported by Hauser, Landon, and Dorn,[31] who had concluded from metallographic observations that prism slip appeared to be a main factor in controlling the local accommodation of strains in the grain-boundary regions.[32] Thus in Fig. 3.3, at 300 K, a basal slip stress of ~0.3 MPa, or greater, operates on average on dislocation pileups within the grain volumes to produce a local shear stress of ~40 MPa in the grain-boundary regions so as to activate the prism slip system.

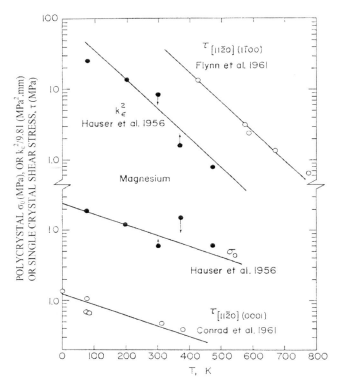

Figure 3.3 Match for polycrystal and single-crystal magnesium materials of the temperature dependencies for H–P parameters σ_0 and k_ε^2 with basal slip system and pyramidal slip system shear stresses, respectively, as adapted from Armstrong.[30]

Prasad, Madhava, and Armstrong[33] showed for compressive stress measurements made on polycrystal zinc that the temperature dependence of $\sigma_{0\varepsilon}$ followed that of τ_0 for the basal (0001)[11–20] slip system and k_ε^2 followed the τ_0 (= τ_c) for the pyramidal (11–23}[11–22] slip system, utilizing for the two cases the single crystal shear stress measurements reported by Wielke[34] and by Soifer and Shteinberg[35], respectively. In this case, somewhat lower $\sigma_{0\varepsilon}$ measurements were obtained at temperatures of 100 K and lower because of the incidence of pileup induced cleavage cracking that was confined within isolated grains. For zinc, the basal and pyramidal shear stresses are quite comparable at 300 K but that for pyramidal slip increases much more strongly as the temperature decreases. Thus, a strong increase in k_ε develops until full cleavage cracking intervenes, and a well-established H–P relation is then observed for the tensile cleavage fracture stress.[36] Mannan and Rodriguez[37] established a similar k_ε^2 dependence matching τ_0 for pyramidal slip as a function of temperature for the H–P behavior of cadmium material.

3.4.2 Application to FCC Polycrystal Deformations

To begin with, it should be mentioned that the deformation properties of aluminum were studied extensively in the early twentieth century, in single-crystal, multicrystal, and polycrystal forms, as were reviewed not so long ago.[29] And, a compilation of the ambient temperature H–P measurements[38, 39] for aluminum provide the lowest H–P microstructural stress intensity, k_ε, values among the metal systems, for example, a value of ~1 MPa. mm$^{1/2}$ applies for aluminum compared with ~5 for an interesting comparison to be made of copper and nickel behaviors, ~10 for magnesium, and ~24 MPa. mm$^{1/2}$ for iron. The reason for the low k_ε value for aluminum[38] is that cross-slip is the locally controlling stress for plastic flow transmission between the polycrystal grains of FCC metals and cross-slip occurs easily in aluminum at room temperature.

For aluminum at 300 K, with $m = 3.1$, $m_s = 2.2$, $G = 25$ GPa, $\tau_c = \tau_{III} = 4.7$ MPa from Bell,[40] $b = 0.29$ nm, and $\alpha = 0.8$, a low value of $k_\varepsilon = 1.2$ MPa.mm$^{1/2}$ is calculated from Eq. 3.8. Further evidence for FCC cross-slip control is provided by the comparison of approximately equal k_ε values being obtained from H–P plots of measurements reported for copper[41] and nickel,[42] including in the latter case, an early result for nanopolycrystal material. Consider in Eq. 3.8 that the shear modulus of nickel is approximately two times greater than the shear modulus of copper, 76 GPa as compared with 40 GPa, but the ambient temperature cross-slip stress, τ_{III}, for nickel is approximately one-half that of copper,[40] ~17 MPa as compared with ~30 MPa, so that the H–P k_ε values are approximately equal; see also later Ref. 65.

Figure 3.4 shows a comparison of the temperature dependencies of τ_{III} and k_ε^2 for copper as obtained from results reported by Mitchell and Thornton[43] and by Carreker and Hibbard,[41] respectively.

The open-triangle point for k_ε^2 at 300 K is the average of estimations made both from the Carreker and Hibbard results and from an H–P plot[6] of other measurements reported by Feltham and Meakin.[44]

Figure 3.5 shows a corresponding comparison on a strain-rate basis at ambient temperatures of the Mitchell and Thornton[43] cross-slip, τ_{III}, measurements and k_ε^2 measurements reported by Meyers, Andrade, and Chokshi.[45] In this case, the Meyers, Andrade, and Chokshi measurement of k_ε^2 at ~10^{-4} s^{-1} is indicated in Fig. 3.5 to be displaced downward so as to be in agreement with the corresponding result indicated in Fig. 3.4.

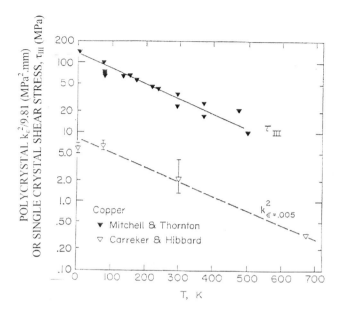

Figure 3.4 A comparison of the temperature dependencies for the cross-slip stress, τ_{III}, and H–P k_ε^2 of copper, as obtained from measurements reported by Mitchell and Thornton[43] and by Carreker and Hibbard,[41] respectively.

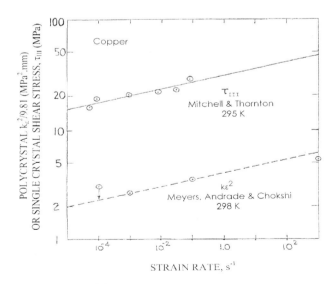

Figure 3.5 A comparison of the strain-rate dependencies for cross-slip, τ_{III}, and H–P k_ε^2 of copper, as obtained from Mitchell and Thornton[43] and Meyers, Andrade, and Choksi[45] measurements.

3.5 THE STRAIN-RATE SENSITIVITY DEPENDENCE ON GRAIN SIZE

The preceding results obtained for a thermally activated dependence both in $\sigma_{0\varepsilon}$ and k_ε for HCP magnesium, zinc, and cadmium and for FCC copper naturally lead to expectation of an H–P-type grain size influence on the strain-rate sensitivity. But why hadn't such an effect been observed in the many H–P research investigations made earlier on iron and steel materials, in which case only σ_{0y} is known to be thermally activated? The main reason appears to be that the high value of $k_y \sim 24$ MPa.mm$^{1/2}$ associated with the yield point behavior is too high for easily discernable thermally activated influence, as is true even for the higher athermal measurements of $k_T \sim 90$ MPa.mm$^{1/2}$ for deformation twinning[46] and $k_C \sim 100$ MPa.mm$^{1/2}$ for cleavage[47] of iron and steel materials. For example, with $k_y = 24$ MPa.mm$^{1/2}$, $m = 2.9$, $m_s = 2.2$, $G = 80$ GPa, $b = 0.27$ nm, and $\alpha = 0.84$, a relatively high value of $\tau_C = 830$ MPa is obtained for Cottrell locking of dislocations at grain boundaries in iron. On the other hand, Fisher and Cottrell[48] have reported lower values of thermally dependent k_y measurements for the special case of quench-aged iron material. The situation for iron and steel was reviewed by Petch[49] in a comprehensive assessment of the theory of grain size influences on the yield point and strain ageing behaviors.

The combination of Eqs. 3.6–3.8 provides for evaluation of V^{*-1} for a polycrystal in terms of both $\sigma_{0\varepsilon}$ and k_ε contributions as[15, 50]

$$V^{*-1} = (mk_BT)^{-1}[\partial\sigma_{0\varepsilon}/\partial\mathbf{ln}(\mathrm{d}\gamma/\mathrm{d}t)]_T +$$

$$(k_\varepsilon/2mk_BT\tau_C)[\partial\tau_C/\partial\mathbf{ln}(\mathrm{d}\gamma/\mathrm{d}t)]_T\ell^{-1/2} \qquad (3.9)$$

A negligible dependence of τ_C on strain-rate (or temperature) for BCC metals removes the grain size dependence in the second term on the right-hand side of the equation. Otherwise, with substitution of the respective grain volume V_0^* and grain-boundary V_C^* quantities,

$$V^{*-1} = V_0^{*-1} + (k_\varepsilon/2m\tau_C V_C^*)\ell^{-1/2} \qquad (3.10)$$

In Eq. 3.10, the product $\tau_C V_C^*$ should be constant, as indicated in Fig. 3.2, so long as the athermal stress component of τ_C is not too large; hence, H–P-type dependence should be observed for the grain size dependence of V^{*-1}.

3.5.1 Strain-rate Sensitivity for Hexagonal Close-Packed Metals

Prasad and Armstrong[50] showed that pioneering measurements reported for cadmium by Risebrough and Teghtsoonian[51] followed Eq. 3.10. And,

Fig. 3.6 demonstrates the application of Eq. 3.10 to a strain-rate change test performed in compression at two-test temperatures for polycrystalline zinc material.[33]

An extent of deformation twinning, favoring an increase in V^*, may have contributed to determining the lower value of V^{*-1} at the larger grain size, as indicated by the upper-pointing arrow. The increased slope value of the V^{*-1} dependence was attributed to the increase in k_ε. By far, the larger increase in V^{*-1} at 77 K is seen to be in the H–P grain-size–dependent term of Eq. 3.10.

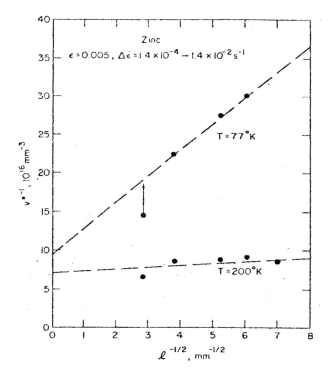

Figure 3.6 H–P-type V^{*-1} dependence of polycrystal zinc measured in compression at 200 K and 77 K, as adapted from Prasad, Madhava, and Armstrong.[33]

In Fig. 3.7, A^{*-1} measurements made by Mannan and Rodriguez,[37] as determined via $A^* = V^*/\mathbf{b}$, on polycrystal cadmium material tested at different temperatures and at different strain values, are plotted on a same H–P-type basis as for Fig. 3.6. The figure shows that the A_0^{*-1} value for cadmium is essentially zero. A quantitative analysis of the slope values both for the selected A^* measurements shown in Fig. 3.7 and for additional measurements obtained in the same investigation confirmed that the relevant activation area was that for pyramidal slip.[52]

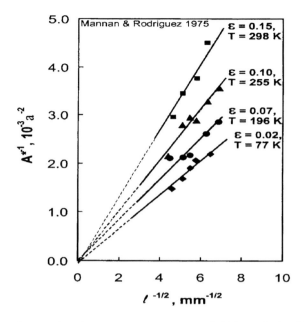

Figure 3.7 Polycrystal cadmium measurements of A^{*-1} made by Rodriguez and Mannan[37] and showing H–P-type dependence on $\ell^{-1/2}$ at different strain values also tested at different temperatures, as adapted from results reported by Rodriguez, Armstrong, and Mannan.[52]

3.5.2 Strain-rate Sensitivity for Face-Centered Cubic Metals

Another reason for the strain-rate sensitivity properties of FCC metals being of particular interest with regard to an influence of the polycrystal grain size, following on from the discussion in Section 3.4.2 of a low H–P k_ε being observed for aluminum, is that the FCC stress–strain properties have a history of being correlated with a flow stress dependence on the square root of dislocation density, beginning from pioneering dislocation model calculations made by Taylor,[25] for example, as related to his description of a (post-cross-slip) parabolic stress–strain curve for deformation results obtained on both single-crystal and polycrystal aluminum materials. Ashby[53] has incorporated a dislocation density–based grain size effect into the Taylor theory, whereby an influence of grain size may be understood, especially relating to the accommodation of plastic strains via geometrically necessary dislocations at grain boundaries. Kocks and Mecking[54] have provided an important review of the subject. Emphasis was given in the review to excellent results that were reported by Narutani and Takamura[55] on measuring a grain size dependence for the stress–strain properties of polycrystal nickel materials while also correlating the material stress–strain measurements with independently determined dislocation density measurements made via calibrated electrical resistivity measurements and with V^* measurements.

In Fig. 3.8, Rodriguez[56] has plotted on an H–P-type V^{*-1} versus $\ell^{-1/2}$ basis the reported Narutani and Takamura measurements made at several values of strain; see Fig. 3.9 of Ref. 55. With a connection between V^* and the dislocation density made through $\rho \sim \mathbf{b}^4/V^{*2}$, Narutani and Takamura obtained a lesser dependence of the flow stress on dislocation density than was obtained separately from the measurements made via electrical resistivity. Rodriguez pointed out that the problem was corrected by employing the (smaller) extrapolated values of V_0^{*-1} obtained on the ordinates axis in Fig. 3.8. Also, Rodriguez determined very much smaller values of V_c^* from the slope values of the V^{*-1} dependencies in Fig. 3.8 and showed the V_c^* values to be comparable with those associated with cross-slip, as had been reported by Puschl.[57] In a further connection with these important nickel results, Armstrong[29] established a quantitative H–P-type relationship between separate strain-hardening results presented for 32- and 91-micron grain size materials. Thus, the Narutani and Takamura results in Fig. 3.8 are taken to be quantitatively compatible with an H–P dislocation pileup model description. And, the two cases of FCC nickel in Fig. 3.8 and HCP cadmium in Fig. 3.7 present themselves as interesting opposite situations in which the strain-rate sensitivity properties, at least at conventional grain sizes, are either mainly determined by the grain volume or the grain-boundary properties, respectively.

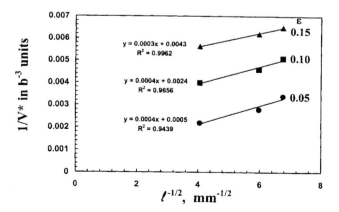

Figure 3.8 Polycrystal measurements of V^{*-1} at 77 K for nickel reported by Narutani and Takamura[55] as expressed on an H–P-type basis by Rodriguez[56].

3.5.3 Strain-Rate Sensitivities at Wider Grain Size Range

The H–P-type V^{-1} dependencies shown in Figs. 3.6–3.8 extend to a conventional smallest grain size of ~20 microns and, as mentioned for Eq. 3.1, this is not an unusual smallest dimension for establishment of grain-size-dependent polycrystal strength properties.

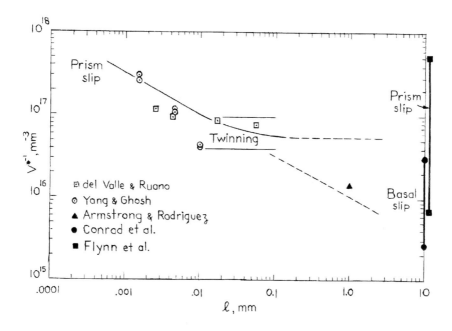

Figure 3.9 V^{*-1} vs. ℓ for magnesium results[59] plotted on a log/log scale and referenced to single-crystal measurements for basal[18] and prism[19] slip systems, as developed from the polycrystal results reported by del Valle and Ruano[60] and Yang and Ghosh.[61]

A more recent effort has dealt with the wider attempt of connecting on a log/log scale, as for the strength evaluations in Fig. 3.1, conventional-grain-size V^{*-1} results with nanopolycrystal grain size results and, at the larger grain size limit, with single-crystal measurements.[2, 58] Figure 3.9 shows one such connection begun for magnesium materials.[59] In the figure, the V^{*-1} dependence includes a combination of single-crystal[18, 19] and polycrystal[60, 61] results extending downward in grain size near to the 1-micron scale. The inclined solid and dashed line has a slope of –1/2 in accordance with prediction from Eq. 3.10. The open-square points of del Valle and Ruano[60] correspond to tensile $\varepsilon = 0.002$ values determined for the magnesium AZ31 alloy. In a further report, the authors determined a $k_\varepsilon = 6.6$ MPa.mm$^{1/2}$ value for the same strain.[62] A minor extent of twinning occurred at the largest grain sizes and became more prominent at larger strains. Yang and Ghosh[61] achieved smaller "ultrafine" grain sizes for the open-circle points shown for the same alloy and found extensive twinning in compressive test results obtained on their larger 10-micron-grain-size material. A value of $k_\varepsilon \sim 4.0$ MPa.mm$^{1/2}$ has been estimated as an effective compressive stress value from their hardness measurements. Deformation twinning is known to be very nearly athermal in both HCP and BCC metals[46] because of the significant local twinning shear strain requirement determining τ_C in k_T; hence an amount of twinning involvement in the deformation behavior contributes to an increase in V^*. And, the experimental k_ε values for the materials in Fig. 3.9 are relatively low compared with a value of ~ 10 MPa.mm$^{1/2}$ that was originally determined for texture-free material by Hauser *et al.*[31] (see Fig. 3.3) and confirmed in more recent measurements reported by Caceres and Blake[63] both for tensile and compression tests, thus confirming an absence of any significant influence of texture.

In Fig. 3.9, the filled-triangle point plotted at $\ell = 1.0$ mm was calculated as a lowest possible value of $V^{*-1} = V_0^{*-1}$ for a single crystal from the linear relation $\tau_0^* V_0^* = 3.1 \times 10^{-20}$ J in Fig. 3.2 and taking $\tau_0^* = 0.4$ MPa from the measurements of Conrad *et al.*[18] that are shown in Fig. 3.3. Furthermore, with $\tau_C = 39$ MPa for prism slip in Fig. 3.3, $m = m_s = 6.5$ (from Ref. 6), $G = 17.5$ GPa, **b** = 0.32 nm, and $\alpha = 0.84$, a value of $k_\varepsilon = 10.6$ MPa.mm$^{1/2}$ is obtained from Eq. 3.8, in agreement with the results reported by Hauser *et al.*[31] and the recent determination by Caceres and Blake[63]. The reduction by half of the computed k_ε to an average $k_\varepsilon \sim 5.3$ MPa.mm$^{1/2}$ for the del Valle *et al.*[62] and Yang and Ghosh[61] results in Fig. 3.9, which are lower because of the material textures, produces an essentially equivalent grain-size-dependent contribution to V^{*-1} from the second V_C^{*-1} term in Eq. 3.10. By comparison of the Yang and Ghosh measurements made at their smallest grain size with the vertical ranges of V^{*-1} shown for the basal slip and prism slip systems at the right-side ordinate axis of Fig. 3.9, it may be seen that the ultrafine polycrystal measurements, being free of deformation twinning, approach the highest prism slip values determined for the Flynn *et al.*[19] single-crystal measurements.

A more pertinent example of investigating the relation of strain-rate sensitivity and grain size at ultrafine grain sizes, this time extending to nanopolycrystal grain size measurements, is shown in Fig. 3.10 as developed by Armstrong and Rodriguez[65] in an H–P assessment of important FCC results originally compiled for combined copper and nickel materials by Asaro and Suresh.[64] First, it should be noted again, as mentioned earlier in Section 3.4.2, that copper and nickel material strength results, which might be thought to be at somewhat different stress levels because of their very different elastic moduli, were in fact deemed to be quite comparable in terms of having approximately the same H–P k_ε values because of the model dislocation pileup result of k_ε being proportional to the square root of the product of G and τ_{III}, whose product has approximately the same value for the two materials. In the log/log representation of Fig. 3.10, the V^{*-1} results determined by Rodriguez[56] from the Narutani and Takamura[55] measurements made for nickel at $\varepsilon = 0.05$ and 77 K are shown at their respective conventional grain sizes. And, the H–P relation from Eq. 3.10 drawn through the Narutani and Takamura points is extended into the nanoscale regime, both for the 77 K temperature of the measurements and, also, for 295 K as estimated through employment of a lowered value of k_ε to match the 295 K temperature of the compiled Asaro and Suresh measurements.[64] The H–P-based connection of V^{*-1} results is thus established by a single H–P-type relationship despite an order of magnitude difference that is shown between $V^* \sim 100\text{--}1000\ b^3$ at conventional grain sizes and $V^* \sim 10\text{--}100\ b^3$ at nanopolycrystal grain sizes.

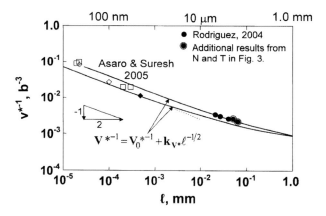

Figure 3.10 A predicted H–P-type V^{*-1} dependence shown on a log/log basis[65] to connect conventional ~10- to 100-micron scale V^{*-1} measurements made at 77 K, as reported for nickel,[55] and compiled copper plus nickel results obtained at 295 K and including nanopolycrystal ~20 nm to 500 nm grain sizes[64]; "N and T" in the figure legend are for Narutani and Takamura,[55] and "Rodriguez, 2004" is the present Ref. 56.

It's also of interest that Asaro and Suresh employed a P–L expression for the stress dependence on strain-rate, in line with Eq. 3.5 earlier but, instead, employing the total stress, σ, in place of τ^*, and a numerical value, $\sqrt{3}$, in place of the Taylor factor, m. Thus, for this case, the strain-rate sensitivity is specified in terms of the exponent, $m^{*\prime}$, as

$$m^{*\prime} = [\partial \mathbf{ln}\tau / \partial \mathbf{ln}(d\gamma/dt)]_T = \sqrt{3}[k_B T/\sigma V^*]_T \qquad (3.11)$$

On such different basis for $m^{*\prime}$, Armstrong and Rodriguez[65] showed that the parameter should be contained within the limits,

$$\sqrt{3}[k_B T/\sigma_0 V_0^*] < m^{*\prime} < \sqrt{3}[k_B T/2m\tau_{III} V_{III}^*] \qquad (3.12)$$

Reasonable correlation was made between the limiting estimations of ineq. 3.12 and the range of the reported measurements. Note, too, that there is an indirect reciprocal relationship of $m^{*\prime}$, above, and m^* in Eq. 3.5, as will be discussed later with respect to Eq. 3.18. Other copper nanopolycrystal measurements[66, 67] have been reported very recently on the same P–L basis as employed by Asaro and Suresh.

3.6 HALL–PETCH STRENGTHS FOR NANOPOLYCRYSTALS

Meyers, Mishra, and Benson[68] have done a comprehensive review of the mechanical properties of nanocrystalline materials, including a compilation of 374 references. Attention was given to experimental observations of V^* decreasing in the nanopolycrystal regime for FCC metals, while for BCC metals, V^* remains constant at the relatively low value observed for conventional-grain-size materials. Conflicting observations were reported for the influence of grain size in creep testing. A considerable number of experimental results were described for the influence of a reduction in grain size on improving material strength properties. The subject is added to here on an H–P dislocation-pileup-model basis and includes consideration of both grain size strengthening in an effective low-temperature condition and grain size weakening at effective higher temperatures.[1, 69, 70]

3.6.1 The Pileup Model Extension for Strength Properties

In Fig. 3.11, a compilation of strength results reported by Conrad[71] are shown on a log/log basis with an extended superposition of H–P curves determined at the $\varepsilon = 0.05$ and 0.20 values as obtained from the careful investigation of Hansen and Ralph.[72]

At 300 K and at the largest grain diameter $\ell (= d) \sim 1.0$ mm, the $\varepsilon = 0.05$ plot of the flow stress is seen to continue to exhibit an influence of the $k_\varepsilon \ell^{-1/2}$ term, whereas, both because of the reduced k_ε at $\varepsilon = 0.20$ and an increased strain hardening in the H–P $\sigma_{0\varepsilon}$ stress intercept, an essentially constant level of the flow stress is seen to be determined. And in the nanopolycrystal regime, the $\varepsilon = 0.05$ determined flow stress exhibits a predicted ($-1/2$) slope value that is not quite achieved for the $\varepsilon = 0.20$ flow stress curve because of the raised $\sigma_{0\varepsilon}$ value. Otherwise, the observation is easily made that the plotted nanopolycrystal stress measurements fall below the conventional H–P grain size extrapolation. And, for the individual studies where sufficient measurements at nanoscale grain sizes were obtained, the lower stresses involve reductions in both $\sigma_{0\varepsilon}$ and k_ε values. The interest here is to investigate the situation of the nanopolycrystal grain size having become sufficiently small that the strength is being determined ultimately by a single dislocation loop overcoming the grain-boundary obstacle.[73,74]

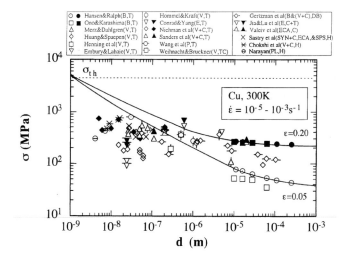

Figure 3.11 A log/log presentation of the compilation by Conrad[71] of grain-size-dependent strength measurements for copper and including extended curves for the H–P dependence as described by Armstrong, Conrad, and Nabarro[58]; $d = \ell$ in the present text.

Li and Liu[74] produced the pioneering computation for expansion of a circular dislocation loop against its own line tension and a grain-boundary obstacle. The resultant stress is expressed as

$$\sigma = m[\tau_0 + (3Gb/4\pi\ell)\{(5/6)(\ln[4\ell/b] - 1) - 1/16\} + \tau_c] \qquad (3.13)$$

Equation 3.13, though being only one of many pileup-type calculations described in an important review article by Li and Chou,[75] is one made in addition to another early calculation[76] anticipating future interest in nanopolycrystal strength properties. That issue and other model pileup applications to material strength properties have been reviewed by

Armstrong,[77] including a role for H–P influences on composite material properties. In Fig. 3.12, the dashed curve shown just above the H–P dependence established by Hansen and Ralph is that computed on the basis of k_ε being employed in Eq. 3.10 to determine τ_C and then that τ_C employed in the evaluation of Eq. 3.13.

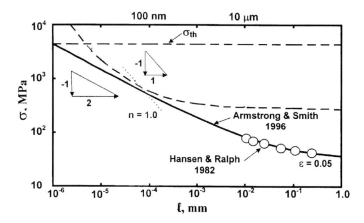

Figure 3.12 Dislocation pileup-based H–P dependence for copper, spanning a range, in grain diameters, from conventional to nanopolycrystal sizes and highlighting the transition to a single-dislocation loop expanding against the grain-boundary obstacle.[65]

At the larger grain sizes shown in Fig. 3.12, the average number, n, of dislocations in the pileups accounts for the separation of the (solid) H–P and (dashed) single-dislocation curves. For example, at $\ell = 100$ microns and with

$$n = 2\alpha k_\varepsilon \ell^{1/2}/\pi m G \mathbf{b} \qquad (3.14)$$

as evaluated from Eq. 7 in Ref. 8 with substitution of $k_\varepsilon \ell^{-1/2}/m$ for the effective shear stress, $\tau_e = (\tau - \tau_{0\varepsilon})$, a number of dislocation loops $n \sim 26$ is spread on average across a typical grain. As the grain size is made smaller, n decreases until at $n = 1.0$ the H–P and dashed curves should meet. This is marked by the dotted line of slope (–1) at ~100 nm where the solid and dashed curves reach a minimum separation rather than intersection, because of the analytic approximation made in the continuum mechanics description on which Eq. 3.14 is based.

The results in Fig. 3.12 bear also on Fig. 3.13, which has been adapted from a graphical description originally developed by Nabarro[78] to illustrate over a large range in grain size the upper-limiting stress dependencies which might be achievable in creep tests,[58] even including an upper-limiting H–P dependence.

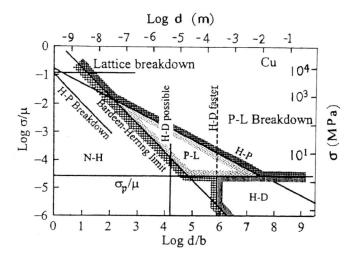

Figure 3.13 The grain size dependence estimated for upper-limiting creep stresses of copper, as adapted from Nabarro[78]; $\mu = G$, $d = \ell$, "N–H" is Nabarro–Herring, σ_p is the Peierls stress, P–L is for power–law-type creep, and "H–D" is for Harper–Dorn-type creep.

Any friction type, $\sigma_{0\varepsilon}$, stresses are neglected in the figure. At stresses of the order of 1.0 MPa or less, below the Peierls–Nabarro stress, σ_p, Nabarro–Herring (N–H) creep should be observed at grain sizes of the order of ~100 microns or less, and Harper–Dorn (H–D) creep would apply at larger grain sizes. A reciprocal grain diameter dependence for a Bardeen–Herring limit to creep is reached at higher stresses after which P–L creep predominates. And, in turn, the range in stress and grain size for P–L creep behavior is limited by H–P dependence that breaks down to the reciprocal $(d/\mathbf{b})^{-1}$ Bardeen–Herring limit at a grain diameter below the same ~100 nm pileup limit described for Fig. 3.12.

Presumably, logarithmic creep, as investigated in detail in the pioneering copper single-crystal study reported by Conrad,[16] and reviewed by Weertman[79] in relation to the other creep-type constitutive equations, would apply at stresses above the limiting H–P dependence shown for P–L creep. In this regime, Armstrong[1,69,70] had reported on the grain size dependence of the plastic strain-rate achieved at constant total stress but incorporating thermal activation in the H–P $\sigma_{0\varepsilon}$ friction stress. The main purpose was to make a connection between effective low-temperature grain-size-strengthening and effective high-temperature (creep-type) grain-size-weakening behaviors.[1] The topic relates to modern interest in an inverse H–P dependence observed at the smallest nanopolycrystal grain sizes.[68, 80]

3.6.2 Transition from Grain Size Strengthening to Weakening

Meyers *et al.*[68] have provided a list of model equations for the occurrence of grain size weakening under creep conditions. For example, the N–H behavior mentioned earlier in connection with Fig. 3.13 was given as

$$(d\varepsilon/dt)_{NH} = (A_{NH}D_L G\mathbf{b}/k_B T)(\mathbf{b}/\ell)^2(\sigma/G) \tag{3.15}$$

In Eq. 3.15, A_{NH} is a constant, D_L is the lattice diffusion coefficient, and the other symbols have already been defined. At constant strain-rate, σ is proportional to ℓ^2, thus being reflective of grain size weakening behavior. Such model equation descriptions relate generally to the grain-boundary regions being weaker either because of enhancement of mass transport by diffusion mechanisms as mentioned earlier or because of additional deformation mechanisms coming into play such as grain-boundary shearing processes or even cracking or for all of these reasons.[1, 68]

Figure 3.14 shows important H–P results compiled for zinc materials by Conrad and Narayan,[80] including their own measurements made with colleagues, and adapted[81] to include a number of dashed H–P lines.[80, 82]

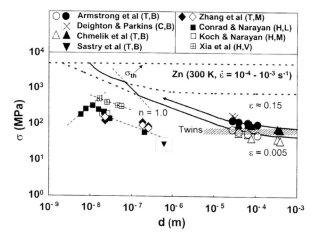

Figure 3.14 Grain-size-strengthening and grain-size-weakening measurements compiled for zinc materials by Conrad and Narayan[80] and connected with several H–P dependencies at nanopolycrystal and larger grain sizes vs. an inverse H–P dependence at the smallest grain sizes;[81] $d = \ell$.

At the smallest nanopolycrystal grain sizes, an inverse H–P dependence is indicated. Beginning from the larger-grain-size end of the figure, parallel H–P dependencies from Ref. 6 at $\varepsilon = 0.005$ and 0.15 are shown to be extended into the nanopolycrystal regime with constant k_ε values. The transition to single-loop expansion against the grain-boundary resistance occurs again at a value of d (= ℓ) of ~100 nm. In the vicinity of this grain size and extending further downward to just larger than 10 nm, several sets of data[82–85] are shown also to follow H–P dependencies but, as mentioned earlier, at indicated lower k_ε values.[81] Below ~10 nm an inverse grain-size-weakening dependence is observed with an indicated P–L dependence of slope 1.0. Conrad and Narayan[80] provided a detailed analysis of the total results, including those obtained with colleagues[83–85]. Particular consideration was given to an important influence at larger grain sizes of deformation twinning, as marked in the figure and especially relating to the twin-free observation of an increased V^{*-1} at smaller grain sizes in the H–P-strengthening regime; see the discussion of twinning relating to the results in Fig. 3.9.

Rodriguez and Armstrong[81] proposed a combination of both grain-size-strengthening and grain-size-weakening explanations for the strain-rate sensitivity results shown for zinc in Fig. 3.15.

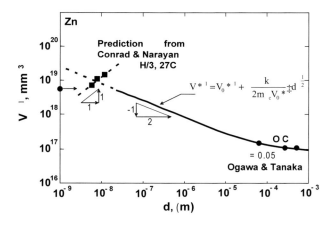

Figure 3.15 A log/log graph of the H–P-type V^{*-1} dependence on grain size extended into the nanopolycrystal regime for comparison with the prediction of both an inverse H–P strain-rate sensitivity dependence on grain size and an upper-limiting value shown on the ordinate scale of $V_c^{*-1} = \mathbf{b}^{-3}$ for pyramidal slip[81]; H is hardness and $d = \ell$.

First, H–P dependence was extended to the nanopolycrystal grain size regime from the excellent conventional polycrystal grain size measurements reported by Ogawa and Tanaka[86] for $\varepsilon = 0.05$ in compression tests made at

273 K. The solid curve for the H–P dependence is seen to end at the level of the (filled circle) V^{*-1} value shown on the ordinate scale with an arrow pointing to the H–P terminus. The point corresponds to calculation of an upper-limiting value of $V^{*-1} = \mathbf{b}^{-3} = 5.7 \times 10^{18}$ mm^{-3} for the $(\mathbf{a} + \mathbf{c})$ Burgers vector for pyramidal slip. At the smallest grain sizes, the several filled-square points have been plotted in accordance with a generalized grain-size-weakening constitutive equation prediction given by Langdon[87] as

$$(d\varepsilon/dt) = (AD_L G\mathbf{b}/k_B T)(\mathbf{b}/d)^p (\sigma/G)^q \qquad (3.16)$$

In Eq. 3.16 that was ascribed to grain-boundary shearing, A, p, and q are experimental constants; see the similarity to Eq. 3.15 for N–H creep with $p = 2$ and $q = 1$. Even for the weakening case of grain-boundary shearing, however, there may be roles to be played by dislocation pileups. Federov, Gutkin, and Ovid'ko[88] have modeled the deformation of the grain-boundary itself shearing against triple points via dislocation pileups that are contained within the boundary interfaces, whereas Conrad[89] has proposed a constitutive equation for grain-boundary strengthening in which case pileups from within the grains force grain-boundary shearing to occur. The model considerations are indicative of progress that is being made in modeling the grain-boundary deformation behavior, especially that for weakening behavior, in proceeding onward from the situation described in Ref. 1 pertaining to "the difficulty of characterizing the structure of grain boundaries and the nature of the events which occur there."

Otherwise, for the zinc grain-size-weakening results of interest in Figs. 3.14 and 3.15, and with $p = q = 1.0$, V^{*-1} is evaluated from Eq. 3.16 as

$$V^{*-1} = \sigma/mk_B T \qquad (3.17)$$

With the several σ values determined from Fig. 3.14, the corresponding filled-square points in Fig. 3.15 were able to be computed. Thus, an interesting consequence of an inverse H–P dependence on grain size for nanopolycrystals leads to prediction of a reversed grain size dependence of the strain-rate sensitivity.[81]

3.7 DISCUSSION

The log/log H–P presentations in Figs. 3.11 and 3.14 are adapted from original versions presented by Conrad[71] and by Conrad and Narayan,[80] as mentioned at the place of each figure introduction. Not shown in each case are the divisions proposed to separate the grain-size-dependent results into separate mechanism-dependent regions, for example, beginning from the largest grain sizes to include (I) a region primarily corresponding to the main influence of the dislocation density; then, at smaller grain size, (II) a region of H–P pileup application; and, finally at the smallest grain size, (III) a region of inverse H–P behavior. In the present description, H–P dependence has been argued to cover the full extent of regions I and II, even including the transition to H–P-type dependence corresponding to expansion of a single-dislocation loop against its own line tension and the grain-boundary obstacle, as employed in the calculation made by Li and Liu.[74] At limiting larger grain sizes, however, for which $\sigma_{0\varepsilon} > k_\varepsilon \ell^{-1/2}$, it may be seen on the graphical log/log basis that the polycrystal flow stress is largely described by the Taylor-type influence of dislocation density that is contained in $\sigma_{0\varepsilon}$ and is associated only with a smaller apparent grain size dependence of σ_ε. Thus, there isn't a large difference between the stage I/II description given by Conrad and that proposed here.

A preference for the strain-rate sensitivity being expressed in terms of V^{*-1}, see Eq. 3.4, or by the related m^* parameter, in Eq. 3.5, is made clear when the alternative $m^{*\prime}$ parameter, see the Eqs. 3.11 and 3.12, is evaluated as

$$m^{*\prime} = mk_B T/\sigma_\varepsilon V^* \tag{3.18}$$

Thus in Eq. 3.18, $m^{*\prime}$ is seen to depend inversely on the product of the total stress, σ_ε, and V^*, as mentioned earlier with regard to Eq. 3.5, and $m\tau^*$ for the thermally activated stress versus $m\tau$ for the sum of grain volume and grain-boundary terms on the right-hand side of Eq. 3.6. The σ_ε and V^* variables, for example, have separate dependencies on strain, for example, as reported[22] for single-crystal niobium measurements. In that case, V^* was found to be independent of strain, as typical of a thermally activated Peierls stress dependence, and $m^{*\prime}$ followed the σ_ε^{-1} dependence on ε. Beyond the limiting cases for $m^{*\prime}$ given in the Eq. 3.12, if $V_0^* > V_c^*$, as is generally true for the FCC case of dislocation intersections versus cross-slip, and $\sigma_{0\varepsilon} < k_\varepsilon \ell^{-1/2}$, as is true at nanopolycrystal grain dimensions, then in the interim small-grain-size regime the grain size dependence may be approximated as

$$m^{*\prime} \sim -(k_B T/2\tau_c V_c^*)\{1 - (\sigma_{0\varepsilon}/k_\varepsilon \ell^{-1/2})\} \tag{3.19}$$

In Eq. 3.19, $m^{*\prime}$ increases as the grain diameter decreases. In several investigations, Della Torre et al.,[90] Ghosh,[91] and Wei[92] have reported on details of experimentally measuring $m^{*\prime}$ and discussed its interpretation.

The reduction shown by the downward displacements of the several dashed H–P lines in Fig. 3.14, thus reflecting a decrease in k_ε for these zinc materials, is not nontypical of the broader number of measurements made of nanopolycrystal behavior for other materials. Armstrong *et al.*[58] suggested that one reason for the observation could be that the grain-boundary regions were more likely to be relatively disordered at nanometric-grain-size dimensions compared with conventional-grain-size materials. The suggestion was also put forward that the unusual observation of deformation twinning being observed in nanopolycrystal aluminum material might be the result of such twin nuclei being already present in the disordered grain-boundary structures. Li[93] has modeled the potential influence of impurity and porosity effects on the equilibrium and nonequilibrium properties of nanopolycrystal grain boundaries. Narayan and Zhu[94] and Wu, Narayan, and Zhu[95] have explained the observation of nanoscale twinning in terms of the partial dislocation structures present at boundaries in nickel and copper–5% germanium alloys, respectively, as made via model dislocation reaction descriptions matching their high-resolution electron microscopy results. Evidence for nanopolycrystal grain boundaries being generally disordered seems to be an accepted observation at grain boundaries in molecular dynamics modeling of nanopolycrystal structures, for example, as described by Weertman *et al.*[96] and by Kadau *et al.*[97]

An interesting opposite consideration that may prove the case for nanoscale boundaries being generally disordered is provided by the role established in nanopolycrystals of relatively perfect twin boundaries contributing to a substantial H–P dependence. Asaro and Suresh[65] counted twin boundaries formed during material production in their reported nanopolycrystal grain size measurements. The reported data lead to determinations of $k_y \sim 3.8$ MPa.mm$^{1/2}$ for the tensile test results and a compressive stress $k_\varepsilon \sim 4.3$ MPa.mm$^{1/2}$ determined for the reported hardness measurements. In follow-up results, Zhu, Samanta, Kim, and Suresh[98] pointed out via atomistic reaction pathway computations applied to the obstacle nature of nano-twin boundaries in copper material that both significant H–P strengthening and high ductility, along with an increased strain-rate sensitivity, were reasonable expectations for well-defined interfacial obstacles to transmission of plastic flow. The consideration of grain-boundary characterizations relates also to boundary-type discriminations made in an early investigation by Wyrzykowski and Grabski[39] for aluminum at ultrafine grain sizes.

Such positive H–P ductility connection mentioned earlier had been proposed already for conventional-grain-size material through the observation that the true fracture strain of magnesium and Armco iron materials followed H–P-type dependence.[99, 100] The result should be valid if the material work hardening is independent of grain size and mechanical instability doesn't intervene. In addition to the consideration of ductility dependence on grain size, there is the natural question of how H–P dependence for the flow stress might also connect with other mechanical properties of nanopolycrystals, in the same manner as was considered earlier for fracture, hardness, creep, and fatigue properties.[1, 27] Zhu et al.[101] have co-edited a recent proceedings on ultrafine-grained material mostly including both methods of producing the materials for engineering applications and evaluation of the resultant properties. Also, a review has been given by Song et al.[102] of higher strength and, especially, higher fracture toughness properties being obtained for ultrafine-grain-size steels of ~1 micron grain size and smaller, as achieved in traditional material compositions. Hwang et al.[103] and, more recently, Bondar[104] have reported on the pronounced shear-banding behavior associated with ultrafine-grain-size low-carbon steels and copper materials, respectively, when subjected to dynamic loading. The results connect with other aspects of high-rate loading behaviors described by Armstrong and Walley[11] for conventional and fine-grained structural materials as well as by Armstrong[105] for a role of dislocation pileups in achieving advantageous deformation and energy release properties of small particle sizes in energetic (explosive) material composites.

3.8 CONCLUSION

The experimental H–P dependence observed for the yield and flow stresses of conventional-grain-size polycrystals, beginning with iron and steel materials, is shown to be useful for assessing nanopolycrystal strength and strain-rate sensitivity behaviors for a larger selection of materials. The experimental H–P dependencies are interpreted in terms of the dislocation pileup model that allows for transition at ultrafine grain size to a single-dislocation loop expanding against the grain-boundary resistance. In contrast to the yield-point-associated, higher, H–P k_ε values observed for BCC metals, those k_ε values for FCC and certain HCP metals are sufficiently low to be thermally activated. In consequence, the reciprocal activation volume characterization of the strain-rate sensitivity shows H–P-type dependence also that, therefore, is greatly enhanced at nanopolycrystal grain sizes. At the smallest grain sizes, there is evidence of the flow stress exhibiting inverse H–P-type dependence. And, the inverse dependence is attributed to the same type of grain-boundary-weakening character that is characteristic of effective high-temperature creep. For this case, reversed grain size dependence is predicted to occur also for the strain-rate sensitivity.

Acknowledgments

Prof. J. C. M. Li is thanked for inviting the present article that has benefitted from his own key research results that were obtained with colleagues and students. Dr. Stephen Walley, University of Cambridge, is thanked for providing very useful reference articles. Dr. Qiuming Wei, University of North Carolina at Charlotte, provided helpful corrections to Eq. 3.11, and Professor C. P. Chang, National Sun Yat-Sen University, Taiwan, provided helpful corrections to Fig. 3.15. Otherwise, a number of results described in the present report were achieved as part of a cooperative research project underway with Prof. Placid Rodriguez, Indian Institute of Technology, Madras, and whose sudden death on August 31, 2008, is very sadly noted.

References

1. R. W. Armstrong (1970), in eds., J. J. Burke and V. Weiss, *Ultrafine–Grain Metals; Sixteenth Sagamore Army Materials Research Conference*, Syracuse University Press, NY, p. 1.
2. T. R. Smith, R. W. Armstrong, P. M. Hazzledine, R. A. Masumura, and C. S. Pande (1995), in eds., M. A. Otooni, R. W. Armstrong, N. J. Grant, and K. Ishizaki, *Grain Size and Mechanical Properties – Fundamentals and Applications*, Materials Research Society, Pittsburgh, PA, **362**, p. 31.
3. E. O. Hall (1951), *Proc. Phys. Soc. Lond.*, **B64**, 747.
4. N. J. Petch (1953), *J. Iron Steel Inst.*, **174**, 25.
5. T. L. Altshuler and J. W. Christian (1967), *Phil. Trans. Roy. Soc. Lond. A*, **261**, 251.
6. R. W. Armstrong, I. Codd, R. M. Douthwaite, and N. J. Petch (1962), *Philos. Mag.*, **7**, 45.
7. J. D. Embury and R. M. Fisher (1966), *Acta Metall.*, **14**, 147.
8. R. W. Armstrong (1983), in ed., T. N. Baker, *Yield, Flow, and Fracture of Polycrystals*, Applied Sci. Publ., London, UK, p. 1.
9. K. Hayashi and H. Etoh (1989), *Mater. Trans. Japan Inst. Met.*, **30**, 925.
10. J. S. C. Jang and C. C. Koch (1990), *Scr. Met.*, **24**, 1599.
11. R. W. Armstrong and S. M. Walley (2008), *Intern. Mater. Rev.*, **53**, 105.
12. E. Orowan (1940), *Proc. Phys. Soc. Lond.*, **B52**, 8.
13. G. Schoeck (1965), *Phys. Stat. Sol.*, **8**, 499.
14. J. C. M. Li (1968), in eds., A. R. Rosenfield, G. T. Hahn, A. L. Bement, Jr., and R. I. Jaffee, *Dislocation Dynamics*, McGraw-Hill Book Co., NY, p. 87.
15. R. W. Armstrong (1973), *(Indian) J. Sci. Indust. Res.*, **32**, 591.
16. H. Conrad (1958), *Acta Metall.*, **6**, 339.
17. H. Conrad and S. Frederick (1962), *Acta Metall.*, **10**, 1013.

18. H. Conrad, R. W. Armstrong, H. Wiedersich, and G. Schoeck (1961), *Philos. Mag.*, **6**, 177.

19. P. W. Flynn, J. Mote, and J. E. Dorn (1961), *Trans. Metall. Soc. – AIME,* **221**, 1148.

20. G. A. Alers, R. W. Armstrong, and J. H. Bechtold (1958), *Trans. Metall. Soc. – AIME,* **212**, 523.

21. H. L. Prekel and H. Conrad (1968), in eds., A. R. Rosenfield, G. T. Hahn, A. L. Bement, Jr., and R. I. Jaffee, *Dislocation Dynamic*, McGraw-Hill Book Co., NY, p. 431.

22. R. E. Reed, H. D. Guberman, and R. W. Armstrong (1970), *Phys. Stat. Sol.*, **37**, 647.

23. F. J. Zerilli and R. W. Armstrong (1987), *J. Appl. Phys.*, **61**, 1816.

24. F. J. Zerilli (2004), *Metall. Mater. Trans. A,* **35A**, 2547.

25. G. I. Taylor (1938), *J. Inst. Met.,* **62**, 307.

26. U. F. Kocks (1958), *Acta Metall.,* **6**, 85.

27. R. W. Armstrong (1970), *Metall. Trans.,* **1**, 1169.

28. A. Seeger (1956), in eds., J. C. Fisher, W. G. Johnston, R. Thomson, and T. Vreeland, Jr., *Dislocations and Mechanical Properties of Crystals*, John Wiley & Sons, NY, p. 243.

29. R. W. Armstrong (2005), in eds., K. H. J. Buschow, R. W. Cahn, M. C. Flemings, E. J. Kramer, S. Mahajan, and P. Veyssiere, *Encyclopedia of Materials: Science and Technology – Updates*, Elsevier Science, Oxford, UK, published online: Elsevier ScienceDirect.

30. R. W. Armstrong (1968), *Acta Metall.*, **16**, 347.

31. F. E. Hauser, P. R. Landon, and J. E. Dorn (1956), *Trans. Metal. Soc. – AIME,* **206**, 589.

32. F. E. Hauser, P. R. Landon, and J. E. Dorn (1956), *Trans. Amer. Soc. Met.,* **48**, 986.

33. Y. V. R. K. Prasad, N. M. Madhava, and R. W. Armstrong (1974), in *Grain Boundaries in Engineering Materials; Fourth Bolton Landing Conf.*, Claitor's Press, Baton Rouge, LA, p. 529.

34. B. Wielke (1973), *Acta Metall.,* **21**, 289.

35. Ya.M. Soifer and V. G. Shteinberg (1972), *Phys. Stat. Sol.,* **10**, K113.

36. G. W. Greenwood and A. G. Quarrell (1954), *J. Inst. Met.,* **82**, 551.

37. S. L. Mannan and P. Rodriguez (1975), *Acta Metall.,* **23**, 221.

38. R. W. Armstrong (1979), in eds., D. W. Borland, L. M. Clarebrough, and A. J. W. Moore, *Physics of Materials; A Festschrift for Dr. Walter Boas on the Occasion of His 75th Birthday*, CSIRO and University of Melbourne, Australia, p. 1.

39. J. W. Wyrzykowski and M. W. Grabski (1986), *Philos. Mag.,* **53**, 505.

40. J. F. Bell (1965), *Philos. Mag.,* **11**, 1135.

41. R. P. Carreker, Jr., and W. R. Hibbard, Jr. (1953), *Acta Metall.,* **1**, 654.

42. G. D. Hughes, S. D. Smith, C. S. Pande, H. R. Johnson, and R. W. Armstrong (1986), *Scr. Metall.,* **20**, 93.

43. T. E. Mitchell and P. R. Thornton (1963), *Philos. Mag.,* **8**, 1127.

44. P. Feltham and J. D. Meakin (1957), *Philos. Mag.,* **2**, 105.

45. M. A. Meyers, U. R. Andrade, and A. H. Chokshi (1995), *Metall. Mater. Trans. A,* **26A**, 2881.

46. R. W. Armstrong and F. J. Zerilli (1999), in eds., S. Ankem and C. S. Pande, *Advances in Twinning,* Metall. Mater. Soc. – AIME, Warrendale, PA, p. 67.

47. R. W. Armstrong (1987), *Eng. Fract. Mech.,* **28**, 529.

48. R. M. Fisher and A. H. Cottrell (1963), in *Relation between Structure and Mechanical Properties,* National Phys. Lab., London, UK, p. 445.

49. N. J. Petch (1990), in eds., J. A. Charles and G. C. Smith, *Advances in Physical Metallurgy; Sir Alan Cottrell's 70th Birthday Meeting,* Inst. Met., London, UK, p. 11.

50. Y. V. R. K. Prasad and R. W. Armstrong (1974), *Philos. Mag.,* **29**, 1421.

51. N. R. Risebrough and E. Teghtsoonian (1967), *Canadian J. Phys.,* **45**, 591.

52. P. Rodriguez, R. W. Armstrong, and S. L. Mannan (2003), *Trans. Indian Inst. Met.,* **56**, 189.

53. M. F. Ashby (1970), *Philos. Mag.,* **21**, 399.

54. U. F. Kocks and H. Mecking (2003), *Prog. Mater. Sci.,* **48**, 171.

55. T. Narutani and J. Takamura (1991), *Acta Metall. Mater.,* **39**, 2037.

56. P. Rodriguez (2004), *Metall. Mater. Trans. A,* **35A**, 2697.

57. W. Puschl (2002), *Prog. Mater. Sci.,* **47**, 415.

58. R. W. Armstrong, H. Conrad, and F. R. N. Nabarro (2004), in eds., I. Ovid'ko, C. S. Pande, R. Krishnamoorti, E. Lavernia, and G. Skandan, *Mechanical Properties of Nanostructured Materials and Nanocomposites,* Mater. Res. Soc., Warrendale, PA, p. 69.

59. R. W. Armstrong and P. Rodriguez (2008), unpublished research.

60. J. A. del Valle and O. A. Ruano (2006), *Scr. Mater.,* **55**, 775.

61. Q. Yang and A. K. Ghosh (2006), *Acta Mater.,* **54**, 5159.

62. J. A. del Valle, F. Carreno, and O. Ruano (2006), *Acta Mater.,* **54**, 4247.

63. C. H. Caceres and A. H. Blake (2007), *Mater. Sci. Eng. A,* **462**, 193.

64. R. J. Asaro and S. Suresh (2005), *Acta Mater.,* **53**, 3369.

65. R. W. Armstrong and P. Rodriguez (2006), *Philos. Mag.,* **86**, 5787.

66. A. S. Khan, B. Farrokh, and L. Takacs (2008), *J. Mater. Sci.,* **43**, 3305.

67. A. Mishra, M. Martin, N. N. Thadhani, B. K. Kad, E. A. Kenik, and M. A. Meyers (2008), *Acta Mater.,* **56**, 2770.

68. M. A. Meyers, A. Mishra, and D. J. Benson (2006), *Prog. Mater. Sci.,* **51**, 427.

69. R. W. Armstrong (1968), in eds., A. R. Rosenfield, G. T. Hahn, A. L. Bement, Jr., and R. I. Jaffee, *Dislocation Dynamics,* McGraw-Hill Book Co., NY, p. 293.

70. R. W. Armstrong (1974), *Canadian Metall. Quart.,* **13**, 187.

71. H. Conrad (2004), *Metall. Mater.Trans. A,* **35A**, 2681.

72. N. Hansen and B. Ralph (1982), *Acta Metall.,* **30**, 411.

73. R. W. Armstrong and T. R. Smith (1996), in eds., C. Suryanarayana, J. Singh and F. H. Froes, *Processing and Properties of Nanocrystalline Materials*, The Metall. Soc. – AIME, Warrendale, PA, p. 345.

74. J. C. M. Li and G. C. T. Liu (1967), *Philos. Mag.,* **15**, 1059.

75. J. C. M. Li and Y. T. Chou (1970), *Metall. Trans.,* **1**, 1145.

76. R. W. Armstrong, Y. T. Chou, R. M. Fisher, and N. Louat (1966), *Philos. Mag.,* **14**, 943.

77. R. W. Armstrong (2005), *Mater. Sci. Eng. A*, **409**, 24.

78. F. R. N. Nabarro (2000), *Soviet Phys. – Sol. State Phys.,* **42**, 1417.

79. J. Weertman (1999), in eds., M. A. Meyers, R. W. Armstrong, and H. O. K. Kirchner, *Mechanics and Materials: Fundamentals and Linkages*, John Wiley & Sons, Inc., NY, Ch. 13, p. 451.

80. H. Conrad and J. Narayan (2002), *Acta Mater.,* **50**, 5067.

81. P. Rodriguez and R. W. Armstrong (2006), *(Indian) Bull. Mater. Sci.,* **29**, 717.

82. Q. Xia, C. Hamilton, K. Rechangel, C. Crowe, and G. Collins (1994), *Mater. Sci. Eng. Forum*, **170**, 147.

83. X. Zhang, H. Wang, R. O. Scattergood, C. Koch, and J. Narayan, private communication in Ref. 80.

84. C. C. Koch and J. Narayan (2001), in eds., D. Farkas *et al.*, *Structure and Mechanical Properties of Nanophase Materials – Theory and Computer Simulation Versus Experiment*, Mater. Res. Soc. Warrendale, PA, **634**, B5.1.1.

85. C. C. Koch, R. O. Scattergood, K. Linga Murty, R. K. Gudura, G. Trichy, and K. V. Rajulapati (2004), in eds., I. Ovid'ko *et al.*, *Mechanical Properties of Nanostructured Materials and Nanocomposites*, Mater. Res. Soc., Warrendale, PA, **791**, p., 51.

86. K. Ogawa and K. Tanaka (1980), *Proc. 23rd Japan Congress on Materials Research*, Soc. Mater. Sci., Japan, p. 39.

87. T. G. Langdon (2006), *J. Mater. Sci.,* **41**, 597.

88. A. A. Federov, M.Yu. Gutkin, and I. A. Ovid'ko (2003), *Acta Mater.,* **51**, 887.

89. H. Conrad (2007), *Nanotechnology* **18**, 325701.

90. F. Della Torre, P. Spatig, R. Schaublin, and M. Victoria (2005), *Acta Mater.,* **53**, 2337.

91. A. K. Ghosh (2006), *Mater. Sci. Eng A*, DOI: 10.1016/j.msea.2006.08.122.

92. Q. Wei (2007), *J. Mater. Sci.,* **42**, 1709.

93. J. C. M. Li (2007), *Appl. Phys. Lett.,* **90**, 041912.

94. J. Narayan and Y. T. Zhu (2008), *Appl. Phys. Lett.,* **92**, 151908.

95. X. L. Wu, J. Narayan, and Y. T. Zhu (2008), *Appl. Phys. Lett.,* **93**, 031910.

96. J. R. Weertman, D. Farkas, K. Hemker, H. Kung, M..Mayo, R. Mitra, and H. Van Swygenhoven (1999), *MRS Bull.,* **24**(2), 44.

97. K. Kadau, T. C. Germann, P. S. Lomdahl, B. L. Holian, D. Kadau, P. Entel, M. Kreth, F. Westerhoff, and D. E. Wolf (2004), *Metall. Mater. Trans. A,* **35A**, 2719.

98. T. Zhu, J. Li, A. Samanta, H. G. Kim, and S. Suresh (2007), *Proc. Nat. Acad. Sci.,* **104**, 3031.

99. R. W. Armstrong (1985), in eds., H. J. McQueen, J.-P. Bailon, J. I. Dickson, J. J. Jonas, and M. G. Akben, *Seventh Intern. Conf. on Strength of Metals and Alloys, ICSMA-CIRMA*, Pergamon Press, Oxford, UK, p. 195.

100. R. W. Armstrong (1997), *Trans. Indian Inst. Met.,* **50**, 521.

101. Y. Zhu, Y. Estrin, T. G. Langdon, X. Liao, T. C. Lowe, Z. Shan, and R. Z. Valiev (2008), *J. Mater. Sci.,* **43**, 7255.

102. R. Song. D. Ponge, D. Raabe, J. G. Speer, and D. K. Matlock (2006), *Mater. Sci. Eng. A,* **441**, 1.

103. B. Hwang, S. Lee, Y. C. Kim, N. J. Kim, and D. H. Shin (2006), *Mater. Sci. Eng. A,* **441** 308.

104. M. P. Bondar (2008), *Comb. Explos. Shock Waves,* **44**, 365.

105. R. W. Armstrong (2009), *Rev. Adv. Mater. Sci.*, **19**, 14.

Chapter 4

A COMPOSITE MODEL OF NANOCRYSTALLINE MATERIALS

George J. Weng

Department of Mechanical and Aerospace Engineering,
Rutgers University, New Brunswick, NJ 08903, USA
E-mail: weng@jove.rutgers.edu

Inspired by the morphology revealed in molecular dynamic simulations, we develop a composite model to study the viscoplastic behavior of nanocrystalline materials. The composite consists of the plastically harder grain interiors serving as inclusions and the plastically softer grain boundaries (or grain-boundary affected zone) serving as the matrix, with the possibility of additional interfacial grain-boundary sliding. The constitutive equations of both phases are represented by a set of power-law, unified theory whereas that of GB sliding is taken to be Newtonian. To address this nonlinear, strain and strain-rate dependent heterogeneous problem, we introduce the methods of secant viscosity and field-fluctuation to build a homogenization scheme, so that the overall stress-strain relations of the nanocrystalline material can be calculated from those of the constituent phases. The conditions without and with grain-boundary sliding are applied to Ni and Cu, respectively, to examine how their stress-strain relations, strain-rate sensitivity, and activation volume change as a function of grain size. The results show that, as the grain size decreases from micrometers all the way down to a few nanometers, both flow stress and strain-rate sensitivity increase and then decrease, whereas the activation volume decreases and then increases. These general trends are found to be consistent with the dislocation theories of Armstrong and Rodriguez[29,30] and the test results of Trelewicz and Schuh.[20]

Mechanical Properties of Nanocrystalline Materials
Edited by James C. M. Li
Copyright © 2011 Pan Stanford Publishing Pte. Ltd.
www.panstanford.com

4.1 INTRODUCTION

The influence of grain size on the yield strength and hardness of polycrystalline materials has long been described by the Hall–Petch relation.[1,2] This equation, given by

$$\sigma_y = \sigma_0 + k_y d^{-1/2}, \tag{4.1}$$

states that the yield strength increases linearly with the inverse of the square root of grain size, d. Such a linear dependence has been explained from the standpoint of dislocation pile-ups of Eshelby *et al.*[3] by Hall and Petch, and by Armstrong *et al.*[4] It can also be explained from the mechanism suggested by Li[5] that grain boundaries serve as dislocation sources, and from the dislocation density models of Conrad[6] and Ashby.[7] This relation holds sufficiently well in the traditional coarse grain range, but it apparently cannot continue to hold as the grain size approaches zero, for the yield strength would approach infinity. This was not an issue until 1984, when it became possible to process nanocrystalline materials by inert gas condensation.[8] Since then many new processing routes have been developed and many experimental and theoretical studies have been carried out to investigate the influence of grain size all the way down to the very fine grained range, even below 10 nm.

Early experimental investigations on nanocrystalline materials by Nieman *et al.*,[9] El-Sherik *et al.*,[10] Gertman *et al.*,[11] Sanders *et al.*,[12,13] and Khan *et al.*,[14] immediately showed a departure of the yield strength from this linear relation, and by Chokshi *et al.*,[15] Lu *et al.*,[16] and Fourgere *et al.*[17] a negative slope in the lower grain size range. This was initially thought to be due to the imperfections and voids in the samples, but subsequent investigations with virtually imperfection-free samples by Schwaiger *et al.*[18] still showed such a departure, and by Conrad and Narayan,[19] and Trelewicz and Schuh[20] a decline in the hardness. It is now generally believed that the departure and decline of the Hall–Petch plot are not artifacts of the specimen imperfections but are genuine characteristics of the nanocrystalline materials. As such, there is a transition of mechanical strength from a positive to a negative slope in the Hall–Petch plot and, associated with such a change, there exists a critical grain size at which the material is at its strongest state.

This in turn brings up the all important question: How to determine this critical grain size and the maximum strength?

Several theories have been put forward to explain the Hall–Petch transition. By considering the equilibrium position of a dislocation pile-up under an external stress, Nieh and Wadsworth[21] arrived at a critical grain size below which the grains could not sustain the pile-up, and thus the occurrence of the Hall–Petch breakdown. This critical grain size was calculated to be about 19.3 nm for copper and 11.2 nm for palladium, both having been reported to exhibit such transition.[15,16] Similar conclusions were also reached by Scattergood and Koch,[22] and by Lian *et al.*[23] Using the ideal shear strength and the Peierls strength as the upper and lower limits, and then taking the geometrical mean as an estimate, Wang *et al.*[24] arrived at a critical grain size of 8.2 nm for copper and 11.6 nm for palladium. Considering GB sliding as the dominant role of deformation, Gutkin *et al.*[25,26] and Ovidko[27] have suggested that the presence of triple junctions as obstacles is the cause of strengthening whereas the deformation-induced migration of GBs and triple junctions is the source of softening. By considering the competition between grain boundary shear and dislocation plasticity, Argon and Yip[28] calculated the grain size at which both mechanisms contribute to the same overall strain-rate to determine the "strongest size," and found it to be around 12.2 nm for Cu. In addition, Armstrong and Rodriguez[29] have demonstrated that the Hall–Petch type dependence for the reciprocal activation volume, which was previously derived from combination of dislocation mechanics-based thermal activation strain-rate analysis and the dislocation pile-up model for coarse grained materials, remains valid into the nanocrystalline regime. In addition Rodriguez and Armstrong[30] have further attributed the Hall–Petch breakdown to the onset of grain-boundary weakening, and predicted that reversal of strain-rate sensitivity and activation volume following the breakdown would occur.

On a separate front, molecular dynamic (MD) simulations have provided significant insights into the morphology and the properties of nanocrystalline materials at low grain size. Along this line Schiøtz *et al.*,[31-34] Swygenhoven *et al.*,[35-37] Yamakov *et al.*,[38,39] Lund and Schuh,[40] and Tomar and Zhou,[41] among others, have made significant contributions. This approach has been able to provide the stress-strain relations of nanocrystalline materials for grain size below 15 nm. These results are particularly valuable because this is also the range during which experimental data are lacking. MD simulations however involve very large number of atoms and are computationally prohibitive with large grain sizes, and they are usually run at very high strain rates, typically between 10^8/s to 10^{13}/s, that remain somewhat unrealistic. So it has its limitations in providing results for nanocrystalline materials over the entire range of grain size, and over the range of strain rates that materials encounter in real life. It is thus desirable to develop an atomically equivalent continuum model that can provide results over the entire range of grain size and strain-rate.

This is the objective of this article. We plan to build a composite model that consists of the plastically harder grain interior (as inclusions) and the plastically softer grain boundaries or grain-boundary affected zone (as matrix). These would reflect the morphology of MD simulations and the observations of in situ transmission electron microscope, respectively. In the former case we will further include the possible interfacial grain-boundary sliding in the analysis. We will then use the developed composite model to study the transition of mechanical properties as the grain size decreases all the way down to a few nanometers.

4.2 RATIONALE FOR A COMPOSITE MODEL, AND SOME PERSPECTIVES

A typical morphology for nanocrystalline materials revealed by MD simulations is reproduced in Fig. 4.1(a), for an average grain size of 5.2 nm.[31] There are two distinct regions: the grain interior and the grain-boundary zone (GB zone), both capable of undergoing plastic flow. The GB zone has a finite thickness and its volume fraction is not negligible. In some cases additional interfacial grain-boundary sliding has been reported.[31-34,40] Another microstructure in which crystallinity was seen to remains sharp all the way to the front of grain boundary has also been observed.[42] In this case the grain boundary has zero thickness and there is no grain-boundary sliding, but there exists a grain-boundary affected zone (GBAZ) on the outer region of the grain as reproduced in Fig. 4.1(b).[18] The thickness of the GBAZ is finite. Both GB and GBAZ are plastically softer than the grain interior. So in the absence of interfacial sliding both morphologies essentially share the same composite model, as depicted in Fig. 4.1(c). We shall further idealize the grain interiors as spherical inclusions. Due to the difference in their mechanical properties, the stress and strain states of these two regions are not the same and these must be determined accordingly. It is the weighted mean of the strain rates from all constituent phases that gives rise to the

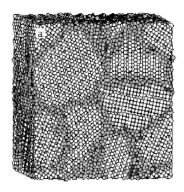

(a)

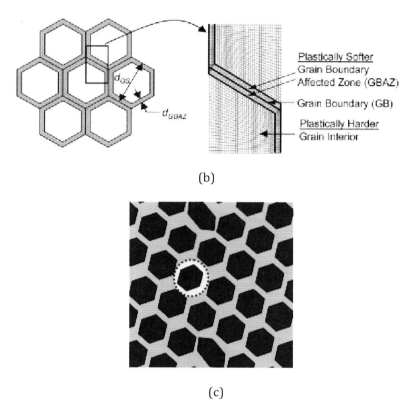

(b)

(c)

Figure 4.1 (a) Morphology of a nanocrystalline Cu by MD simulation (Schiøtz *et al.*[31]), (b) plastically harder grain interior and plastically softer GBAZ (Schwaiger *et al.*[18]), and (c) idealization by a 2-phase composite.

overall strain-rate of the nanocrystalline material. The strain-rate in each phase, in turn, must be determined from its own constitutive equations and calculated at its current level of stress and strain. It is in this spirit that the composite model is to be developed.

We now let the grain size d representing the grain diameter, and t the thickness of the GB zone or GBAZ, and further denote these two regions as phase 1 and phase 0, respectively. The volume fraction of the interior phase (phase 1) can be calculated approximately from

$$c_1 = \left(\frac{d}{d+t}\right)^3, \quad \text{or} \quad c_1 = \left(\frac{d-t}{d}\right)^3, \tag{4.2}$$

depending on Fig. 4.1(a) or (b), with $c_0 = 1 - c_1$. In (a), the GB thickness is about 2–3 lattice spacing, fairly independent of the grain size.[43,44] Conrad and Narayan[19] have suggested $t \approx 3b$ (where b is the Burgers vector). The GBAZ thickness in (b) was reported to be about 7–10 lattice spacings.[18] For instance using the first definition and taking $t = 1$ nm, the variations of c_1 and c_0 as a function of d is shown in Fig. 4.2. It is evident that even though the volume fraction of the GB zone is small at $d \geq 100$ nm, it can reach 25% at $d = 10$ nm. Since the inter-connected GB or GBAZ is plastically softer than the grain interior, its presence can have a profound effect on the overall properties of the material.

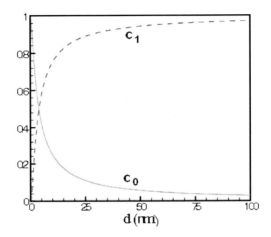

Figure 4.2 Variations of the volume concentrations, c_1 and c_0, of the grain interior and GB zone as the grain size decreases.

In retrospect, the advantage of adopting a composite concept to investigate the nanocrystalline behavior has long been recognized. This could date back to at least as early as the beginning of MD simulations. Earlier works on the rate-independent behavior include Carsley *et al.*,[45] who used the Hall–Petch hardness for the bulk phase and a constant strength for the grain-boundary phase to calculate the grain size dependence of hardness for nickel, iron, and copper. In addition, Wang *et al.*[24] considered a unit cell consisting of crystalline and inter-crystalline phase (including the grain boundary, triple line, and quadruple node), and adopted different strength for each region. They also reported significant deviation from the Hall–Petch relation as the grain size decreases. In the rate-dependent context, Kim *et al.*[46] adopted a similar unit cell to study the dependence of the stress-strain relations on the strain-rate and grain size. The strain-rate of inclusions (crystallites) was taken to be a superposition of a unified viscoplastic flow, lattice diffusion (Nabarro-Herring creep, with a d^{-2}-dependence), and grain-boundary diffusion (Coble creep, a d^{-3}-

dependence), whereas that of the grain-boundary phase was treated as ideally plastic (no work-hardening) but with a d^{-2}-dependence. Taking the grain-boundary thickness to be 1 nm, they calculated the stress-strain relations of copper at three different grain sizes: 1,000 nm, 100 nm, and 10 nm, and, in each case, considered two strain rates: 10^{-5}/s and 10^{-3}/s. Their calculated results showed an increase of flow stress in the stress-strain curves from 1,000 nm to 100 nm, but a decrease from 100 nm to 10 nm. This feature bears the significance of grain size hardening and then grain-size softening. Other composite studies have been carried out by Benson *et al.*[47] and Fu *et al.*[48] These works have an inner core representing the grain interior and an outer shell representing the work-hardened layer near the grain boundary. In contrast to most composite studies their outer shell was harder than the grain interior, and the thickness of the outer shell was assumed to be proportional to $d^{1/2}$. Then applying the mixture rule for the composite, they found that the flow stress initially increased in the Hall–Petch fashion but eventually reached a constant, asymptotic value. No stress-strain relations were reported, however. One may refer to the extensive review of Meyers *et al.*[49] for other composite models. Some other overviews—not necessarily related to the composite approach—can also be found in Dao *et al.*,[50] Ramesh,[51] and Cherkaoui and Capolungo.[52]

These composite studies have shed valuable insights into the separate contributions of the interior phase and the external one toward the overall hardness or strength of the materials. But despite the difference in the properties of both phases the computations have invariably been carried out by the simple mixture rule, that is, they all assumed that the stresses (or strains) in the grain interior and GB zone are equal, given by the externally applied stress (or strain). Thus, despite the difference in their properties, the difference in their stress (or strain) state under an external one has been ignored. It is evident that in a composite material the harder phase must carry a higher stress and undergo a lower strain as compared to the softer phase, with their weighted mean giving rise to the externally applied one. The mixture rule neglects such basic stress (or strain) distribution, and the calculated results in essence represent the lower (or upper) bounds for the flow stress of the composite. In such procedure the compatibility (or equilibrium) condition at the phase boundary is invariably violated.

Recognizing the need to account for the stress and strain heterogeneity among the constituent phases, and to use such stress and strain state to determine the response of the constituent phases, Jiang and Weng[53–55] started to introduce the principle of micromechanics in composites to model the behavior of nanocrystalline materials. In these studies a series of homogenization schemes have been introduced to determine the overall elastoplastic response of nanocrystalline metals, nanocrystalline ceramics, and nanocrystalline polycrystals as the grain size decreases from the traditional coarse to the nano-meter range. The formulation was based on the microgeometries in Fig. 4.1(a) and (c), with the volume fractions of grains and GB zone varying according to the first part of Eq. 4.2. These studies focussed on the rate-independent plastic behavior. The behavior of grain interior was taken to follow the Hall–Petch relation, whereas the grain boundary was treated as an amorphous phase with a pressure-dependent flow stress. One of the key elements introduced in the development of the homogenization schemes was the secant moduli of the constituent phases. Many fundamental properties, such as the departure from the Hall–Petch relation in the yield strength and its transition to a negative slope, the tension-compression asymmetry, influence of porosity, grain-boundary thickness, multi-phase ceramics, among others, have been examined. Some of these studies–and others–have been recently summarized in Weng.[56]

In the meantime several other investigations have also been undertaken. These include the self-consistent approach of Capolungo *et al.*[57–60] without grain-boundary sliding, the finite-element computations of Wei and Anand[61] and Wei *et al.*[62] with grain-boundary sliding and cavitation but without the finite GB zone, and the computational schemes of Jerusalem *et al.*[63] and Jerusalem and Radovitzky[64] under very high strain-rate and shock loading. Warner *et al.*[65] have also developed an atomic-based continuum model, and Joshi *et al.*[66] and Joshi and Ramesh[67] have extended the secant-moduli formulation suggested in Weng[68] to study the plastic behavior of bimodal metals with simultaneous presence of ultrafine grains and coarse grains, and that of materials with hierarchical microstructure. These continuum studies have greatly contributed to our current understanding of the subject.

In this article we shall focus on the rate-dependent behavior. Such a focus has the merit that plastic deformation above 0 K is fundamentally a stress-assisted thermally activated process,[69] and it also has a long historical standing from the perspective of dislocation mechanics.[70–73] Most contemporary measurements also adopted the strain-rate test,[74–81] and a related article can also be found from Armstrong[82] in this book. To treat the rate-dependent issue, we shall extend the concept of secant moduli in rate-independent plasticity to secant viscosity in rate-dependent viscoplasticity. This concept was originally proposed in Li and Weng[83–85] to study the time-dependent creep and stress-strain relations of particle-reinforced composite materials. It allows one to account for the nonlinear dependence of strain-rate of a constituent phase on its current stress and plastic strain. Then by means of a unified transition from elasticity to viscoelasticity through the Laplace transform,[86] and then the transition from viscoelasticity to viscoplasticity by the replacement of the linear viscosity by the secant viscosity, the overall viscoplastic response of a particle-reinforced composite could be determined. This rate-dependent approach draws from its analogy to the rate-independent one which has its origins in Talbot and Willis,[87] Tandon and Weng,[88] Ponte Castaneda,[89] Qiu and Weng,[90] Suquet,[91] and Hu,[92] among others. Both are based on the concept of a linear comparison composite. The key difference lies in the introduction of an intermediate viscoelastic state in the rate-dependent case, and the use of the secant viscosities instead of secant moduli to characterize the current viscoplastic state of the constituent phases.

4.3 CONSTITUTIVE EQUATIONS OF THE GRAIN INTERIOR, GBAZ, GB ZONE, AND INTERFACIAL GRAIN-BOUNDARY SLIDING

Before we proceed to build the composite model, let us first provide the constitutive equations for each constituent phase.

At a given state of deformation the total strain-rate is the sum

$$\dot{\varepsilon}_{ij} = \dot{\varepsilon}_{ij}^e + \dot{\varepsilon}_{ij}^{vp}, \tag{4.3}$$

where the elastic rate, $\dot{\varepsilon}_{ij}^e$, is linearly related to the stress-rate, and the viscoplastic rate, $\dot{\varepsilon}_{ij}^{vp}$, is taken to follow the power-law, unified theory

$$\dot{\varepsilon}_e^{vp} = \dot{\varepsilon}_0^{vp} \cdot \left(\frac{\sigma_e}{s}\right)^n, \tag{4.4}$$

where $\dot{\varepsilon}_e^{vp}$ is the effective viscoplastic strain-rate and σ_e von Mises' effective stress, defined as usual by

$$\sigma_e = \left(\frac{3}{2}\sigma_j'\sigma_j'\right)^{1/2}, \quad \dot{\varepsilon}_e^{vp} = \left(\frac{2}{3}\dot{\varepsilon}_{ij}^{vp}\dot{\varepsilon}_{ij}^{vp}\right)^{1/2}, \tag{4.5}$$

in terms of the second invariant of the deviatoric stress, σ_{ij}', and strain-rate, $\dot{\varepsilon}_{ij}^{vp}$. The components of the viscoplastic strain-rate then follow from the Prandtl-Reuss relation, as

$$\dot{\varepsilon}_{ij}^{vp} = \frac{3}{2}\frac{\dot{\varepsilon}_e^{vp}}{\sigma_e}\sigma_{ij}'. \tag{4.6}$$

The drag stress, s in Eq. 4.4, which represents the hardening state of the phase, grows through the competition of strain hardening and dynamic recovery and can be written as

$$\dot{s} = h \cdot \left(1 - \frac{s}{s_*} \right) \dot{\varepsilon}_e^{vp}, \tag{4.7}$$

so that, upon integration, it depends on the current strain as

$$s = s_* - (s_* - s_0) \exp\left(-\frac{\varepsilon_e^{vp}}{s_*/h} \right), \tag{4.8}$$

where s_0 represents the initial hardening state and s_* the final saturation state. This applies to the gain interior, and GBAZ in Fig. 4.1(b).

For the GB zone in Fig. 4.1(a), we shall adopt a pressure-dependent constitutive equation to reflect its somewhat disordered atomic state. Such a disordered state is close to that of an amorphous phase and the presence of pressure sensitivity would lead to the observed tension-compression asymmetry in nanocrystalline materials.[13,40,53-55] In the spirit of Drucker's[93] pressure-dependent yield function, we write it as

$$s = \lambda p + s_* - (s_* - s_0) \exp\left(-\frac{\varepsilon_e^{vp}}{s_*/h} \right), \tag{4.9}$$

where p is the hydrostatic pressure given by $p = -(1/3)\,\sigma_{kk}$, and λ the pressure-dependent constants. In the rate-independent context Drucker's constitutive equation has proven to be valuable to the modeling of metallic glasses.[94]

Furthermore, the initial and saturation stresses of the grain interior, s_0 and s_*, are grain size dependent, taken to obey the Hall–Petch relation as

$$s_0^{(g)} = s_0^\infty + kd^{-1/2}, \quad s_*^{(g)} = a \cdot s_0^\infty. \tag{4.10}$$

So in addition to h and n, it has three constants: the frictional stress, s_0^∞, the Hall–Petch constant, k, and the coefficient, a, whereas GBAZ has two: the initial strength $s_0^{(gb)}$ and the saturation strength, $s_*^{(gb)}$, and the GB zone has the additional λ. There are of course the Young's modulus, E, and Poisson's ratio, ν (or bulk modulus, κ and shear modulus, μ) that must be supplied for their respective elastic properties.

The constitutive equation of grain-boundary sliding is taken to be Newtonian. In the 3-D setting the stress and velocity jumps can be written as

$$[\sigma] \cdot n = 0, \quad \text{and} \quad \alpha \cdot [v] = \sigma \cdot n, \tag{4.11}$$

where v is the velocity vector and n the unit outward normal vector from the grain interior surface, and the square brackets [.] represent the jump of the said quantity. This constitutive equation implies a traction continuity and a velocity jump across the interface, with a sliding viscosity tensor, α. This tensor has three distinct components, one normal, α_n, and two tangential, α_s and α_t; it can be written as

$$\alpha_{ij} = \alpha_n \, n_i n_j + \alpha_s \, s_i s_j + \alpha_t \, t_i t_j, \tag{4.12}$$

in terms of n_i and the two orthogonal unit tangential vectors, s_i and t_i. When all $\alpha_i \to \infty$, the interface recovers to perfect bonding and there is no interfacial grain-boundary sliding or cavitation, and when $\alpha_i \to 0$, the interface is completely lubricant. For our GB sliding problem, $\alpha_n \to \infty$, and the two shear viscosities are equal, with $\alpha_t = \alpha_s$.

4.3.1 Strain-Rate Sensitivity, *m*, and Activation Volume, *v**, of the Grain Interior and GB (or GBAZ) Zone

In addition to yield strength, two widely considered properties are strain-rate sensitivity and activation volume of nanocrystalline materials. At this point it is useful to see how these two parameters are linked with the unified constitutive equations given above for each constituent phase. Wei *et al.*[77] have pointed out that there are two commonly adopted definitions of strain-rate sensitivity: one is the physically based *S*, and the other the engineering-based *m*. In terms of the shear stress, τ, and shear strain-rate, $\dot\gamma$, and invoking von Mises' relations $\tau = \sigma_e/\sqrt{3}$ and $\dot\gamma = \sqrt{3}\dot\varepsilon_e$, these two constants can be calculated from

$$S = \frac{\partial \tau}{\partial \ln (\dot\gamma^{vp})} = \frac{1}{\sqrt{3}} \frac{\partial \sigma_e}{\partial \ln (\dot\varepsilon_e^{vp})}, \quad m = \frac{\partial \ln (\sigma_e)}{\partial \ln (\dot\varepsilon_e^{vp})}. \tag{4.13}$$

Each of these expressions has its own merit. The physically based *S* has the virtue that, since for a stress-assisted, thermally activated process, the shear strain-rate can be written in terms of the Arrhenius function, it is directly related to the activation volume, *v**, such that

$$v^* = \frac{k_B T}{S} = \frac{\sqrt{3} k_B T}{\sigma_e m}, \tag{4.14}$$

where k_B is the Boltzmann constant and *T* the absolute temperature. The strain-rate sensitivity, *m*, however, covers a wider range of stress and is the favored parameter in engineering applications. For a simple power-law material such as in steady-state creep, $m = 1/n$. So for a low *n*-value such as under Newtonian flow or Coble creep, it has the high rate sensitivity of $m = 1$, and for the common dislocation climb-plus-glide, $n = 3$–7, and thus $m = 0.14$–0.33. Under most dynamic testing the value of *m* is much smaller, typically below 0.06.

Now return to the power-law unified constitutive equations. Since $S = m\sigma_e/\sqrt{3}$, the physically based *S* will increase with σ_e. Thus *S* measured at higher flow stress will be higher than that measured at lower flow stress. Due to their reciprocal relation, activation volume *v** will have exactly the opposite trend. This also implies that, for either the grain interior or the GB zone, its strain-rate sensitivity, S, will increase with deformation owning to strain hardening, and it will also increase with strain-rate which tends to lead to higher flow stress. Their activation volume will decrease with both deformation and strain-rate. For the grain interior there is the additional factor of grain size dependence in Eq. 4.10, and thus its S will increase and its v^* will decrease with the Hall–Petch type relation. This trend is consistent with Armstrong and Rodriguez's[29] demonstration that, during the Hall–Petch strengthening, one can expect

$$\frac{1}{v^*} = \frac{1}{v_0} + k_{v^*} \cdot d^{-1/2}.$$ (4.15)

4.3.2 The Secant Viscosity of the Individual Phase

The constitutive behaviors of grain interior and grain boundary phases are nonlinear, rate-dependent, and capable of strain-hardening. To address this complex issue we shall adopt the concept of secant viscosities to build the homogenization scheme. To this end it is helpful to observe the following two characteristics:

(i) The decomposition of total strain-rate into the elastic and viscoplastic strain rates in Eq. 4.3 calls for a Maxwell-type linear viscoelastic comparison phases.

(ii) At a given stage of deformation, the shear viscosity, say η_r of the r-th Maxwell phase, will represent the *secant* viscosity, η_r^s, of the real, nonlinear phase, which is defined through the deviatoric relations

$$\dot{\varepsilon}_{ij}^{vp(r)} = \frac{1}{2\eta_r^s} \sigma_{ij}^{'(r)}, \quad \text{or} \quad \dot{\varepsilon}_e^{vp(r)} = \frac{1}{3\eta_r^s} \sigma_e^{(r)},$$ (4.16)

where, in view of Eq. 4.4,

$$\eta_1^s = s^{(g)}/3\dot{\varepsilon}_e^{vp(g)} \cdot \left(\frac{\dot{\varepsilon}_e^{vp(g)}}{\dot{\varepsilon}_0^{vp(g)}}\right)^{1/n^{(g)}}, \quad \eta_0^s = \frac{s^{(gb)}}{3\dot{\varepsilon}_e^{vp(g)}} \cdot \left(\frac{\dot{\varepsilon}_e^{vp(gb)}}{\dot{\varepsilon}_0^{vp(gb)}}\right)^{1/n^{(gb)}},$$ (4.17)

for the grain interior (phase 1) and GB zone (phase 0). Unlike the ordinary viscosity in a linear viscoelastic material these secant viscosities will continue to increase due to strain hardening, and, in the case of grain interior, it further depends on the grain size.

4.4 ELASTIC PROPERTIES AND INTERNAL STRESS OF THE NANOCRYSTALLINE MATERIAL

Since the elastic properties of the grain boundary may be softer or harder than those of the grain interior.[95–98] It is useful to provide a formula for the effective moduli of the nanocrystalline material. This is a classic topic in the micromechanics of composites, and several models, such as the Mori-Tanaka approach,[99-102] self-consistent method,[103,104] generalized self-consistent scheme,[105] differential scheme,[106,107] and the double-inclusion method,[108] can be called for to give an estimate. For the microgeometry depicted in Fig. 4.1(c), both Mori-Tanaka and generalized self-consistent scheme are quite suitable. To avoid additional complications in the field fluctuation and Laplace transform that we will encounter later, the former one will be adopted, and this lead to[99]

$$\kappa_c = \kappa_0 \left[1 + \frac{c_1(\kappa_1 - \kappa_0)}{c_0\alpha_0(\kappa_1 - \kappa_0) + \kappa_0} \right],$$

$$\mu_c = \mu_0 \left[1 + \frac{c_1(\mu_1 - \mu_0)}{c_0\beta_0(\mu_1 - \mu_0) + \mu_0} \right], \tag{4.18}$$

where

$$\alpha_0 = \frac{(1 + \nu_0)}{3(1 - \nu_0)} = \frac{3\kappa_0}{3\kappa_0 + 4\mu_0} \quad \beta_0 = \frac{2(4 - 5\nu_0)}{15(1 - \nu_0)} = \frac{6}{5}\frac{\kappa_0 + 2\mu_0}{3\kappa_0 + 4\mu_0}, \tag{4.19}$$

and κ_r and μ_r are the bulk and shear moduli of the r-th phase, while ν_0 is the Poisson's ratio of the matrix (GBAZ or GB zone). This pair of moduli is known to coincide with the Hashin-Shtrikman[109] and Walpole[110] lower bounds if the matrix is the softer phase, and with their upper bounds if the matrix is the harder one. They will never violate the bounds.

Due to the difference in their elastic constants, the average stresses of the grain interior and GB zone are also different. In terms of the applied stress, $\overline{\sigma}_{ij}$, their hydrostatic and deviatoric components can be written as[99]

$$\sigma_{kk}^{(1)} = \frac{\kappa_1}{(c_1 + c_0\alpha_0)(\kappa_1 - \kappa_0) + \kappa_0}\overline{\sigma}_{kk},$$

$$\sigma_{ij}^{\prime(1)} = \frac{\mu_1}{(c_1 + c_0\beta_0)(\mu_1 - \mu_0) + \mu_0}\overline{\sigma}_{ij}^{\prime}, \tag{4.20}$$

and

$$\sigma_{kk}^{(0)} = \frac{\alpha_0(\kappa_1 - \kappa_0) + \kappa_0}{(c_1 + c_0\alpha_0)(\kappa_1 - \kappa_0) + \kappa_0}\overline{\sigma}_{kk},$$

$$\sigma_{kk}^{\prime(0)} = \frac{\beta_0(\mu_1 - \mu_0) + \mu_0}{(c_1 + c_0\beta_0)(\mu_1 - \mu_0) + \mu_0}\overline{\sigma}_{ij}^{\prime}, \tag{4.21}$$

for the grain interior and GB zone, respectively. (An over-bar implies that it is the averaged, composite term.)

4.5 NANOCRYSTALLINE MATERIALS WITHOUT INTERFACIAL GRAIN-BOUNDARY SLIDING

In this section we consider the ordinary two-phase composite without interfacial grain-boundary sliding. This applies to the GBAZ microgeometry as shown in Fig. 4.1(b) or the GB zone microgeometry as shown in Fig. 4.1(a) but without the additional interfacial GB sliding. The case with the additional GB sliding will be taken up in Section 4.6. With such an assumption a simple homogenization scheme without going through the Laplace transform could be established. This will be presented first.

4.5.1 The Secant Viscosity of the Nanocrystalline Materials

By analogy to Eq. 4.18, a simple secant-viscosity approach can be used to estimate the effective secant viscosity of the nanocrystalline material. When the deformation is dominated by the inelastic strain-rate, the effective shear viscosity can be written as[111]

$$\eta_c = \eta_0 \left[1 + \frac{c_1(\eta_1 - \eta_0)}{(2/5) \cdot c_0(\eta_1 - \eta_0) + \eta_0} \right], \tag{4.22}$$

where the coefficient 2/5 comes from the β_0-term in the viscous context under the condition of plastic incompressibility (by setting $v_0 = 1/2$). Since both the grain interior and GB zone are taken to be plastically incompressible, the dilatational viscosity of the composite approaches infinity.

Extending this linear effective viscosity to the nonlinear context, the effective secant viscosity can be written similarly as

$$\eta_c^s = \eta_0^s \left[1 + \frac{c_1(\eta_1^s - \eta_0^s)}{(2/5) \cdot c_0(\eta_1^s - \eta_0^s) + \eta_0^s} \right], \tag{4.23}$$

so that it could be used in

$$\bar{\sigma}_{ij}' = 2\eta_c^s \bar{\bar{\varepsilon}}_{ij}', \tag{4.24}$$

to calculate the current stress of the nanocrystalline material under a constant strain-rate loading, $\bar{\bar{\varepsilon}}_{ij} = const$. The dilatational stress is simply related to the dilatational strain through the elastic bulk modulus κ_c in Eq. 4.18, leading to the rate, $\dot{\bar{\sigma}}_{kk} = 3\kappa_c \dot{\bar{\varepsilon}}_{kk}$.

Equation (4.24) is seemingly linear, but it is dependent on the plastic state of the grain interior and GB zone and thus is highly nonlinear. It is evident from Eq. 4.23 that the effective secant viscosity depends on η_1^s and η_0^s, which, in view of Eq. 4.17, further depend on the drag stresses, $s^{(g)}$ and $s^{(gb)}$ and plastic strain rates, $\dot{\varepsilon}_e^{vp(g)}$ and $\dot{\varepsilon}_e^{vp(gb)}$. The drag stresses are functions of the plastic strain, ε_e^{vp} through Eq. 4.8 and, in the case of grain interior, it also depends on the grain size, d, in Eq. 4.10. The grain size dependence also enters into Eq. 4.24 explicitly through c_1 and c_0 in Eq. 4.23, as written in Eq. 4.2. When fully executed, a wealth of mechanical properties of nanocrystalline materials can be extracted from this simple equation.

4.5.2 The Field-Fluctuation Approach to Connect the Strain Rates of the Grain Interior and GB Zone to the External, Applied Strain-Rate

A key step in the application of Eq. 4.23 is the determination of the secant viscosities, η_1^s and η_0^s, under a given applied strain-rate, $\dot{\bar{\varepsilon}}_{ij}$. This in turn requires the determination of the effective strain-rate, $\dot{\bar{\varepsilon}}_e^{vp(g)}$ and $\dot{\bar{\varepsilon}}_e^{vp(gb)}$, and their respective drag stresses, $s^{(g)}$ and $s^{(gb)}$, which are functions of $\varepsilon_e^{vp(g)}$ and $\varepsilon_e^{vp(gb)}$. Thus a fundamental problem is the determination of the individual strain rates at a given applied external rate, $\dot{\bar{\varepsilon}}_e^{vp}$. This can be accomplished through application of the field-fluctuation method.

Such a method was originally developed for the elastic composites by Bobeth and Diener[112] and Kreher and Pompe.[113] Subsequently it was extended to the rate-independent plasticity by Suquet[91] and Hu,[92] and to the rate-dependent viscoplastic behavior by Li and Weng.[84] The novelty of field fluctuation in a composite material is that, under the same boundary condition –whether it is under a constant stress for the creep deformation, a constant strain for a relaxation process, or a constant strain-rate for the calculation of stress-strain curves–a change in a material constant of a particular phase will result in a change of the overall energy that is solely dependent on this particular change. Based on such an observation some internal-to-external connections can be readily established.

For the present problem the relevant energy term is the overall work rate, which can be written globally as

$$\dot{U} = \bar{\sigma}_{ij}\dot{\bar{\varepsilon}}_{ij} = \bar{\sigma}_{ij}\left(\dot{\bar{\varepsilon}}_{ij}^e + \dot{\bar{\varepsilon}}_{ij}^{vp}\right) = \dot{U}^e + \dot{U}^{vp},$$

(4.25)

for the composite, and it can also be written locally in terms of the contributions from the grain interior and GB zone, as

$$\dot{U} = c_1\left(\dot{U}_1^e + \dot{U}_1^{vp}\right) + c_0\left(\dot{U}_0^e + \dot{U}_0^{vp}\right).$$

(4.26)

In Eq. 4.25 the total work rate is decomposed into the elastic and viscoplastic components, whereas in Eq. 4.26 such decompositions have been further written for both constituent phases. These two expressions must be equal for consistency. Then recognizing that, for the composite,

$$\dot{U}^{vp} = 3h_c^s \cdot (\dot{\bar{e}}_e^{vp})^2,$$

(4.27)

and for the individual phases,

$$\dot{U}_1^{vp} = 3h_1^s \cdot (\dot{e}_e^{vp(1)})^2, \text{ and } \dot{U}_0^{vp} = 3h_0^s \cdot (\dot{e}_e^{vp(0)})^2,$$

(4.28)

the equality will take the form

$$c_1[\dot{U}_1^e + 3h_1^s \cdot (\dot{e}_e^{vp(1)})^2] + c_0[\dot{U}_0^e + 3h_0^s \cdot (\dot{e}_e^{vp(0)})^2] = \dot{U}^e + 3h_c^s \cdot (\dot{\bar{e}}_e^{vp})^2$$

(4.29)

Since this relation holds for any field, we may take the applied strain-rate, $\dot{\bar{\varepsilon}}_e^{vp}$, fixed. We may also keep η_c^s fixed, and vary only η_1^s which in turn will also cause the effective viscosity, η_c^s, to change. This is tantamount to taking the partial derivative of Eq. 4.29 with respect to η_1^s. This in turn leads to $3c_1 \cdot (\dot{e}_e^{vp(1)})^2 = 3(\partial h^s / \partial h_1^s) \cdot (\dot{\bar{e}}_e^{vp})$. After carrying out such a procedure again for η_0^s, we arrive at the connections between the internal strain rates of the grain interior and GB zone with the external, applied strain-rate:

$$\dot{e}_e^{vp(1)} = \left(\frac{1}{c_1} \frac{\partial h_c^s}{\partial h_1^s} \right)^{1/2} \dot{\bar{e}}_e^{vp}, \quad \dot{e}_e^{vp(0)} = \left(\frac{1}{c_0} \frac{\partial h_c^s}{\partial h_0^s} \right)^{1/2} \dot{\bar{e}}_e^{vp}.$$

(4.30)

Making use of the effective viscosity in Eq. 4.23 we find

$$\frac{\partial h_c^s}{\partial h_1^s} = \frac{c_1 R^2}{\left[(2/5) \cdot c_0(1-R) + R \right]^2},$$

$$\frac{\partial \eta_c^s}{\partial \eta_0^s} = 1 + \frac{c_1(1-R)}{(2/5) \cdot c_0(1-R) + R} - \frac{c_1 R}{[(2/5) \, c_0 \, (1-R) + R]^2},$$

(4.31)

where R is the ratio, $R = \eta_0^s / \eta_1^s$.

The stress-strain curves then can be calculated from Eq. 4.24 under constant strain-rate: $\dot{\bar{\varepsilon}}_{ij}$ = *const*. Due to the continuous change of the secant viscosities, η_1^s and η_0^s, in the course of deformation, this must be done incrementally, with the continuous update of these quantities and the effective secant viscosity, η_c^s of the composite. Through the relation in Eq. 4.2 between d and c_1, the grain size dependence and strain-rate sensitivity of the nanocrystalline materials can be obtained.

4.6 NANOCRYSTALLINE MATERIALS WITH ADDITIONAL INTERFACIAL GRAIN-BOUNDARY SLIDING

In this case there are three different sources of deformation: grain interior, GB zone, and interfacial grain-boundary sliding. Deformation in the grain interior is dislocation-mediated, whereas that of the GB zone comes from the uncorrelated events in which a few atoms or a few tens of atoms slide with respect to each other.[32] Interfacial grain-boundary sliding usually involves the diffusion process and is relatively slow, but since these three regions deform independently, it is not the rate-controlling mechanism under normal or high rate loading. As in the preceding section we shall start from the elastic state, but this time we will use it to build an intermediate viscoelastic state and then translate it into the viscoplastic state. Due to the presence of interfacial sliding the combined inclusion and the imperfect interface could become softer than the outer GB zone. As such, the theory needs to be able to capture the overall dilatational compressibility. This was not an issue in the preceding section because the grain interior was harder than the outer one and any plastic compressibility would have been insignificant.[90] To ensure such an outcome the Laplace transform technique will be used to build the viscoelastic state from the elastic one through the correspondence principle.

4.6.1 Elastic Properties of Nanocrystalline Materials with Additional Interfacial Imperfections

The issue of interfacial grain-boundary sliding has its basis in an elastic imperfect interface problem. Unlike the GB zone, this interface has zero thickness. As with Eqs. 4.11 and 4.12, the deformation takes place through the jump of displacement across the interface. In the elastic context this can be similarly written as

$$[\sigma] \cdot n = 0, \quad \text{and} \quad \alpha^{(e)} \cdot [u] = \sigma \cdot n, \tag{4.32}$$

where $\alpha^{(e)}$ is the elastic counterpart of α, with

$$\alpha_{ij}^{(e)} = \alpha_n^{(e)} n_i n_j + \alpha_s^{(e)} s_i s_j + \alpha_t^{(e)} t_i t_j, \tag{4.33}$$

with entirely analogous interpretations.

Determination of the effective moduli of a 2-phase composite with an interfacial imperfection has been a subject of considerable study in the past. These include Benveniste's[114] work with interfacial shear jump and Hashin's[115] work with normal and shear jumps. Within the framework of nanocrystalline materials, Nan *et al.*,[116] Wang *et al.*,[117] and Sharma and Ganti[118] have considered the influence of grain-boundary sliding and interphase to estimate the effective elastic moduli. At this point it is also fitting to note that there has been considerable amount of work dealing with the elastic moduli of composites with nano-sized particles. This line of approach has included the effects of surface energy and surface stress, and significant contributions have been made by Sharma and Ganti,[119] Sharma and Wheeler,[120] Duan *et al.*,[121] Huang and Wang,[122] Huang and Sun,[123] and Chen *et al.*,[124,125] among others. The issue of surface energy can be quite significant when there is a sharp interface between two distinctly different atomic structures, but since the grain interior and grain boundary share the same type of atoms the issue is of a secondary nature in nanocrystalline materials.

Now let us return to the problem of imperfect interface. Most recently Duan *et al.*[126] made use of the concept of an "equivalent inclusion"–which represents the combination of the real inclusion and the imperfect interface–and the elastic energy equivalence to estimate its moduli. The concept of elastic energy equivalence is a powerful and elegant one; it has also been adopted by Shen and Li[127,128] in their interphase study. This approach has the merit of giving explicit expressions for both effective bulk and shear moduli of the equivalent inclusion, as

$$\frac{\kappa_e}{\kappa_1} = \frac{m_r \mu_0}{3\kappa_1 + m_r \mu_0},$$

$$\frac{\mu_e}{\mu_1} = \frac{24 M m_\theta + m_r \left(16M + m_\theta N\right)}{80 g_3 M + 4 g_3 m_\theta \left[10\left(7 - \nu_1\right) + M\right] + m_r \left[2 g_3 \left(140 - 80\nu_1 + 3M\right) + m_\theta N\right]},$$

$$(4.34)$$

where

$$M = g_3 \left(7 + 5\nu_1\right), \quad N = 5\left(28 - 40\nu_1 + M\right), \quad g_3 = \mu_1/\mu_0,$$

$$m_r = \alpha_n^{(e)} R_1/\mu_0, \quad m_\theta = \alpha_s^{(e)} R_1/\mu_0, \qquad (4.35)$$

and $R_1 = d/2$ is the radius of the grain size. Because $\alpha_n \to \infty$ in our problem, we have $m_r \to \infty$, and this in turn leads to

$$\kappa_e = \kappa_1,$$

$$\mu_e = \mu_1 \frac{\left(16M + m_\theta N\right)}{\left[2 g_3 \left(140 - 80\nu_1 + 3M\right) + m_\theta N\right]}. \qquad (4.36)$$

Since the interface has zero thickness, the volume concentration of the equivalent inclusion remains as c_1. The effective moduli of the nanocrystalline material now can be calculated from Eq. 4.18 as

$$\kappa_c = \kappa_0 \left[1 + \frac{c_1(\kappa_e - \kappa_0)}{c_0\alpha_0(\kappa_e - \kappa_0) + \kappa_0} \right]$$

$$\mu_c = \mu_0 \left[1 + \frac{c_1(\mu_e - \mu_0)}{c_0\beta_0(\mu_e - \mu_0) + \mu_0} \right] \tag{4.37}$$

The stress distribution can be written similarly as in Eqs. (4.20) and (4.21), with phase 1 replaced by phase "*e*", the equivalent inclusion phase.

4.6.2 Viscoelastic State Through the Correspondence Principle

To be able to capture the overall dilatational compressibility due to the weakening effect of the interfacial grain-boundary sliding, a unified scheme based on the transition from elasticity to viscoelasticity, and then to viscoplasticity, will be adopted now. This transition scheme was first developed in Li and Weng[85] to study the viscoplastic behavior of particle-reinforced metal-matrix composites. This approach first makes use of the correspondence principle between elasticity and viscoelasticity through the Laplace transform, and then replaces the Maxwell viscosity of the constituent phases by their respective secant viscosity. Introduction of the intermediate viscoelastic state makes this approach a bit more involved, but it can deliver the overall compressibility even when both constituent phases are plastically incompressible.

Through the correspondence principle, the elastic state in Eq. 4.37 can be transformed to the Laplace domain, to yield the effective moduli in the "transformed domain' (TD), as

$$\kappa^{TD} = \kappa_0^{TD} \left[1 + \frac{c_1(\kappa_1 - \kappa_0^{TD})}{c_0\alpha_0^{TD}(\kappa_1 - \kappa_0^{TD}) + \kappa_0^{TD}} \right],$$

$$\mu^{TD} = \mu_0^{TD} \left[1 + \frac{c_1(\mu_e^{TD} - \mu_0^{TD})}{c_0\beta_0^{TD}(\mu_e^{TD} - \mu_0^{TD}) + \mu_0^{TD}} \right], \tag{4.38}$$

with

$$\alpha_0^{TD} = \frac{3\kappa_0}{3\kappa_0 + 4\mu_0^{TD}}, \quad \text{and} \quad \beta_0^{TD} = \frac{6}{5}\frac{\kappa_0 + 2\mu_0^{TD}}{3\kappa_0 + 4\mu_0^{TD}}. \tag{4.39}$$

where κ_e^{TD} and μ_e^{TD} of the equivalent inclusion follow from Eq. 4.36. In light of the decomposition of total strain-rate into the elastic and plastic components in Eq. 4.3, both phase 1 and phase 0 are taken to be of the Maxwell type in its deviatoric response, but remain elastic in its dilatational response. In this spirit, the "transformed" shear moduli of the constituent phases carry the forms

$$\mu_r^{TD} = \frac{\mu_r s}{s + T_r}, \quad \text{with} \quad T_r = \frac{\mu_r}{\eta_r}, r = 1, 0, \tag{4.40}$$

where η_r is the shear viscosity of the r-th phase which will replace the secant shear viscosity η_r^s later, and s is the usual Laplace parameter. In addition, the overall stress-strain relation of the composite in the transformed domain can be written through

$$\hat{\bar{\varepsilon}}_{kk} = 3\kappa_c^{TD} \hat{\bar{\varepsilon}}_{kk}, \quad \text{and} \quad \hat{\bar{\sigma}}'_{ij} = 2\mu_c^{TD} \hat{\bar{\varepsilon}}'_{ij}, \tag{4.41}$$

where the hat $^\wedge$ on top of the overall stress and strain signifies the "transformed" state.

In the absence of interfacial sliding, this approach has been used to study the influence of inclusion shape on the time-dependent creep, strain-rate sensitivity, relaxation behavior, and complex moduli of a class of two-phase viscoelastic composites.[129,130]

4.6.3 Viscoplastic State Through the Replacement of Maxwell Viscosity by the Secant Viscosity of the Constituent Phases

With the linear viscosity of the constituent phases continuously replaced by their secant viscosity, the Laplace inversion of Eq. 4.41 can yield a variety of information for the viscoplastic behavior of nanocrystalline materials under various boundary conditions. For instance for the study of time-dependent creep under a constant stress, one has $\hat{\bar{\sigma}}_{ij} = (1/s) \cdot \bar{\sigma}_{ij}$, and for the study of stress-strain relations under a constant strain-rate, one has $\hat{\bar{\varepsilon}}_{ij} = (1/s^2) \cdot \dot{\bar{\varepsilon}}_{ij}$. The overall strain or stress can be obtained in turn from the Laplace inversion of $\hat{\bar{\varepsilon}}_{ij}$ or $\hat{\bar{\sigma}}_{ij}$, symbolically as $\bar{\varepsilon}_{ij}(t) = L^{-1} \{ \hat{\bar{\varepsilon}}_{ij} \}$ or $\bar{\sigma}_{ij}(t) = L^{-1} \{ \hat{\bar{\sigma}}_{ij} \}$. Barai and Weng[131] have applied this scheme to study the time-dependent creep of nanocrystalline materials in the absence of interfacial sliding. Grain boundary may contain porosity and/or impurity, and Li[132] has found that their presence could have a significant influence on the yield strength of nanocrystalline materials. In this case a modified GB zone which combines the original GB and porosity or impurity could be introduced with the help of Eq. 4.18, and further used to study their effects. By considering the simultaneous presence of grains and porosity inside the matrix, Barai and Weng[133] have also made use of the Laplace inversion to study the coupling behavior of grain size, porosity, flow stress, and strain-rate of nanocrystalline materials, also without the interfacial sliding.

Now with the specific objective of determining the stress-strain relations of nanocrystalline materials under a constant strain-rate loading, Eq. 4.41 can be inverted back to the real space, in the form

$$\bar{\sigma}_{kk}(t) = 3\eta_\kappa(t)\dot{\bar{\varepsilon}}_{kk}, \qquad \bar{\sigma}'_{ij}(t) = 2\eta_\mu \dot{\bar{\varepsilon}}'_{ij},\qquad (4.42)$$

where $\eta_\kappa = L^{-1}\left(\kappa_c^{TD}/s^2\right)$ and $\eta_\mu = L^{-1}(\mu_c^{TD}/s^2)$. These two constants represent the overall dilatational and shear viscosities of the viscoelastic composite. For the system with interfacial sliding it can be shown that[134]

$$\eta_\kappa(t) = \kappa_0\{(1 + A_2)t + A_1\,[1 - \exp(-\tau_1 t)]\},$$
$$\eta_\mu(t) = \eta_0\left(1 - \exp(-T_0 t)\right) + A + B\exp(-T_0 t) + C\exp(s_1 t)$$
$$+ D\exp(s_2 t) + E\exp(s_3 t) + F\exp(s_4 t),\qquad (4.43)$$

where parameters $A_1, A_2, ...,$ etc. depend on the elastic and viscosity constants, and volume concentrations of the constituent phases. A complete list of these constants can be found in Barai and Weng.[134] It is through c_1 and Eq. 4.10 that the effect of grain size, d, enters into Eq. 4.43.

By replacing the Maxwell viscosities which appear in the constants in Eq. 4.43 by the corresponding secant viscosities of the constituent phases provided by 4.17, Eq. 4.42 in turn leads to the dependence of the overall stress on the applied strain-rate through the effective *secant* "bulk" and "shear" viscosities as

$$\bar{\sigma}_{kk}(t) = 3\eta_\kappa^s \dot{\bar{\varepsilon}}_{kk}, \qquad \bar{\sigma}'_{ij}(t) = 2\eta_\mu^s \dot{\bar{\varepsilon}}'_{ij}.\qquad (4.44)$$

Since η_κ^s does not go to infinity, the material is plastically compressible even if the individual phases are plastically incompressible.

Recognizing that the material is now plastically compressible, the field fluctuation principle can be shown to lead to

$$\dot{\varepsilon}_e^{vp(1)} = \left[\frac{1}{c_1}\left(\frac{1}{3}\frac{\partial\eta_\kappa^s}{\partial\eta_1^s}(\dot{\bar{\varepsilon}}_{kk})^2 + \frac{\partial\eta_\mu^s}{\partial\eta_1^s}\dot{\bar{\varepsilon}}_e^2\right)\right]^{1/2},$$

$$\dot{\varepsilon}_e^{vp(0)} = \left[\frac{1}{c_0}\left(\frac{1}{3}\frac{\partial\eta_\kappa^s}{\partial\eta_0^s}(\dot{\bar{\varepsilon}}_{kk})^2 + \frac{\partial\eta_\mu^s}{\partial\eta_0^s}\dot{\bar{\varepsilon}}_e^2\right)\right]^{1/2},\qquad (4.45)$$

where $\dot{\bar{\varepsilon}}_e^2 = (2/3)\dot{\bar{\varepsilon}}'_{ij}\dot{\bar{\varepsilon}}'_{ij}$. No such relation is needed for the interfacial grain-boundary sliding, for it is Newtonian with a constant viscosity.

The stress-strain relations then can be computed incrementally from Eq. 4.44, with continuous update of the secant viscosities of the constituent phases as explained in the previous section.

This marks the end of the development of the two computational schemes with the composite model.

4.7 STRESS-STRAIN RELATIONS, HALL–PETCH PLOTS, STRAIN-RATE SENSITIVITY, AND ACTIVATION VOLUME OF NANOCRYSTALLINE NICKEL AND COPPER

We now show some results calculated from these two schemes. The first case is for an electrodeposited nickel which has been tested by Schwaiger et al.,[18] and the second one is for a very fine-grained nanocrystalline copper under very high strain-rate, obtained through MD simulations by Schiøtz et al.[32] These two cases respectively correspond to the microgeometries of Fig. 4.1a,b without and with the additional interfacial sliding, and the results have been obtained by the methods presented in Sections 4.5 and 4.6.

4.7.1 Strain-Rate Sensitivity of Electrodeposited Ni

We first examine the influence of grain size on the stress-strain relation of Ni under the strain-rate of 3×10^{-4}/s. The material constants used in the calculations are listed in Table 4.1. The results at three different grain sizes: $d = 10\,\mu$m, 320 nm, and 40 nm, are shown in Fig. 4.3, which also includes the

Table 4.1 Material constants used in the calculations of Ni

	Grain interior	**Grain boundary affected zone**
E(GPa)	200	200
v	0.3	0.3
s_0^∞(MPa)	4.03	—
k (GPa $\sqrt{nm}$)	15.8	—
a	2.63	—
s_0(MPa)	—	360
s_*	—	820
h(GPa)	5.6	20
n	200	55
$\dot{\varepsilon}_0(10^{-4}$/hr)	1.0	1.0

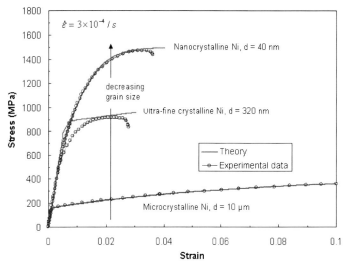

Figure 4.3 Stress-strain relations of an electrodeposited Ni with d = 40 nm, 320 nm, and 10 μm (theory from Li and Weng[111] and experiment from Schwaiger *et al.*[18]).

experimental data of Schwaiger *et al.*[18] These grain sizes cover the range from the traditional coarse grain, to ultra-fine grain, and then to the nano–meter range. Significant grain size hardening is evident from both experiment and theory. This confirms that the Hall–Petch type hardening persists down to 40 nm for Ni. The response of the nanocrystalline Ni under three different orders of strain rates are shown in Fig. 4.4, where (a) is the test data of Schwaiger *et al.*[18] and (b) is the calculated curves. At d = 40 nm, both results indicate a noticeable increase of flow stress with increasing strain-rate. We have also made a similar calculation for the ultra-fine Ni at d = 320 nm. The calculated 1% yield stress for both the ultra-fine and nanocrystalline Ni over these strain rates are shown in Fig. 4.5, along with the test data. These results indicate that there is little change in the flow stress with d = 320 nm, but a noticeable increase with d = 40 nm, pointing to a greater strain-rate sensitivity of nanocrystalline nickel.

The data in Fig. 4.5 can be used to calculate the strain-rate sensitivity parameters, S and m, and the activation volume v^*, of the composite through the definitions of Eqs. (4.13) and (4.14). We found that: (i) at d = 320 nm, m = 0.008, S = 6.70 MPa, and v^*/b^3 = 64.33; and (ii) at d = 40 nm, m = 0.016, S = 20.99 MPa, and v^*/b^3 = 20.03, with the Burgers' vector of Ni at b = 0.2489 nm (its lattice constant is 0.3520 nm). To put these calculated results in perspective, we have superimposed these two sets of computed results into

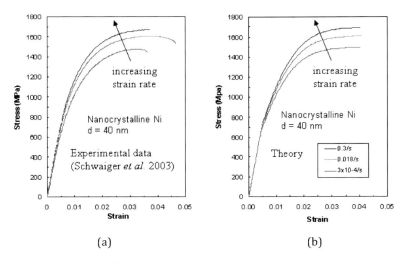

(a)   (b)

Figure 4.4 Nanocrystalline Ni under three orders of strain-rate: (a) data from Schwaiger *et al.*,[18] and (b) theory from Li and Weng.[111]

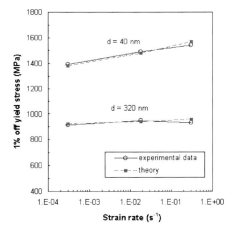

Figure 4.5 Strain-rate dependence of the flow stress for an electrodeposited Ni (data from Schwaiger *et al.*[18] and theory from Li and Weng[111]).

the data compiled by Wei[135] for the strain-rate sensitivity parameter m, in Fig. 4.6, and by Armstrong[81] for the activation volume v^*, in Fig. 4.7, both in (pink) x-signs. These data points are seen to be in line with the published data, and with the known trend that the strain-rate sensitivity increases with decreasing grain size whereas the activation volume decreases. The covered range of grain size down to 40 nm here is still on the ascending side of the Hall–Petch plot, so the Hall–Petch breakdown has not reached yet.   The

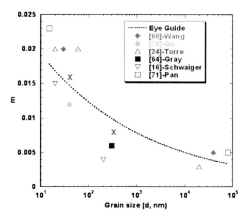

Figure 4.6 Calculated strain-rate sensitivity parameter, *m*, for Ni from the theoretical curves in Fig. 5 (*m* = 0.16 and 0.008 for *d* = 40 nm and 320 nm, respectively, in (pink) x signs; remaining part of the figure reproduced from Wei[135]).

calculated trends by the composite model are also consistent with the dislocation theories of Armstrong and Rodriguez,[29] including the reciprocal relation of $1/v^*$ written in Eq. 4.15 as presented in Fig. 4.7.

4.7.2 Strain-Rate Sensitivity of Cu Under Very High Strain-Rate

We now show the calculated results for Cu under the very high strain rates of 10^8–10^9/s commonly encountered in MD simulations. This was done through the theory developed in Section 4.6 with the additional interfacial grain-boundary sliding.

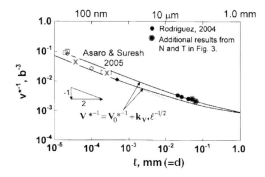

Figure 4.7 Calculated activation volume, v^*/b^3, for Ni from the theoretical curves in Fig. 4.5 (v^*/b^3 = 20.03 and 64.33 for *d* = 40 nm and 320 nm, respectively, in [pink] x signs; remaining part of the figure reproduced from Armstrong[82]).

The material constants used are listed in Table 4.2. Since we are not concerned with the tension-compression asymmetry here the pressure-

Table 4.2 Material constants used in the calculation of Cu

	Grain interior	GB zone	Grain-boundary sliding
E(GPa)	145	50	—
v	0.3	0.3	—
s_0^∞(MPa)	5.03	—	—
k(GPa $\sqrt{nm}$)	7.9	—	—
a	1.5	—	—
s_0(GPa)	—	1.3	—
s_*(GPa)	—	5.0	—
λ	—	0	—
h(MPa)	5.0	9.0	—
n	50	220	—
$\dot{\varepsilon}_0(10^{-4}$/hr)	5.0	5.0	—
α_s(GPa/nm·sec)	—	—	12.0

dependent constant λ for the GB zone was taken to be zero. We first calculated the stress-strain relations of Cu at d = 5.21 nm, 4.13 nm, and 3.28 nm, all under the strain-rate of 5 × 10^8/s. The calculated results, together with the MD simulations of Schiøtz *et al.*[32] are shown in Fig. 4.8. The response of Cu under the different strain rates of 2.5 × 10^8/s, 5 × 10^8/s, and 1×10^9/s for d = 4.13 nm are shown in Fig. 4.9. It is seen that the composite model is able to capture the MD simulated results sufficiently well. The material still exhibits the positive strain-rate effect, but at this low level of grain size there is evidence of grain size softening. The material in essence has already passed over the maximum strength state, down to the side of negative slope in the Hall–Petch plot.

The developed model is capable of predicting the stress-strain relations of Cu over a wider range of grain size. In order to shed some light on the transition of yield strength as a function of grain size, we have extended the calculations upward to 45 nm. The results for the flow stress at 6% strain are illustrated in Fig. 4.10, under the same strain-rate of 5 × 10^8/s. Three different values of interfacial sliding viscosity, α_s, have been adopted in computations. The top curve corresponds to perfect bonding between the grain and GB zone, without the additional grain-boundary sliding, whereas the other two are with increasing level of sliding activity. This set of curves vividly shows the transition from a positive to a negative slope of the Hall–Petch plot. It is

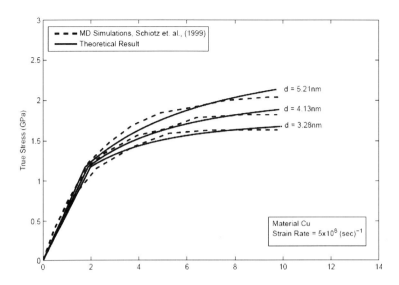

Figure 4.8 Stress-strain relations of nanocrystalline Cu at *d* = 5.21 nm, 4.13 nm and 3.28 μm (theory from Barai and Weng[134]; MD simulations from Schiøtz *et al.*[32]).

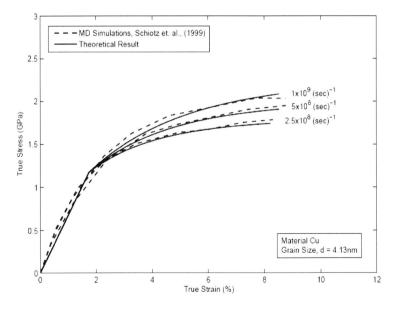

Figure 4.9 Very fine-grained nanocrystalline Cu under very high rate loading (theory from Barai and Weng[134]; MD simulations from Schiøtz *et al.*[32]).

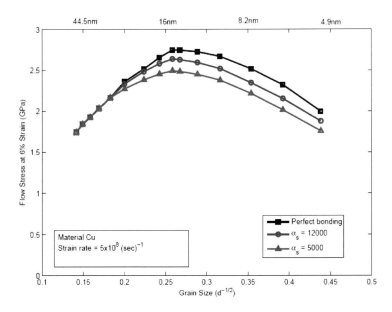

Figure 4.10 Hall–Petch plot for Cu under the high strain-rate of 5×10^8/s with various GB sliding parameter α_s (curves from Barai and Weng[134]).

also evident that GB sliding would lower the yield strength of the material, but it does not alter the critical grain size at which the maximum yield strength occurs. For this Cu, the calculated critical grain size is around 14–15 nm. This compares with Argon and Yip's[28] calculated value of 12.2 nm, and Schiøtz and Jacobsen's[33] simulated value of 15 nm. They all lie between the 12–15 nm range. The three maximum strengths obtained here are about 2.3 GPa, 2.5 GPa, and 2.7 GPa, respectively, with increasing level of sliding viscosity. On the other hand Schiøtz and Jacobsen[33] gave 2.3 GPa for the maximum strength, while Argon and Yip gave a normalized value (with respect to μ) of about 0.028.

To evaluate the strain-rate sensitivity parameters S and m, and the activation volume, v^*, for Cu, we have calculated its flow stress at the same three strain rates from 40 nm down to 2 nm. The results are shown in Fig. 4.11. The choice of this range of grain size was inspired by an experimental investigation on the hardness of Ni-W alloys by Trelewicz and Schuh[20] from $d = 200$ nm down to only 3 nm. The slope of each curve immediately gives the strain-rate sensitivity parameter, S, through $S = \left(1/\sqrt{3}\right)[\partial\sigma/\partial(\ln \dot{\varepsilon})]$ and another strain-rate sensitivity parameter, m, at each grain size. A close

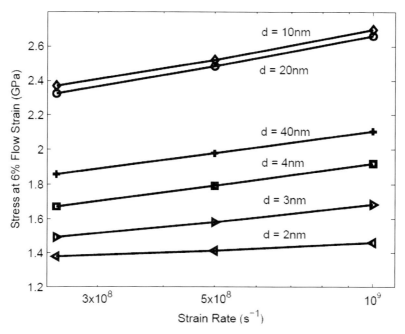

Figure 4.11 Model calculation for the flow stress, strain-rate, and grain size coupling of nanocrystalline Cu (curves from Barai and Weng[134]).

examination of these slopes indicates that it increases from d = 40 nm to 20 nm, but decreases from 10 nm all the way down to 2 nm. Since the activation volume, v^*, is inversely proportional to S at a constant temperature, it will decrease and then increase also.

To see the transition of these properties more closely, we first calculated the strain-rate sensitivity parameter, m, from $m = \sqrt{3}S/\sigma$, as a function of grain size. The results for d = 40 nm, 20 nm, and 10 nm are 0.0515, 0.0558, and 0.0534, respectively. For direct comparison, these three data points have been superimposed onto Wei's[135] compiled data for Cu in Fig. 4.12, in (pink) x-signs. They appear on the high end of the data set, but are not totally out of bound. The higher rate sensitivity is likely to be due to the extremely high strain rates of 10^8–10^9/s used in MD simulations. These three data points also indicate an increase of the strain-rate sensitivity from 40 nm to 20 nm, and then a decrease from 20 nm to 10 nm. There exists a critical grain size at which the strain-rate sensitivity m is at its maximum, and this occurs around d = 10–20 nm.

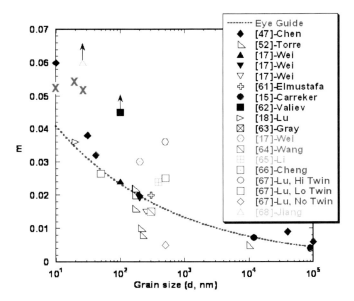

Figure 4.12 Calculated strain-rate sensitivity parameter, *m*, for Cu with *d* = 10, 20, and 40 nm, in (pink) x signs (0.0534, 0.0558, and 0.0515, respectively; remaining part of the figure reproduced from Wei[135]).

We then used $v^* = \sqrt{3}k_BT/[\partial\sigma/\partial(\ln \dot{\varepsilon})]$ and the slopes in Fig. 4.11 to calculate the activation volume for each grain size, with the Burgers vector *b* = 0.2556 nm. The results are shown in Fig. 4.13(a). The trend is strikingly similar to the test data of Ni-W alloy presented by Trelewicz and Schuh,[20] as reproduced in Fig. 4.13(b). Both sets of results indicate a decrease and then increase of the activation volume as the grain size decreases from 40 nm to 2–3 nm. The minimum value occurs around *d* = 10–20 nm in both plots.

A closer examination on the magnitudes of these two figures, however, indicates that the activation volume shown in Fig. 4.13(a) is somewhat lower than that in Fig. 4.13(b). As was with the higher *m*-values shown in Fig. 4.12, these lower *v**-values are likely to be due to the much higher strain rates involves here. In an extensive investigation on the high strain-rate properties of metals and alloys and after examining the test data of Follansbee *et al.*[136] and Swegle and Grady,[137] both Armstrong and Walley[138] and Armstrong *et al.*[139] have pointed out that the normalized activation volume, v^*/b^3, of Cu could decrease from 1,000 to the order of unity as the strain-rate goes beyond 10^5/s. Under such high rate loading–especially under the shock loading–they further suggested that the rate dependence of flow stress could change from the drag-controlled mobile dislocation motion to the dislocation nucleation process, and this in turn could lead to the strong upturn of the flow stress and the substantially lower activation volume. A reproduction of their observation on the activation area, A^*, is shown in Fig. 4.13(c), with $v^*=A^*b$. Besides the difference in the v^*/b^3 values, it is also noted that Fig. 4.13(a) indicates a grain size weakening exponent of about -0.5 for v^* whereas Fig. 4.13(b) shows a weakening exponent of about -1.0. The latter is in better agreement with the $+1.0$ exponent for the reciprocal v^* for Zn (data from Conrad and Narayan[19]), as reported by Armstrong[82] (see his Fig. 4.15 in this book). These differences are likely to be caused by the extremely high rate of deformation involved in MD simulations, but further investigations are needed.

The calculated increase and then decrease of the strain-rate sensitivity parameter, m, in Fig. 4.12, and the decrease and then increase of the activation volume, v^*, in Fig. 4.13 (a) and (b), are significant. At the vanishing grain size limit, Trelewicz and Schuh[20] have concluded that the nanocrystalline material is approaching the amorphous limit, which has very low strain-rate sensitivity and very high activation volume. Our composite model also suggests that, at such low grain size limit, the volume fraction of the nanocrystalline material is dominated by the GB zone which is treated as amorphous.[53-55] The GB zone has a substantially higher n-value than the grain interior (220 *vs.* 50; see Table 4.2), and thus it has lower strain-rate sensitivity and higher activation volume. The transitions of m and v^* displayed here are also the reflections of the increase and then decrease of the Hall–Petch plot of the flow stress predicted by Armstrong and Rodriguez[29] and Rodriguez and Armstrong.[30] We have demonstrated, quantitatively, that such a trend indeed exists.

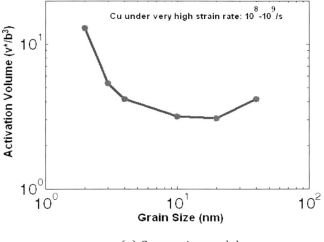

(a) Composite model

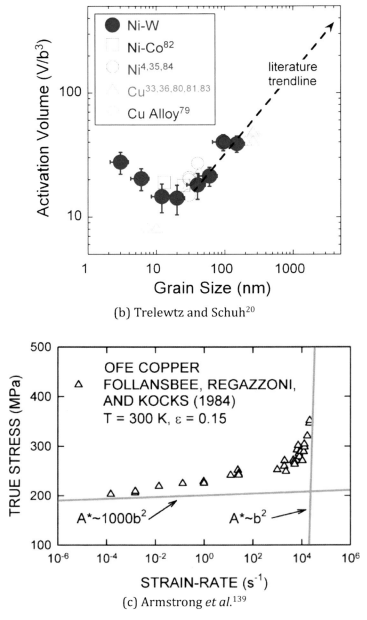

(b) Trelewtz and Schuh[20]

(c) Armstrong *et al.*[139]

Figure 4.13 Dependence of the activation volume, *v** as a function of grain size and strain-rate: (a) calculated from the theory under very high rate, (b) test data (taken from Trelewicz and Schuh[20]), and (c) decrease of activation volume to b^3 under very high strain-rate (taken from Armstrong *et al.*[139]).

4.8. CONCLUDING REMARKS

In this article we have developed a micromechanics-based composite model to study the mechanical properties of nanocrystalline materials. Two specific morphologies have been considered. The first one involves a plastically harder grain interior and a plastically softer grain-boundary affected zone (GBAZ) as depicted in Fig. 4.1(b). In this case the grain boundary has zero thickness and there is no grain-boundary sliding. The second morphology involves a harder grain-interior phase as inclusions and a softer grain-boundary phase (the GB zone) as the matrix, with the possibility of interfacial grain-boundary sliding. This is represented in Fig. 4.1(a). From the composite standpoint, the first case is in effect a special case of the second one without grain-boundary sliding, but the fact that there is absolutely no interfacial sliding has allowed the problem to be formulated by the much simpler direct secant viscosity analogy. The one with the additional interfacial sliding has to go through the Laplace transform and inversion process so the overall plastic compressibility of the composite can be preserved. But in both cases the viscoplastic properties of the inclusions and the matrix are described by a set of power-law, unified theory that is capable of strain hardening. The drag stress of the grain interior is further taken to be grain size dependent, and the GB zone is modeled as an amorphous phase with a pressure-dependent property. The interfacial grain-boundary sliding is modeled as a Newtonian process.

Both composite problems are nonlinear, heterogeneous, time, and strain-rate dependent. To address these complicated problems we have made use of the idea of transition from elasticity to viscoelasticity, and then from viscoelasticity to viscoplasticity in a composite material, and introduced the secant viscosity of the constituent phases into the formulations. In this process a field-fluctuation approach is also introduced to determine the link between the local strain rates of the constituent phases with the applied strain-rate. This makes it possible to calculate the deformation of the individual phase from its own local stress and strain state, and to determine the overall behavior of the nanocrystalline material from the properties of its constituent phases at a given grain size.

These two schemes have been applied to study the properties of an electrodeposited Ni and a MD simulated Cu, respectively. The calculated results for Ni from d = 320 nm down to 40 nm show significant grain size hardening and an increase of strain-rate sensitivity and decrease of the activation volume. This grain size range appears to be on the ascending side of the Hall–Petch plot, and the obtained trends are found to be consistent with the theory of Armstrong and Rodriguez.[29] Quantitatively the calculated strain-rate sensitivity parameter, *m*, and activation volume, *v**, are found to be in line with the data compiled by Wei[135] and Armstrong,[82] respectively. The calculated results for Cu from 40 nm down to 2 nm are associated with the very high strain-rates of 10^8–10^9/s, and we have uncovered some rather interesting new features that are possibly related to these unusually high rates at the extremely low end of grain size. From 40 nm down to 20 nm, the results show the positive side of the Hall–Petch plot for the flow stress, and the strain-rate sensitivity, *m*, continues to increase whereas the activation volume, *v**, continues to decrease. But from 20 nm down to 10 nm, and further down to 2 nm, the calculated results show a negative slope of the Hall–Petch plot for the flow stress, a decrease in the strain-rate sensitivity, and an increase in the activation volume. The trends of decrease and then increase of the activation volume, and the increase and then decrease of the strain-rate sensitivity, are consistent with those extracted from the indentation test data of Trelewicz and Schuh,[20] and with the dislocation models of Armstrong and Rodriguez[29] and Rodriguez and Armstrong.[30] But we also found that the magnitude of the calculated *v** is noticeably lower than the indentation data. We believe that this is likely due to the much higher strain-rate involved in our case. Under such conditions, Armstrong and Walley[138] and Armstrong *et al.*[139] have pointed out that the activation volume could substantially decrease, to the order of b^3 as the rate goes beyond 10^5/s. Although their observations were based on the properties of coarse-grained Cu and Fe, it is quite possible that it also applies to nanocrystalline materials.

In closing, we believe that this continuum composite approach has shed significant insights into the strain-rate sensitivity of nanocrystalline materials. The theory has been formulated with full account of stress and strain heterogeneity of the constituent phases, and most of the calculated results are in line with the measured data. As the mechanical properties of nanocrystalline materials involve both continuum mechanics and materials science, full understanding of this subject will require considerations from both perspectives. While recognizably not easy, the linkage between the two remains to be better established.

Acknowledgments

I am indebted to Professor Ronald W. Armstrong; without his initiative and encouragement this article would have never come to being. In the course of this writing I have also benefited from his many insightful observations on the strain-rate sensitivity of nanocrystalline materials. I also thank

Professor Jacqueline Li of the City College of New York for her long-standing collaboration on the rate-dependent viscoplastic behavior of composites and nanocrystalline materials. Mr. Pallab Barai, a graduate student at Rutgers University, has also made significant contributions to nanocrystalline materials at his young age, leading to several joint publications. The invitation of Professor J.C.M. Li to write this chapter is also greatly appreciated. Finally, this work was supported by the NSF Engineering Directorate, Mechanics and Structure of Materials Program, under grant CMS-0510409.

References

1. E.O. Hall, *Proc. Phys. Soc. Lond.*, **B64**, 747 (1951).
2. N.J. Petch, *J. Iron Steel Inst.*, **174**, 25 (1953).
3. J.D. Eshelby, F.C. Frank and F.R.N. Nabarro, *Philos. Mag.*, **42**, 351 (1951).
4. R.W. Armstrong, I. Codd, R.M. Douthwaite and N.J. Petch, *Philos. Mag.*, **7**, 45 (1962).
5. J.C.M. Li, *Trans. Metall. Soc*. A.I.M.E., **227**, 239 (1963).
6. H. Conrad, *Acta Metall.*, **11**, 75 (1963).
7. M.F. Ashby, *Philos. Mag.*, **21**, 399 (1970).
8. R. Birringer, H. Gleiter, H.P. Klein and P. Marquardt, *Phys. Lett.*, **102A**, 365 (1984).
9. G.W., Nieman, J.R. Weertman and R.W. Siegel, *Scripta Metall.*, **23**, 2013 (1989).
10. A.M. El-Sherik, U. Erb, G. Palumbo and K.T. Aust, *Scripta Metall. Mater.*, **27**, 1185 (1992).
11. V.Y. Gertman, M. Hoffmann, H. Gleiter and R. Birringer, *Acta Metall. Mater.*, **42**, 3539 (1994).
12. D.G. Sanders, J.A. Eastman and J.R. Weertman, *Acta Mater.*, **45**, 4019 (1997).
13. D.G. Sanders, C.J. Youngdahl and J.R. Weertman, *Mater. Sci. Eng.*, **A234**, 77 (1997).
14. A.S. Khan, H. Zhang and L. Takacs, *Int. J. Plasticity*, **16**, 1459 (2000).
15. A.H. Chokshi, A. Rosen, J. Karch and H. Gleiter, *Scripta Metall.* **23**, 1679 (1989).
16. K. Lu, W.D. Wei and J.T. Wang, *Scripts Metall. Mater.*, **24**, 2319 (1990).
17. G.E. Fougere, J.R. Weertman, R.W. Siegel and S. Kim, *Scripta Metall. Mater.*, **26**, 1879 (1992).
18. R. Schwaiger, B. Moser, M. Dao, N. Chollacoop and S. Suresh, *Acta Mater.*, **51**, 5159 (2003).
19. H. Conrad and J. Narayan, *Appl. Phys. Lett.*, **81**, 2241 (2002); *Acta Mater.*, **50**, 5067 (2002).
20. J.R. Trelewicz and C.A. Schuh, *Acta Mater.*, **55**, 5948 (2007).

21. T.G. Nieh and J. Wadsworth, *Scripta Metall.*, **25**, 955 (1991).
22. R.O. Scattergood and C.C. Koch, *Scripta Metall.*, **27**, 1195 (1992).
23. J. Lian, B. Baudelet and A.A. Nazarov, *Mater. Sci. Eng.*, **A172**, 23 (1993).
24. N. Wang, Z. Wang, K.T. Aust and U. Erb, *Acta Metall. Mater.*, **43**, 519 (1995).
25. M.Yu. Gutkin, I.A. Ovid'ko, and C.S. Pande, *Philos. Mag.*, **21**, 847 (2004).
26. M.Yu. Gutkin, I.A. Ovid'ko, and N.V. Skiba, *Acta Mater.*, **52**, 1711 (2004).
27. I.A. Ovid'ko, *Rev. Adv. Mater. Sci.*, **10**, 89 (2005).
28. A.S. Argon and S. Yip, *Philos. Mag. Lett.*, **86**, 713 (2006).
29. R.W. Armstrong and P. Rodriguez, *Philos. Mag.*, **86**, 5787 (2006).
30. P. Rodriguez and R.W. Armstrong, *Bull. Mater. Sci.*, **29**, 717 (2006).
31. J. Schiøtz, F.D. Di Tolla and K.W. Jacobsen, *Nature*, **391**, 561 (1998).
32. J. Schiøtz, T. Vegge, F.D. Di Tolla and K.W. Jacobsen, *Phys. Rev. B*, **60**, 971 (1999).
33. J. Schiøtz and K.W. Jacobsen, *Science*, **301**, 1357 (2003).
34. J. Schiøtz, *Scripta Mater.*, **51**, 837 (2004).
35. H.V. Swygenhoven and A. Caro, *Appl. Phys. Lett.*, **71**, 1652 (1997).
36. H.V. Swygenhoven, M. Spaczer, A. Caro and D. Farkas, *Phys. Rev. B*, **60**, 22 (1999).
37. H.V. Swygenhoven, A. Caro and D. Farkas, *Mater. Sci. Eng. A*, **309–310**, 440 (2001).
38. V. Yamakov, D. Wolf, S.R. Phillpot and H. Gleiter, *Acta Mater.*, **50**, 61 (2002).
39. V. Yamakov, D. Wolf, S.R. Phillpot, A.K. Mukherjee and H. Gleiter, *Philos. Mag. Lett.*, **83**, 385 (2003).
40. A.C. Lund and C.A. Schuh, *Acta Mater.*, **53**, 3193 (2005).
41. V. Tomar and M. Zhou, *J. Mech. Phys. Sol.*, **55**, 1053 (2007).
42. K.S. Kumar, S. Suresh, M.F. Chisholm, J.A. Horton and P. Wang, *Acta Mater.*, **51**, 387 (2003).
43. S. Takeuchi, *Scripta Mater.*, **44**, 1483 (2001).
44. J. Schiøtz, Private communication (2003).
45. J.E. Carsley, J. Ning, W.M. Milligan, S.A. Hackney and E.C. Aifantis, *Nanostruct. Mater.*, **5**, 441 (1995).
46. H.S. Kim, Y. Estrin and M.B. Bush, *Acta Mater.*, **48**, 493 (2000).
47. D.J. Benson, H.-H. Fu and M.A. Meyers, *Mater. Sci. Eng.*, **A319–A321**, 854 (2001).
48. H. Fu, D.J. Benson and M.A. Meyers, *Acta Mater.*, **49**, 2567 (2001).
49. M.A. Meyers, A. Mishra and D.J. Benson, *Prog. Mater. Sci.*, **51**, 427 (2006).
50. M. Dao, L. Lu, R.J. Asaro, J.T.M. De Hosson and E. Ma, *Acta Mater.*, **55**, 4041 (2007).
51. K.T. Ramesh, *Nanomaterials: Mechanics and Mechanisms* (Springer, NY, 2009).

52. M. Cherkaoui and L. Capolungo, *Atomistic and Continuum Modeling of Nanocrystalline Materials* (Springer, NY, 2009).

53. B. Jiang and G.J. Weng, *Metall. Mater. Trans.*, **34A**, 765 (2003).

54. B. Jiang and G.J. Weng, *Int. J. Plasticity*, **20**, 2007 (2004).

55. B. Jiang and G.J. Weng, *J. Mech. Phys. Sol.*, **52**, 1125 (2004).

56. G.J. Weng, *Rev. Adv. Mater. Sci.*, **19**, 41 (2009).

57. L. Capolungo, M. Cherkaoui and J. Qu, *J. Eng. Mater. Tech.*, **127**, 400 (2005).

58. L. Capolungo, C. Joachim, M. Cherkaoui and J. Qu, *Int. J. Plasticity*, **21**, 67 (2005).

59. L. Capolungo, M. Cherkaoui and J. Qu, *Int. J. of Plasticity*, **23**, 561 (2007).

60. L. Capolungo, D.E. Spearot, M. Cherkaoui, D.L. McDowell, J. Qu and K.I. Jacob, *J. Mech. Phys. Sol.*, **55**, 2300 (2007).

61. Y.J. Wei and L. Anand, *J. Mech. Phys. Sol.*, **52**, 2587 (2004).

62. Y.J. Wei, C. Su and L. Anand, *Acta Mater.*, **54**, 3177 (2006).

63. A. Jerusalem, L. Stainier and R. Radovitzky, *Philos. Mag.*, **87**, 2541 (2007).

64. A. Jerusalem and R. Radovitzky, *Modell. Simul. Mater. Sci. Eng.*, **17**, 025001 (2009).

65. D.H. Warner, F. Sansoz and J.F. Molinari, *Int. J. Plasticity*, **22**, 754 (2006).

66. S.P. Joshi, K.T. Ramesh, B.Q. Han and E.J. Lavernia, *Metall. Mater. Trans.*, **37A**, 2397 (2006).

67. S.P. Joshi and K.T. Ramesh, *Script Mater.*, **57**, 877 (2007).

68. G.J. Weng, *J. Mech. Phys. Sol.*, **38**, 419 (1990).

69. U.F. Kocks, A.S. Argon and M.F. Ashby, *Progr. Mater. Sci.*, **19**, 1 (1975).

70. E. Orowan, *Proc. Phys. Soc. Lond.*, **B52**, 8 (1940).

71. G. Schoeck, *Phys. Stat. Sol.*, **8**, 499 (1965).

72. J.C.M. Li, in *Dislocation Dynamics*, Eds. A.R. Rosenfield, G.T. Hahn, A.L. Bement, Jr., and R.I. Jaffee (McGraw-Hill Book, NY, 1968), p. 87.

73. R.W. Armstrong, *(Indian) J. Sci. Indust. Res.*, **32**, 591 (1973).

74. F.D. Torre, H. Van Swygenhoven and M. Victoria, *Acta Mater.*, **50**, 3957 (2002).

75. K.S. Kumar, H. Van Swygenhoven and S. Suresh, *Acta Mater.*, **51**, 5743 (2003).

76. D. Jia, K.T. Ramesh and E. Ma, *Acta Mater.*, **51**, 3495 (2003).

77. Q. Wei, S. Cheng, K.T. Ramesh and E. Ma, *Mater. Sci. Eng.*, **A381**, 71 (2004).

78. S.P. Joshi and K.T. Ramesh, *Mater. Sci. Eng.*, **A493**, 65 (2008).

79. A.S. Khan, B. Farrokh and L. Takacs, *Mater. Sci. Eng.*, **A489**, 77 (2008).

80. A.S. Khan, B. Farrokh and L. Takacs, *J. Mater. Sci.*, **43**, 3305 (2008).

81. B. Farrokh and A.S. Khan, *Int. J. Plasticity*, **25**, 715 (2009).

82. R.W. Armstrong, in *Mechanical Properties of Nanocrystalline Materials*, Ed. J.C.M. Li (Pan Stanford Publishing, C/o World Scientific Publishing Co., Hackensack, NJ, 2009), p. 61.

83. J. Li and G.J. Weng, *J. Mech. Phys. Sol.*, **45**, 1069 (1997).

84. J. Li and G.J. Weng, *Acta Mech.*, **125**, 141 (1997).

85. J. Li and G.J. Weng, *Int. J. Plasticity*, **14**, 193 (1998).

86. Z. Hashin, *J. Appl. Mech.*, **32**, 630 (1965).

87. D.R.S. Talbot and J.R. Willis, *IMA J. Appl. Math.*, **35**, 39 (1985).

88. G.P. Tandon and G.J. Weng, *J. App. Mech.*, **55**, 126 (1988).

89. P. Ponte Castañeda, *J. Mech. Phys. Sol.*, **39**, 45 (1991).

90. Y.P. Qiu and G.J. Weng, *J. Appl. Mech.*, **59**, 261 (1992).

91. P. Suquet, C.R. *Acad. Des Sci.*, **320**, Ser. IIb, 563 (1995).

92. G.K. Hu, *Int. J. Plasticity*, **12**, 439 (1996).

93. D.C. Drucker, *Q. Appl. Math.*, **7**, 411 (1950).

94. P.E. Donovan, *Acta Metall.*, **37**, 445(1989).

95. M.D. Kluge, D. Wolf, J.F. Lutsko and S.R. Phillpot, *J. Appl. Phys.*, **67**, 2370 (1990).

96. D. Wolf and M. Kluge, *Scripta Metall.*, **24**, 907 (1990).

97. J.L. Bassani, V. Vitek and I. Alber, *Acta Metall. Mater.*, **40**, S307 (1992).

98. I. Alber, J.L. Bassani, M. Khantha, V. Vitek and G.J. Wang, *Trans. R. Soc. Lond.*, **A339**, 555 (1992).

99. G.J. Weng, *Int. J. Eng. Sci.*, **22**, 845 (1984).

100. Y. Benveniste, *Mech. Mater.*, **6**, 147 (1987).

101. G.J. Weng, *Int. J. Eng. Sci.*, **28**, 1111 (1990).

102. G.J. Weng, *Int. J. Eng. Sci.*, **30**, 83 (1992).

103. R. Hill, *J. Mech. Phys. Sol.*, **13**, 213 (1965).

104. B. Budiansky, *J. Mech. Phys. Sol.*, **13**, 223 (1965).

105. R.M. Christensen and K. H. Lo, *J. Mech. Phys. Sol.*, **27**, 315 (1979).

106. R. Roscoe, *British J. Appl. Phys.*, **3**, 267 (1952).

107. A.N. Norris, *Mech. Mater.*, **4**, 1 (1985).

108. M. Hori and S. Nemat-Nasser, *Mech. Mater.*, **14**, 189 (1993).

109. Z. Hashin and S. Shtrikman, *J. Mech. Phys. Sol.*, **11**, 127 (963).

110. L.J. Walpole, *J. Mech. Phys. Sol.*, **14**, 151, 289 (1966), **17**, 235 (1969).

111. J. Li and G.J. Weng, *Int. J. Plasticity*, **23**, 2115 (2007).

112. M. Bobeth and G. Diener, *J. Mech. Phys. Sol.*, **34**, 1(1986)

113. W. Kreher and W. Pompe, *Internal Stress in Heterogeneous Solids* (Akademie, Berlin, 1989).

114. Y. Benveniste, *Mech. Mater.*, **4**, 197 (1985).

115. Z. Hashin, *J. Mech. Phys. Sol.*, **39**, 745 (1991).

116. C.W. Nan, X.P. Li, K.F. Cai and J.H. Tong, *J. Mater. Sci. Lett.*, **17**, 1917 (1998).

117. G.F. Wang, X.Q. Feng, S.W. Yu and C.W. Nan, *Mater. Sci. Eng.*, **A363**, 1 (2003).

118. P. Sharma and S. Ganti, *J. Mater. Res.*, **18**, 1823 (2003).

119. P. Sharma and S. Ganti, *J. Appl. Mech.*, **71**, 663 (2004).

120. P. Sharma and L. Wheeler, *J. Appl. Mech.*, **74**, 447 (2007).

121. H.L. Duan, J. Wang, Z.P. Huang and B.L. Karihaloo, *J. Mech. Phys. Sol.*, **53**, 1574 (2005).

122. Z.P. Huang and J. Wang, *Acta Mech.*, **182**, 195 (2006).

123. Z.P. Huang and L. Sun, *Acta Mech.*, **190**, 151 (2007).

124. T. Chen, G.J. Dvorak, and C.C. Yu, *Acta Mech.*, **188**, 39 (2007).

125. H. Chen, G.K. Hu and Z.P. Huang, *Int. J. Solids Struct.*, **44**, 8106 (2007).

126. H.L. Duan, X. Yi, Z.P. Huang and J. Wang, *Mech. Mater.*, **39**, 81 (2007).

127. L. Shen and J. Li, *Int. J. Solids Struct.*, **40**, 1393 (2003).

128. L. Shen and J. Li, *Proc. R. Soc.*, **A147**, 1 (2005).

129. Y.M. Wang and G.J. Weng, *J. Appl. Mech.*, **59**, 510 (1992).

130. J. Li and G.J. Weng, *J. Eng. Mat. Tech.*, **116**, 495 (1994).

131. P. Barai and G.J. Weng, *Acta Mech.*, **195**, 327 (2008).

132. J.C.M. Li, *Appl. Phys. Lett.*, **90**, 041912 (2007).

133. P. Barai and G.J. Weng, *Int. J. Plasticity*, **24**, 1380 (2008).

134. P. Barai and G.J. Weng, *Int. J. Plasticity*, **25,** 2410 (2009).

135. Q. Wei, *J. Mater. Sci.*, **42**, 1709 (2007).

136. P.S. Follansbee, G. Regazzoni and U.F. Kocks, in *Mechanical Properties of Materials at High Strain-rate*, Ed. J. Harding (Inst. Physics, London, 1984), p. 71.

137. J.W. Swegle and D.E. Grady, *J. Appl. Phys.*, **58**, 692 (1985).

138. R.W. Armstrong and S.M. Walley, *Int. Mater. Rev.*, **53**, 105 (2008).

139. R.W. Armstrong, W. Arnold and F.J. Zerilli, *J. Appl. Phys.*, **105**, 023511 (2009).

Chapter 5

THE EFFECT OF MICROSTRUCTURAL FEATURES ON THE MECHANICAL PROPERTIES OF NANOCRYSTALLINE METALS

Zbigniew Pakiela, Malgorzata Lewandowska, and Krzysztof J. Kurzydlowski

Warsaw University of Technology, Faculty of Materials Science and Engineering, Woloska 141, 02-507 Warsaw, Poland

E-mail: zpakiela@inmat.pw.edu.pl, malew@inmat.pw.edu.pl, and kjk@inmat.pw.edu.pl

5.1 INTRODUCTION

In recent years nanocrystalline metallic materials (NCMs) have been the focus of much intensive interdisciplinary research. The research interest has been motivated by the exceptional properties currently possessed and the perceived potential properties and performance of products produced from such materials. Currently, NCMs already are of high technical significance, and the prognosis for the next few years indicates that they will continue to be of increasing importance.

Nanometric elements of microstructure have been used to formulate the properties of materials for a long time. For example, the precipitation of nanometric particles of metastable phases in aluminum produces a significant increase in strength. The rapid growth of interest in NCMs has been observed since the late '80s, particularly after publication of a review paper by H. Gleiter,[1] who is one of the pioneers of research on NCMs.

Mechanical Properties of Nanocrystalline Materials
Edited by James C. M. Li
Copyright © 2011 Pan Stanford Publishing Pte. Ltd.
www.panstanford.com

In the most general and widely recognized classification, NCMs can be divided into 1-D (nanotubes, nanowires, nanofibres), 2-D (nanolayers), and 3-D (bulk materials and nanopowders). Recent progress has been particularly impressive in the development of 1- and 2-D nanomaterials. In the case of bulk NCMs, development has been much slower because of the difficulties in manufacturing these materials and obtaining reproducible microstructures and properties.

For industrial applications, the requirements for NCMs, particularly with regard to mechanical properties, are very demanding. Very high strength combined with good plasticity and fracture toughness is an essential requirement. Extensive research is conducted by numerous teams around the world, but currently the achieved properties are not satisfactory. High hardness and strength can be obtained relatively easily. Unfortunately the enhanced strength and hardness are obtained only at the expense of ductility and fracture toughness. However, experimental and theoretical analyses indicate that the combination of high strength with good ductility and fracture toughness is possible.

The aim of this chapter is to consider the current views on the relevant microstructural features of NCMs and to discuss the possible ways of enhancing their mechanical properties by optimization of the microstructure.

5.2 MICROSTRUCTURAL FEATURES OF NANOCRYSTALLINE METALS

NCMs have been the focus of increasing research interest because grain size refinement is one of the most promising ways to improve some mechanical properties. This, in turn, instigated the development of methods to induce grain refinement. Various techniques have been proposed to refine the grain size of polycrystalline metals to less than 100 nm, which is considered the upper limit for the grain size of nanomaterials. Many of the proposed methods are based on the concept of severe plastic deformation, SPD. It has been demonstrated that SPD enables micrograined aggregates to be transformed to nanosized structure by the accumulation and rearrangement, or annihilation, of the crystal lattice defects, primarily dislocations. Experimental observations have shown[2, 3] that for such a transformation to take place, a large degree of plastic deformation is required, which usually exceeds the maximum strain achievable by the conventional methods of forming. As a result, special deformation methods have been invented and employed, including equal channel angular pressing (ECAP),[4] high-pressure torsion,[5] cyclic extrusion compression,[6] and multiaxial forging.[7] More recently, hydrostatic extrusion, HE, has also been used.[8,9] As SPD methods currently dominate the fabrication routes for bulk NCMs, a review of their microstructural features obtained is given later. Examples of metals and alloys processed by HE are used to illustrate the general considerations.

5.2.1 Diversity of Nanostructures

A number of technically pure metals and alloys have been processed by SPD to demonstrate the potential for achieving grain size refinement. It must be stressed, however, that SPD processing brings about various refined microstructures which are dependent on the materials being processed and the technological parameters of the process used. In particular the differences are related to the density of dislocations in the refined grains, the distribution function of the grain boundary structure/properties, and the grain geometry. Some of the most characteristic microstructures obtained by HE processing are illustrated in Fig. 5.1.

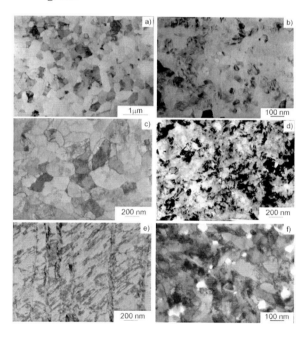

Figure 5.1 Microstructures of HE materials deformed to a total true strain of ~4: (a) aluminum, (b) the 2017 aluminum alloy, (c) copper, (d) titanium, (e) austenitic stainless steel, and (f) Eurofer 97 steel.[9]

In the case of pure aluminum (Fig. 5.1a), equiaxed grains, almost free of dislocations, are observed, whereas in pure copper (Fig. 5.1c) a dislocation cell structure is present after the same true strain. Grain refinement is much more efficient in the case of aluminum alloys (Fig. 5.1b) and titanium (Fig. 5.1d). However, in both cases the grains show a significant dislocation density. Hydrostatically extruded stainless steel (Fig. 5.1e) exhibits a very different microstructure, which possesses a high fraction of twin boundaries.

5.2.2 Grain Size

As the primary objective of SPD processing is to reduce the grain size, its quantitative evaluation is of major importance. In order to quantify grain size, various parameters can be used, including the equivalent diameter, d_2, defined as the diameter of a circle with the same surface area, or the mean intercept length, l. The equivalent diameter, d_2, is measured for the population of individual grains/subgrains revealed in transmission electron microscopy (TEM) observations. The grain size is then quantified by its average value, $E(d_2)$. The mean intercept length is determined for the entire population with no individual grains being measured. Stereological relationships, see, for example, Kurzydlowski and Ralph,[10] show that the value of $E(l)$ can be used to estimate the specific surface area of grain boundaries, Sv, using the simple formula Sv = 2/E(l).

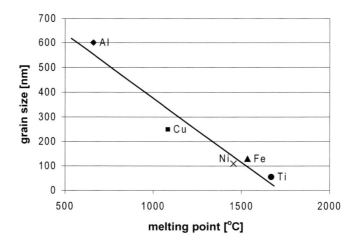

Figure 5.2 Average grain diameters of HE processed to a true strain of ~4 for pure metals, plotted as a function of their melting temperature.[9]

The average grain/subgrain size achieved in a number of HE-processed metals is shown in Fig. 5.2. It can be seen that for pure metals, the smallest grain size was obtained in technically pure titanium. This indicates that the efficiency of SPD at room temperature increases with the melting temperature of the materials being processed, as shown in Fig. 5.2, which plots grain size against melting temperature. This is fully understandable in view of the thermally activated processes of recovery and recrystallization, which could take place either during extrusion or soon after its completion. Thus it can be assumed that the efficient processing of low-melting-point materials may require the billets to be cooled prior to processing and/or the extruded material to be rapidly cooled at the die exit.

The efficiency of SPD refinement is much higher in the case of alloys, in particular aluminum alloys, where the presence of alloying elements strongly retards the rate of recovery. As a result, the grain size is usually smaller in alloys with a higher content of alloying elements, as illustrated for aluminum alloys in Fig. 5.3.

There are a number of factors which influence the grain size in bulk NCMs obtained by SPD. These factors include the applied strain, the strain-rate, the temperature, and the deformation path. In general, all factors which enhance the accumulation of defects, such as large plastic strain and low temperature, are beneficial for grain refinement. However, in order to transform the deformation structure into a nanocrystalline state, some limited thermally activated recovery processes and/or dynamic or postdynamic recrystallization must take place. Thus, the final microstructure results from the balance betw +een two opposite mechanisms, defect accumulation, and structural rearrangement, which leads to their partial annihilation.

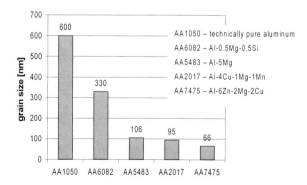

Figure 5.3 Mean equivalent diameters of various aluminum alloys processed by hydrostatic extrusion.

Systematic studies of the final grain size as a function of the total strain accumulated indicate that in the case of HE, the most rapid grain refinement takes place at relatively low strain.[11] At larger strains, the changes of grain size occur more slowly and approach a limiting value for a given material. A further substantial decrease in the grain size can be achieved by combining HE with other SPD techniques, for example, with ECAP, as has been demonstrated for copper and nickel.[12]

In summary, it should be noted that bulk nanorefined forms of a given alloy/metal may exhibit differences in the actual grain size, as expressed by the average value of the equivalent diameter or the mean intercept length/ specific surface area of grain boundaries.

5.2.3 Grain Size Diversity

Polycrystalline metals are stochastic populations of grains characterized by differences in their individual size and shape. NCMs obtained by SPD, in particular, exhibit a considerable degree of grain size variation. This is due to the stochastic nature of the formation and growth of nanograins during the process of plastic deformation. Generally, NCMs produced by low-temperature consolidation of nanopowders, which can be sieved to obtain particles of uniform size, can exhibit a much lower grain size variation.

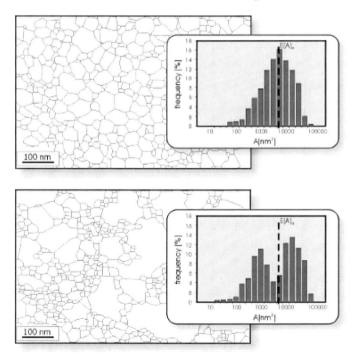

Figure 5.4 Structures possessing the same average grain size and the respective grain size distributions: (a) highly uniform and (b) highly nonuniform.

Grain size diversity has very important consequences for the properties of NCMs. It can be easily demonstrated that profoundly different grain structures may have the same average grain size, as is indicated in Fig. 5.4. As a result, these structures have different flow strength[13] despite possessing the same average grain size.

The grain size diversity can be quantified in terms of the equivalent diameter coefficient of variation, $CV(d_2)$. This parameter, defined as the ratio of the standard deviation to the mean value, is circa 0.2 for fairly uniform grain structures (for more information see Kurzydlowski and Ralph[10]).

Generally, metals and alloys processed by SPD are prone to large nonhomogeneity of the grain size. For HE-processed metals, the coefficient of variation of grain size ranges from 0.2 to 0.5, depending on the material and processing parameters—see Fig. 5.5. However, it is not easy to compare the grain size diversities of materials processed by different techniques as available data are relatively scarce. Nevertheless, it must be emphasized that the bulk NCMs currently available exhibit a considerable variation in grain size, with some grains exceeding the 100 nm threshold for nanomaterials.

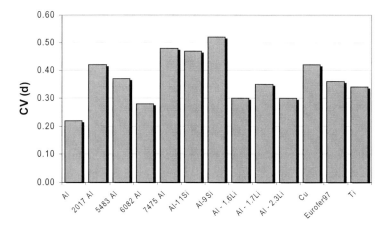

Figure 5.5 Values of the CV parameter for metallic materials processed by hydrostatic extrusion.

5.2.4 Grain Boundary Characteristics

Grain boundaries are one of the most important microstructural elements possessed by polycrystalline materials. Their role is even more crucial in the case of NCMs as their surface area per unit volume is substantially greater than in conventionally structured polycrystalline materials. It can be calculated that in a material possessing grains with a mean intercept length of 100 nm, there are, on average, 200 cm^2 of grain boundaries in each cubic millimeter of the metal and that 5% of the atoms are located at the grain boundaries. When the grain size is reduced to 5 nm the atoms located at the grain boundaries increases to 50%.[14] These figures rationalize the notion that the properties of NCMs are largely governed by grain boundaries.

The grain boundaries in metallic materials differ in their geometry and structure, which in turn influences the properties of the particular material. At the current stage of development of the theory of the effects of grain boundaries, their misorientation angles are the key parameter affecting their structure and properties. The misorientation angle is defined as the minimum angle of rotation which transforms one crystal lattice into that of the adjacent crystal. On the basis of the value of the misorientation angle, the grain boundaries in polycrystalline materials can be divided into three groups: (i) low angle, where the misorientation is less than 5°; (ii) medium angle, having a misorientation angle between 5 and 15°; (iii) high angle, where the misorientation is greater than 15°.

Misorientation angles can be calculated from the Kikuchi line patterns obtained either by TEM or by scanning electron microscopy (SEM) using the electron backscatter diffraction (EBSD) system. In the case of SPD-processed materials, which usually contain a large fraction of low-angle grain boundaries, the distribution function of misorientation angles is the key microstructural factor for understanding the properties. If low-angle grain boundaries are dominant, the microstructure is classified as a "deformation structure." When the processed material contains primarily high-angle grain boundaries, depending on the average grain size, it is called ultrafine grained or nanocrystalline. It has been established that the unique properties of NCMs result from the presence of high-angle grain boundaries.[15] However, as shown in Fig. 5.6, the typical distribution function of misorientation angles for SPD metals is relatively wide. The fraction of high-angle grain boundaries often exceeds 60%, and the grain boundary character distribution function frequently has a bimodal character.[16,17]

The changes in the fraction of boundary types in pure aluminum, as a function of the true strain induced by HE processing, are illustrated in Fig. 5.7. From this graph one can conclude that as the true strain increases, there is a sharp increase in the fraction of high-angle grain boundaries accompanied by a significant decrease in the fraction of low-angle grain boundaries. At a true strain of about 4, the majority of boundaries, some 60%, are high angle and only about 20% are low-angle boundaries. A discussion about the formation mechanism of high-angle grain boundary in HE-processed materials is given elsewhere.[18]

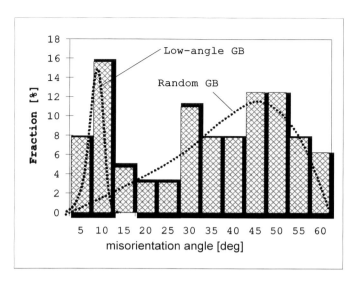

Figure 5.6 Distribution of grain boundary misorientation in SPD material.[16]

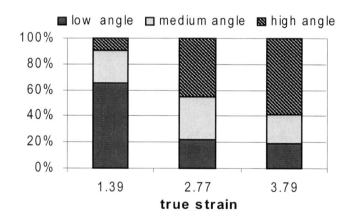

Figure 5.7 The fraction of grain boundary types in pure aluminum subjected to HE as a function of true strain.

5.2.5 Second-Phase Particles

Metallic materials, including those which are considered single phased, usually contain second-phase particles, primarily intermetallic inclusions resulting from the presence of impurities and/or precipitates which form during any heat treatment. The role of these particles on nanograin refinement was studied in experiments carried out on 2017 and 7475 aluminum alloys.[11,19]

5.2.5.1 The impact of second-phase particles on grain refinement

One of the most important aspects of the presence of second-phase particles in materials intended for SPD processing is their influence on the process of grain refinement. This influence of primary intermetallic particles and secondary precipitates is discussed later.

TEM observations have revealed that relatively large primary intermetallic particles alter the evolution of the microstructure during HE processing, as shown in Fig. 5.8. Away from the particles, the microstructure consists of elongated grains enveloped by boundaries nearly parallel to the extrusion direction, but in the vicinity of the intermetallic particles, equiaxed and very fine grains are visible. This implies that the relatively large particles promote the process of grain refinement. It should, however, be noted that their influence is limited to the neighboring zone of thickness comparable to that of the size of the particle. These observations are in good agreement with those for an aluminum alloy containing a large volume fraction of intermetallic particles.[20] For such an alloy, the process of grain refinement is much more rapid than in a single-phase alloy.

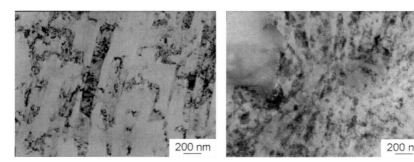

Figure 5.8 The microstructure of hydrostatically extruded the 2017 aluminum alloy (a) in a region without particles and (b) in the vicinity of a particle.[19]

The role of precipitates in grain refinement is very different. The presence of fine precipitates reduces the degree of grain refinement. If the 2017 aluminum alloy is subjected to HE immediately after water quenching, a well-developed microstructure with a grain size of 90 nm is formed, as shown in Fig. 5.9a. Measurement of the misorientation angle revealed that a significant fraction of the grain boundaries were high-angle boundaries (see Fig. 5.9b). When the same alloy was processed after ageing, a less well-developed structure with a higher grain size of about 150 nm was obtained, as shown in Fig. 5.9c, with a lower number of high-angle grain boundaries (Fig. 5.9d). This indicates that precipitates delay the formation of high-angle boundaries.

The reason for this might be that the strain energy transferred to the sample during deformation is used both for the grain refinement process and for fracture/fragmentation of the precipitates.

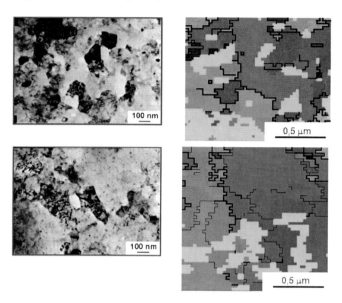

Figure 5.9 The microstructures (a and c) and orientation maps (b and d) of the 2017 aluminum alloy processed immediately after water quenching (a and b) and after ageing (c and d)[19]. Thick, black lines in orientation maps correspond to high-angle, whereas thin to low-angle grain boundaries.

These observations were confirmed by studies on hydrostatically extruded 7475 aluminum alloy samples[9,21] treated to produce one of the following combinations of inclusions and precipitates:

(1) low-density inclusions (solution-heat-treated samples)

(2) low-density inclusions and low-density large precipitates (annealed samples)

(3) low-density inclusions and high-density small precipitates (solution-heat-treated and aged)

The finest grain size, with a mean value of ~60 nm, was obtained in the annealed material, sample (2), as illustrated in Fig. 5.10a. When processed immediately after water quenching, a well-developed microstructure with a grain size of 90 nm was produced (Fig. 5.10b). After ageing, HE processing produced a microstructure containing a high density of dislocations with much less-developed, larger grains (Fig. 5.10c).

Figure 5.10 Microstructures of the HE-processed 7475 aluminum alloy after pretreatment: (a) solution heat treated and water quenched, (b) annealed, and (c) solution heat treated and aged to show the influence of second-phase particles on grain refinement (for details see the earlier text).

These findings indicate that relatively coarse precipitates, which resist plastic deformation, accelerate grain refinement, whereas fine precipitates inhibit the formation of a nanostructure. This results are in good agreement with those reported by Murayama *et al.*[22] and Zhu.[23]

5.2.5.2 The Stability of Second-Phase Particles During Processing

5.2.5.2.1 *Precipitates*

Precipitates are one of the microstructural elements which strengthen aluminum alloys. They form during ageing of solution-heat-treated and water-quenched samples. Age-hardened aluminum alloys exhibit the highest mechanical strength among all aluminum alloys.

The behavior of precipitates during conventional plastic deformation of aluminum alloys is quite well known. Small and coherent precipitates are sheared by moving dislocations. This may lead to the microscopic localization of plastic deformation in the form of shear bands[24] and to changes in their shape and eventually to dissolution.[25] One can expect that similar processes will take place during SPD processing; however, different kinetics apply, as described later.

The ageing process in the 2017 aluminum alloy brings about the formation of platelike θ' precipitates (Fig. 5.11a) of length approximately 200 nm and width 20 nm along the {100} planes. Hydrostatic extrusion to a true strain of 1.8 results in the fragmentation of the precipitates as a result of shearing induced by moving dislocations. With increasing deformation, up to a true strain of 3.8, the precipitates change shape and form small spherical particles with an estimated diameter of 10 nm (Fig. 5.11b). This indicates that precipitates are strongly refined during the process of plastic deformation. It was noted that this refinement took place uniformly throughout the entire volume of the processed material, with no shear banding being visible.

The transformation mechanism causing changes in the size and shape of precipitates is not yet known. It seems that the first stage of transformation, that is, fragmentation, involves interactions between the precipitates and the dislocations, whereas the second stage, the shape change, requires diffusion and thus requires thermal activation. The driving force for dissolution and/or shape change is the strong need to decrease the surface area of the precipitate–matrix interfaces. It may be appreciated that the necessary thermal activation for this process may be provided by adiabatic heating, which occurs during plastic deformation at very high strain rates, as encountered during hydrostatic extrusion. In the case of aluminum alloys, the temperature rise can be estimated at some 200°C, which is sufficient to initiate diffusional transformation.

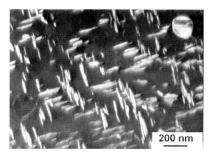 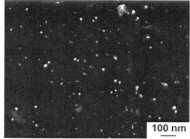

Figure 5.11 TEM micrographs of precipitates in the 2017 aluminum alloy aged at 200°C for 3 h (a) before and (b) after hydrostatic extrusion.[19]

5.2.5.2.2 *Intermetallic particles*

Commercial alloys always contain a small amount of impurities, such as Fe and Si in aluminum, which form primary intermetallic particles during the solidification process. The particles are relatively large, several microns in diameter, and are noncoherent with the matrix. They are thermodynamically stable over a wide range of temperature and are considered stable or only to undergo negligible changes during SPD. However, the observations of microstructural evolution occurring during the SPD of cast alloys, which contain a large number of intermetallic particles, indicate that the particles undergo significant changes in their size and spatial arrangement.[26,27] This suggests that similar changes can take place in alloys which contain a smaller quantity of intermetallic particles. However, this phenomenon has not yet been examined in detail.

In the as-received state, the 2017 aluminum alloy contains relatively large primary intermetallic particles, as shown in Fig. 5.12a, with a chemical composition corresponding to the $(FeCuMn)_3Si_2Al_{15}$ phase. The average equivalent diameter of these particles is $1.4 \propto m$, and their spatial distribution is nonhomogeneous with the small particles existing in clusters. These clusters effectively correspond to a eutectic structure, as the intermetallic compounds usually solidify at a fixed temperature.

Transverse section   Longitudinal section

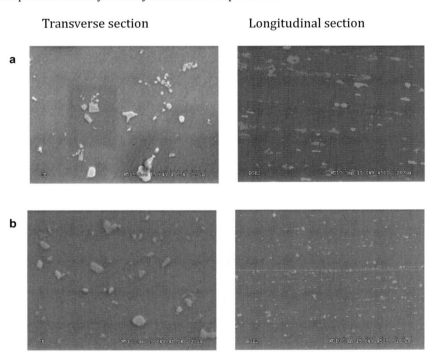

Figure 5.12 SEM images of primary intermetallic particles (a) before and (b) after hydrostatic extrusion. Left column for transverse and right column longitudinal section.[19]

Hydrostatic extrusion results in significant changes in both the geometry of the intermetallic particles and their spatial arrangement (see Fig. 5.12b). These changes have been quantified by measuring their equivalent diameter, d_2 and their shape by determining the shape factors: d_{max}/d_2 and $p/\pi d_2$, to describe the particle elongation and interface boundary waviness, respectively. The spatial distribution was determined using the SKIZ tessellation method, in which the variation coefficient of the areas of the surrounding zone of influence for each particle is analyzed.[28] The values of these parameters, before and after HE, are summarized in Table 5.1.

The data in Table 5.1 clearly show that hydrostatic extrusion causes a slight reduction in the size of the primary intermetallic particles, which, however, remains at the micrometer level. More significant changes in the particle shape occur, as they become more regular—as confirmed by the value of both shape factors, which are substantially lower after HE. Additionally the distribution of the particles becomes more homogenous. It should be noted that the changes in particle geometry and dispersion influence the mechanical properties of the HE-processed material. In particular, it may be expected that hydrostatically extruded aluminum alloys exhibit enhanced fracture toughness due to the changes in shape, size, and distribution of the intermetallic particles.

Table 5.1 Stereological parameters of primary intermetallic particles before and after hydrostatic extrusion.

	Parameter	Before hydrostatic extrusion	After hydrostatic extrusion
Particle size	Mean equivalent diameter, d_2 $(\propto$m)	1.4	1.24
	Equivalent diameter variation coefficient, $CV(d_2)$	0.82	0.64
Particle shape	d_{max}/d_2	1.41	1.34
	$p/\pi d_2$	1.23	1.18
Spatial distribution	SKIZ cell area variation coefficient, $CV(A_{SKIZ})$	1.14	0.75

5.3 MECHANICAL PROPERTIES OF NANOCRYSTALLINE METALS

5.3.1 Strength

Grain size is a prime microstructural parameter affecting the mechanical strength of NCMs. The Hall–Petch equation, one of the most important empirical relationships describing the influence of microstructure on the mechanical properties of metals, predicts a linear dependence of the yield strength/hardness on the inverse square root of the average grain size, $1/(d)^{1/2}$ at low temperatures. This implies that a particularly significant increase in the mechanical strength will be obtained as the grain size is reduced to the nanorange.

The efficiency of strengthening by grain refinement has already been established for a number of metals and alloys.[21,29-30-31-32-33-34-35] Table 5.2 lists the yield strength of a number of alloys in their microcrystalline and nanostructured condition. It is apparent that NCMs frequently possess a tensile strength two or three times greater than that of the alloy in its conventional microcrystalline state. The very high absolute values of the flow stress for various metals should be noted. The yield strength greater than 1,000 MPa obtained for technically pure nano-titanium is typical to that of titanium alloys, and a strength of 500 MPa for a single-phase aluminum alloy in the 5xxx series is similar to that of age-hardenable alloys, which are considered the strongest aluminum alloys.

Table 5.2 Mechanical properties of nanostructured alloys and their microcrystalline counterparts.

Materials	Yield strength (MPa)		Reference
	Microcrystalline	Nanostructured	
AA1050	70	200	[29]
AA1100	40	170	[30]
AA 3004	95	360	[30]
AA6082	(T6) 370	437	[31]
AA6082	112	283	[32]
AA5182	395	430	[31]
AA5483	215	490	[32]
AA2017	350	570	[29]
AA2024	115	330	[30]
AA7075	200	470	[30]
AA7475	230	650	[21]
Al-Zn	375	750	[33]
Ti	345	1060	[34]
Ti-6Al-4V	900	1040	[35]

The strength of NCMs at the temperature of $\sim 0.3\ T_m$, where T_m is the melting point, is usually analyzed by plotting the flow stress against $d^{-1/2}$, as shown for aluminum alloys in Fig. 5.13. An extrapolation of the Hall–Petch relationship to the nanometer range shows that technically pure aluminum may exhibit a yield strength of 500 MPa if its grain size is reduced to 25 nm (see Fig. 5.13). However, the experimental data presented in Fig. 5.13 shows that such a value has not yet been attained for pure aluminum. This could be because to date the minimum grain size obtained for pure aluminum processed by various SPD methods has not been less than 300 nm.

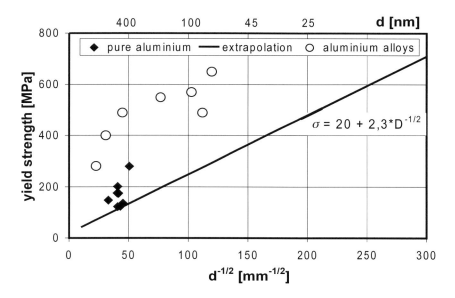

Figure 5.13 Plot of the Hall–Petch relationship for SPD-processed aluminum alloys.[36]

It is suggested that for each metal or alloy, there be a limit to the degree of grain size refinement. This limit depends on the chemical composition, the melting temperature, and the presence of second-phase particles and is controlled by the rate of accumulation, annihilation, and rearrangement of defects which are generated during the processing and by the concurrent thermally activated phenomena of recovery and grain growth.[16]

However, experimentally obtained yield strength values for aluminum processed by SPD are higher than those predicted by extrapolation of the Hall–Petch relationship. This is because the grain refinement achieved by SPD is accompanied by a significant increase in the dislocation density in the grains. It follows that the dislocation and grain boundary strengthening combine to give a value higher than that predicted by the Hall–Petch relationship. For a first approximation, the increase in the flow stress due to dislocation strengthening is accounted for by an increase in the so-called friction stress of the Hall–Petch equation, which is proportional to the square root of the dislocation density. It should be noted that if the variations in the dislocation density are not taken into account, the data points on Hall–Petch plots are aligned along lines with a fictitious slope, which may be very different from the true value, which is obtained for the specimens with constant density of dislocations.

Some experimental results and numerical modeling indicate that there is also an additional factor limiting the grain size strengthening, which is exhibited as a negative Hall–Petch slope.[37] Such a change in the slope of the Hall–Petch diagram is expected when the grain boundaries deform at a stress lower than that required for deformation of the actual grains. It should be realized, in this context, that the resistance to plastic deformation of the grain interiors is expected to increase as the value of $1/d$ increases and is likely to exceed the resistance of grain boundaries if the grain size is sufficient. It must be also appreciated that the grain boundaries for materials with small grain sizes account for a significant volume of the polycrystalline material[38]— a situation which could be considered a grain boundary matrix composite strengthened by nondeformable grain interiors. In such a case further grain refinement does not increase the mechanical strength of the material and is counterproductive. Current models of grain boundary strengthening in the nanorange predict that the inverse Hall–Petch behavior for aluminum alloys should occur with grain sizes in the range of 10–15 nm.[39] If these sizes could be obtained it would result in the strength limit of some aluminum alloys being as high as about 1,000 MPa.

The mechanical strength is also affected by the variation in grain size, as demonstrated experimentally and by *finite-element modeling*[13,40]. The results have shown that by reducing the grain size homogeneity in copper from $CV = 0.07$ to 0.41, the yield strength decreases by about 100 MPa for an average grain size of ~ 30 nm.[40] This shows that grain size homogeneity, which varies considerably in the currently available bulk NCMs, is also a limiting factor for strengthening.

As discussed earlier in this chapter and illustrated in Fig. 5.13, the strength of NCMs can be further increased by the presence of solute atoms and/or precipitates.[36] In the case of nanocrystalline aluminum alloys, a significant improvement in the mechanical strength of the 2xxx and 7xxx series can be anticipated. Figure 5.14 shows the yield strengths of the 7475 aluminum alloy strengthened by (a) precipitates (grain size of 70 $\propto$m), (b) grain boundaries (grain size of 70 nm), and (c) the combined effect (grain size of 70 nm + precipitates). It is clearly seen that the yield strength of about 500 MPa is increased by about 200 MPa by precipitate hardening.

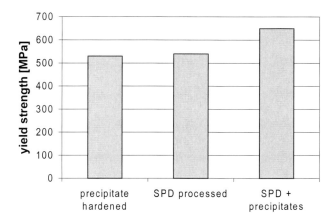

Figure 5.14 Yield strengths of the 7475 aluminum alloy in different microstructural states.[36]

5.3.2 Ductility

Currently NCMs usually exhibit high tensile strength, but it is achieved at the expense of ductility and toughness. Numerous publications indicate that elongation to failure for NCMs usually does not exceed a few percent.[41]

The main reason for the low ductility is the low strain-hardening rate of NCMs. As a result; the various flaws induced by the technological processes, such as nanopores and nanocracks, cause localization of the plastic deformation and therefore premature failure. The low strain-hardening rate of NCMs results from the extremely low rate of accumulation of dislocations in the grains. As already discussed, for a grain size below a certain critical value, dislocation accumulation is impossible.

In the case of submicrocrystalline materials, the reasons for low ductility are similar to those which apply with NCMs. However, in these cases the processing is less severe and the defects can be eliminated. Additionally, the strain-hardening rate is, on average, higher as the larger grains may be able to accumulate a limited density of dislocations.

The potential for increasing the ductility of NCMs is related to an improvement in the processing technologies aiming at eliminating isolated flaws and/or redesigning the microstructure to achieve a higher strain-hardening rate.

An increase of the strain-hardening coefficient, which in turn increases the ductility, can be achieved by deformation at a lower temperature or at a high deformation rate. However, this does not solve the problem entirely, as engineering materials operate in various conditions, not just at a low temperature and a high deformation rate. Another method is based on the concept of bimodal grain size distribution,[42] in which a considerable volume fraction of relatively bigger, strain-hardened grains are featured. However, bimodal-sized polycrystals exhibit a decrease in yield and tensile strength due to an increase in the average grains size. A new method proposed for improving the strain-hardening rate is to fabricate NCMs with grain boundaries which harden during deformation. Theoretical models have been developed to rationalize this concept, but currently there has been no success in fabricating a material possessing a grain boundary population capable of strain hardening. Among the processes which affect the behavior of grain boundaries, the segregation and de-segregation of foreign atoms should be considered. The results obtained in conventional materials and the results of computer modeling indicate that segregation of foreign atoms at grain boundaries can increase their shear strength. However, in literature, no quantitative evaluations could be found which quantifies the influence of segregation on the dynamics of grain boundaries sliding.[43]

In the search for methods to increase the ductility of NCMs, use could be made of the concepts already proved successful in reducing the brittleness of ceramics, which suffer from localized deformation in the form of cracks. In particular nanocrystalline metal–metal composites exhibit increased strength and a reduced tendency for localization of deformation. The manufacturing methods of such composites are described by Embury and Sinclair[44] and Lu *et al.*[45] Some results obtained by the authors of this chapter indicate that a small addition of microcrystalline Co to nanocrystalline Fe can results in an increase in both strength and uniform elongation.[46]

An efficient method of increasing the ductility of NCMs utilizes the effect of mechanical twinning. Karimpoor *et al.*[47] have demonstrated that for nanocrystalline nickel and cobalt, much larger elongations to rupture can be achieved due to mechanical twinning, which occurs more easily in metals with hexagonal close-packed (HCP) rather than face-centered-cubic (FCC) lattices. Tadmor and Bernstein[48] have shown that the tendency for twinning in an FCC lattice depends strongly on the value of the so-called unstable twinning energy. In principle this energy can be estimated by computer modeling and can be modified by appropriate alloying additions. However, to date this concept has not been utilized in a practical manner.

5.3.3 Fracture Toughness

Early analyses of fracture toughness indicated that a decrease in the grain size to a nanometric size will reduce the size of cracks in polycrystalline materials. The results obtained for NCMs fabricated by various methods indicate that these expectations are not very realistic. Micrometer-sized defects are observed in materials produced by powder consolidations and chemical, electrochemical, and physical deposition, as well as many materials produced by severe plastic deformation.

The characteristic feature of failure for many NCMs is ductile fracture. Such fracture is observed both in materials with significant ductility, for example, in an Al–1.7%Fe alloy fabricated by the physical vapor deposition (PVD) method,[49] which cracked at 6% elongation, and in materials which cracked at no measurable plastic deformation. Another example is ductile fracture observed in a Ni–W alloy fabricated by the electrodeposition method, which cracked at a plastic strain much lower than 1%.[50] Also a Ni–19%Fe alloy fabricated by the powders metallurgy method, which cracked at no measurable plastic deformation,[51] revealed ductile fracture.

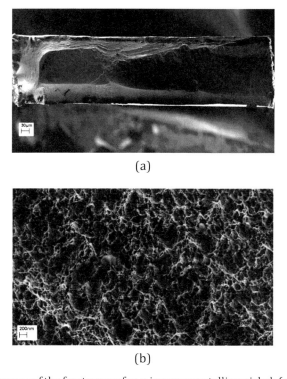

(a)

(b)

Figure 5.15 Images of the fracture surfaces in nanocrystalline nickel: (a) macroscopic and (b) under high magnification.[52]

Similar results were obtained by present authors for nanocrystalline Ni, fabricated by the electrodeposition method.[52] The material has a mean grain size of about 20 nm and was characterized by ductile fracture, even though it cracked with no macroscopically measurable plastic deformation. As shown in Fig. 5.15, at a macroscopic scale, the fracture surface looks like that for brittle materials. However, at high magnification ductile fracture is clearly revealed. This means that despite the fact that nanocrystalline Ni preserved plasticity at a microscopic scale, the capacity to absorb the energy of cracking is very limited. One of the reasons is the low ability of the material to deform by grain boundary sliding. Slight modification of the processing procedure for the material results in a significant improvement in the ductility.

Figure 5.16 presents the tensile test curve for a sample of nanocrystalline Ni with the same grain size (20 nm) as that discussed earlier but with much greater plasticity, possessing a strain to failure of about 7% compared with the negligible plasticity on the sample shown in Fig. 5.15. It can be supposed that modification of the electrodeposition process improved the material's ability to undergo grain boundary sliding, which is considered responsible for the higher plasticity of the material.

Observations of natural composites, for example, nacre shells, show that they have a structure which simulates a composite-like approach to increasing the fracture toughness of engineering materials. The high fracture toughness of these natural materials results from their specific nanocrystalline structure, in which plates of brittle material (aragonite) are separated with very thin layers of deformable proteins. The small size of the plates ensures that the stress concentration does not exceed their theoretical strength. Also the protein layers prevent the stresses from being transmitted to the neighboring plates.[53]

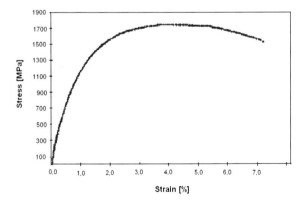

Figure 5.16 Tensile curve of nanocrystalline Ni with relatively high plasticity of mean grain size 20 nm produced using the modified electrodeposition processing procedure.[52]

This concept has been applied for production of metallic nanocomposites which consist of nanocrystalline plates of intermetallic phases embedded in an aluminum matrix.[54] The preliminary results indicate that application of this concept may lead to the production of material possessing high strength and fracture toughness.

5.3.4 Fatigue Strength

One of the most important mechanical properties of engineering materials is the fatigue strength. However, despite the importance of fatigue properties, publications focussing on the fatigue of NCMs are scarce and are often restricted to the effect of nanosurface layers, which are known to improve fatigue strength. There are even fewer publications concerning the fatigue strength of bulk NCMs. Research into the fatigue properties of such metals has been mainly carried out on samples obtained by SPD, with the majority of papers concerning high cycle fatigue (HCF). Only a few works considered LCF and the development of fatigue cracks in NCMs.[54]

The basic conclusion which can be drawn from the available experimental data is that NCMs are characterized by enhanced HCF strength and decreased LCF strength with an increased rate of crack propagation.[55]

Fatigue phenomena are currently analyzed in terms of the sequential process shown in Fig. 5.17.[56] Microdeformation in the same local areas of the material is the stage which usually occurs for 50% to 90% of the lifetime of components operating in fatigue conditions. Therefore, a reduction in the density of microdeformations in the material should increase the fatigue resistance. The reductions in the density of sites for microdeformation can be achieved by an increase in yield stress and an improvement in the homogeneity of he microstructure.

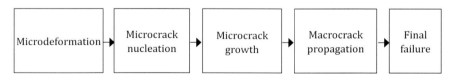

Figure 5.17 Schematic presentation of the stages leading to fatigue failure.

A way of meeting these requirements is to produce a nanostructured material exhibiting a higher yield stress. However, it does not always increase the degree of homogeneity, and the discontinuities and microcracks induced by the process of nanorefinement can be a problem. An example of a flaw induced by nanorefinement is illustrated in Fig. 5.18. This shows an image of the fracture surface of austenitic steel after high-pressure torsion. Metallic materials containing such flaws have low fatigue strength, despite the possession of a nanorefined grain structure.

To increase the strength of a material intended to operate in conditions of HCF can result in its LCF strength being reduced. The reduction in the LCF properties is accompanied by an increase in the rate of fatigue crack propagation. Various authors have shown that during LCF, significant deformation weakening occurs[57-59] and localization of deformation occurs in the form of bands, which in turn lead to a decrease in the fatigue strength.

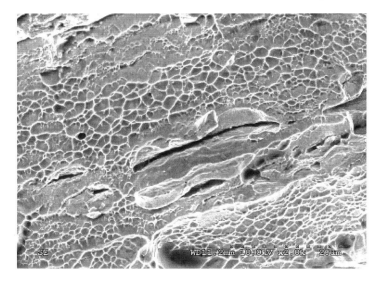

Figure 5.18 An example of microcrack produced in austenitic steel processed by high-pressure torsion.[60]

Although the fracture initiation process can be delayed by a higher yield stress, the crack propagation rate cannot. Materials with lower plasticity have a lower capacity to diffuse energy at the crack tip. As a result the fracture toughness decreases and the rate of crack propagation increases.

5.4 DEFORMATION MECHANISMS OF NANOCRYSTALLINE METALS

Although research on NCMs is conducted in numerous laboratories around the world, the criteria used to define such materials are still not entirely precise. Commonly a material in which the size of the elements, crystallites, grains, blocks, layers, fibers, etc., does not exceed 100 nm, at least in one direction, is considered a nanomaterial. This definition covers 3-D nanoaggregates of equiaxial grains and nanolayers or nanofibers. However, it should be recognized the critical size of structure elements in this definition, 100 nm, has no physical basis and is, therefore, totally discretional.

A different definition of NCMs takes into consideration the percentage of atoms to the total number of atoms or the volume of grain, or crystallite, boundaries to the volume of the sample of material. If this ratio is significant, it is assumed that the material is nanocrystalline. For example a value of 10% gives, in effect, the critical size of crystallites at a level of 30–50 nm, depending on the grain boundary thickness, which is usually in the range of 0.5–1 nm.

Another definition refers to the correlation of nanocrystallite size with the characteristic physical dimensions. In the case of plastic deformation it can be, for example, the critical distance between obstacles for dislocations.

In general, the definitions of NCMs introduce a requirement that the material should be characterized by new or enhanced properties due to the refinement of the microstructure, which are unattainable with conventional materials. Such a definition would allow metals with crystallite dimensions larger than 100 nm[61] to be considered NCMs if they are characterized by unique properties. However, materials with a grain size lower than 1 μm but higher than 100 nm are more often referred to as being ultrafine grained.

Thus, the critical size of structural elements, which are employed to distinguish NCMs, remains an open issue. In the case of the mechanical properties of metallic materials, the criteria enabling the qualification of materials as being nanocrystalline can be based on the operating mechanisms of the deformation process. Applying this criterion, metallic materials can be divided into four groups: amorphous, nanocrystalline, ultrafine grained (submicrocrystalline), and microcrystalline (conventional).

Grain sizes and boundary values separating these four groups of materials are shown in Fig. 5.19. The standard values of the critical grain sizes are given in parentheses.

The role of dislocations in the deformation process of crystalline materials is well understood. In the case of NCMs, especially with a grain size above a certain value, the dislocation theory developed for conventional materials begins to have limited application. With a decrease in the grain size there is a decrease in the density of dislocations inside the grains and the sources at the grain boundaries become of key importance.

In materials which possess a grain size less than 30–70 nm, which can depend on the particular material, dislocations are not accumulated in the grains. Therefore the dislocation density does not increase during deformation. Molecular dynamics simulation indicates that a reduction of the grain size increases the relative importance of the emission of partial dislocations from the grain boundaries and the related stacking fault formations.[62] The rate of deformation is thus controlled by the generation of partial dislocations from grain boundaries.

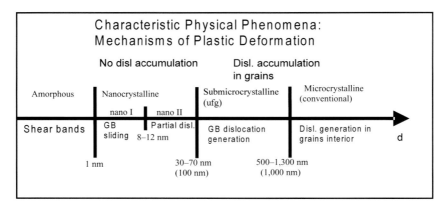

Figure 5.19 The ranges of crystallite sizes determined on the basis of experimental data and computer modeling. The conventional values are given in parentheses. *Abbreviation*: Disl, dislocation.

In materials with even smaller grains, no dislocation activity is observed and deformation occurs by the sliding of grain boundaries. The critical grain size, below which dislocation activity is not observed, is different for each material, the applied strain-rate, and the test temperature. It also depends on factors such as the stacking fault energy. Computer modeling by molecular dynamics showed that the value for copper is 8 nm and for nickel 12 nm.[63] The conclusion has been drawn that the critical grain size increases with an increase in the stacking fault energy.

The question of whether there is a physically rational sharp borderline between conventional (microcrystalline) and submicrocrystalline materials is not totally resolved. Some authors suggest that such a borderline does exist and that it is defined in terms of the number of dislocations generated at the grain boundaries and inside the grains. In the case of conventional materials, plastic deformation depends on the dislocation sources inside the grains and in submicrocrystalline materials—on the dislocation sources at grain boundaries. As a result, submicrocrystalline materials reveal different properties, including the capacity for superplastic deformation at relatively low temperatures at high deformation rates.

NCMs often contain a large number of deformation twins. Computer simulations indicate that in fine-grained materials, twinning can be the dominant mechanism of plastic deformation. In micrograined materials, twinning occurs for materials with low stacking fault energies, at low temperature, or under a high level of material hardening. In the case of NCMs twinning occurs much more easily. The proof is the twinning that can occur in nanocrystalline aluminum. In coarse-grained aluminum twin formation is so rare that until recently it was believed that it could not occur. In nanocrystalline aluminum, however, twinning occurs fairly easily, as has been demonstrated by computer modeling and experimental methods.[64,65]

5.5 SUMMARY

A profound improvement of many mechanical properties resulting from the nanostructuring of metals and the increasing body of knowledge of the microstructural factors that influence these properties open up the prospects for their broader application in innovative industries such as aerospace, surface transport, and biomedical implants. A particular advantage is in applications for vehicles for air and ground transportation. The possible weight reduction facilitated by the significantly enhanced strength of nanostructured metals and alloys may contribute to a reduction in CO_2 emission. Nano-titanium can also successfully be employed for highly loaded bioimplants, which currently are manufactured from a Ti–Al–V alloy, because pure microcrystalline titanium is not sufficiently strong.

However, the applications of nanocrystalline metals and alloys face serious limitations—primarily the cost. NCMs are, and probably will for a long time be, much more expensive than their conventional microcrystalline equivalents. Another major limitation is the size of available nanocrystaline billets as the currently available fabrication methods have severe size restrictions. The only methods currently available to produce relatively large-sized elements with a highly refined structure are based on SPD techniques. However, attempts to achieve larger-sized elements using SPD have not been very successful to date. Sub-micron-grained materials are fairly easily achieved. Refinement of the grain size to the range of 300–500 nm brings about a significant increase in the material strength and maintains an acceptable level of ductility for many applications. The potentials are particularly promising for Al alloys, which, in addition to the increased strength exhibited by the ultrafine-grained structure, possess the so-called fast superplasticity[66] which occurs at deformation rates above 10^{-1}.

References

1. H. Gleiter (1989), *Prog. Mater. Sci.*, 33:323.
2. R. Z. Valiev, R. K. Islamgaliev, and I. V. Alexandrov (2000), *Prog. Mat. Sci.*, 45:103.

3. D. A. Hughes and N. Hansen (2000), *Acta Mat.*, 48:2985.

4. R. Z. Valiev and T. G. Langdon (2006), *Prog. Mat. Sci.*, 51:881.

5. G. Sakai, Z. Horita, and T. G. Langdon (2005), *Mater. Sci. Eng. A*, 393:344.

6. M Richert, Q. Liu, and N. Hansen (1999), *Mat. Sci. Eng. A*, 260:275.

7. B. Cherukuri, T. S. Nedkova, and R. Srinivasan (2005), *Mat. Sci. Eng. A*, 410–411:394.

8. K. J. Kurzydłowski (2006), *Mat. Sci. Forum*, 503–504:341.

9. M. Lewandowska and K. J. Kurzydłowski (2008), *J. Mat. Sci.*, 43:7299–7306.

10. K. J. Kurzydlowski and B. Ralph (1995), *Quantitative Description of Materials Microstructure*, CRC Press, Boca Raton.

11. M. Lewandowska (2006), *J. Microsc.*, 224:34–37.

12. M. Kulczyk, W. Pachla, A. Mazur, M. Sus-Ryszkowska, N. Krasilnikov, and K. J. Kurzydlowski (2007). *Mat. Sci.—Poland*, 25:991.

13. K. J. Kurzydlowski and J. J. Bucki (1993), *Acta Metal Mater.*, 42:3141–3146.

14. H. Gleiter (1995), *Nanostruc. Mater.*, 6:3–14.

15. R. Z. Valiev, Y. Estrin, Z. Horita, T. G. Langdon, M. Zehetbauer, and Y. T. Zhu (2006), *JOM*, 58:33–39.

16. M. Lewandowska, T. Wejrzanowski, and K. J. Kurzydłowski (2008), *J. Mat. Sci.*, 43:7495–7500.

17. M. Ferry, N. E. Hamilton, F. J. Humphreys (2005). *Acta Mat.*, 53:1097.

18. M. Lewandowska (2006), *Sol. Stat. Phenom.*, 114:109–116.

19. M. Lewandowska (2006), *Arch. Metall. Mater.*, 51:569–574.

20. P. J. Apps, J. R. Bowen, and P. B. Prangnell (2003), *Acta Mater.*, 51:2811–2922.

21. K. Wawer, M. Lewandowska, M. Zehetbauer, and K. J. Kurzydlowski, *Kovove Materialy*, in press.

22. P. J. Apps, J. R. Bowen, P. B. Prangnell (2003), *Acta Mat.*, 51:2811.

23. M. Murayama, Z. Horita, and K. Hono (2001), *Acta Mat.*, 49:21.

24. A.W. Zhu (1998), *Acta Mater.*, 46:3211–3220.

25. Y. Brechet, F. Louchet, C. Marchionni, and J.-L. Verger-Gaugry (1987), *Phil. Mag.*, 38:353–360.

26. C. Xu, M. Furukawa, Z. Horita, and T. G. Langdon (2003), *Acta Mater.*, 51:6139–6149.

27. A. Ma, K. Suzuki, Y. Nishida, N. Saito, I. Shigematsu, M. Takagi, H. Iwata, A. Watazu, and T. Amura (2005), *Acta Mater.*, 53:211–220.

28. W. A. Spitzig and J. F. Kelly (1985), *Metallography*, 18:235–261.

29. M. Lewandowska, H. Garbacz, W. Pachla, A. Mazur, and K. J. Kurzydłowski (2005), *Mater. Sci.—Poland*, 23:279–286.

30. Z. Horita, T. Fujinami, M. Nemoto, and T. G. Langdon (2001), *J. Mat. Sci. Tech.*, 117:288–292.

31. H. J. Roven, H. Nesboe, J. C. Werenskiold, and T. Seibert (2005), *Mater. Sci. Eng. A*, 410–411:426–429.

32. P. Widlicki (2009), PhD Thesis, Warsaw University of Technology, Warsaw.

33. K. Islamgaliev, N. F. Yunusova, I. N. Sabirov, A. V. Serueeva, and R. Z. Valiev (2001), *Mat. Sci. Eng. A*, 319–321:877–881.

34. H. Garbacz, M. Lewandowska, W. Pachla, and K. J. Kurzydłowski (2006), *J. Microsc.*, 223:272–274.

35. G. G. Yapici, I. Karaman, and Z.-P. Luo (2006), *Acta Mater.*, 54:3755–3771.

36. M. Lewandowska and K. J. Kurzydlowski (2009), *Mater. Sci. Forum*, 618–619: 405–410.

37. J. Schiotz, F. D. di Tolla, and K. W. Jacobsen (1998), *Nature*, 391:561–563.

38. H. S. Kim and Y. Estrin (2005), *Acta Mater.*, 53:765–772.

39. R. Dobosz, T. Wejrzanowski, and K. J. Kurzydlowski (2009), *Comp. Meth. Mater. Sci.*, 9(1).

40. R. Dobosz (2009), PhD Thesis, Warsaw University of Technology, Warsaw.

41. C. C. Koch (2003), *Scripta Mater.*, 49:657–662.

42. Y. Wang, M. Chen, F. Zhou, and E. Ma (2002), *Nature*, 419:912–915.

43. V. Yamakov, E. Saether, D. Phillips, and E. H. Glaessgen, *Stress Distribution During Deformation of Polycrystalline Aluminum by Molecular-Dynamics and Finite-Element Modeling*, Paper for the Special Session on Nanostructured Materials at the 45th AIAA/ASME/ASCE/AHS/ASC.

44. J. D. Embury and C. W. Sinclair (2001), *Mater. Sci. Eng. A*, 319–321:37–45.

45. L. Lu, M. O. Lai, and L. Froyen (2005), *J. Alloys Compd.*, 387:260–264.

46. Z. Pakiela *et al.* Warsaw University of Technology, internal report.

47. A. A. Karimpoor, U. Erb, K. T. Aust, and G. Palumbo (2003), *Scripta Mater.*, 49:651–656.

48. E. B. Tadmor and N. Bernstein (2004), *J. Mech. Phys. Sol.*, 52:2507–2519.

49. T. Mukai, S. Suresh, K. Kita, H. Sasaki, N. Kobayashi, K. Higashi, and A. Inoue (2003), *Scripta Mater.*, 51:4197.

50. H. Iwasaki, K. Higashi, and T. G. Nieh (2004), *Scripta Mater.*, 50:398.

51. X. Y. Qin, S. H. Cheong, and J. S. Lee (2003), *Mater. Sci. Eng. A*, 363:62.

52. Z. Pakiela, Mikrostrukturalne uwarunkowania właściwości mechanicznych nanokrystalicznych metali (Microstructural Factors Affecting Mechanical Properties of Nanocrystalline Metals), Oficyna Wydawnicza Politechniki Warszawskiej, Warszawa 2008 (in polisch).

53. Baohua Ji and Huajian Gao (2004), *Mater. Sci. Eng. A*, 366:96–103.

54. Z. Pakieła, M. Suś-Ryszkowska, A. Drużycka-Wiencek, K. Sikorski, and K. J. Kurzydłowski (2004), *Inżynieria Materiałowa*, nr 3(140):407–410.

55. T. Hanlon, Y. N. Kwon, and S. Suresh (2003), *Scripta Mater.*, 49:675.

56. S. Suresh (2003), *Fatigue of Materials*, Cambridge University Press, 2nd ed., Cambridge.

57. S. R. Agnew and J. R. Weertman (1998), *Mater. Sci. Eng. A*, 224:145.

58. A. Vinogradov, V. Patlan, Y. Suzuki, K. Kitagawa, and V. I. Kopylov (2002), *Acta Mater.*, 50:1639.

59. V. Patlan, A. Vinogradov, K. Higashi, and K. Kitagawa (2001), *Mater. Sci. Eng. A*, 300:171.

60. A. Drużycka-Wiencek (2004), PhD Thesis, Warsaw University of Technology, Warsaw.

61. S. P. Baker (2001), *Mater. Sci. Eng. A*, 319–321:16.

62. H. Van Swygenhoven,and J. R. Weertman (2006), *Mater. Today*, 9:24–31.

63. H. Van Swygenhoven, M. Spaczer, and A. Caro (1999), *Acta Mater.*, 47:3117.

64. X. Z. Liao, F. Zhou, E. J. Lavernia, D. W. He, and Y. T. Zhu (2003), *Appl. Phys. Lett.*, 83:5062–5064.

65. Y. T. Zhu, X. Z. Liao, S. G. Srinivasan, Y. H. Zhao, M. I. Baskes, F. Zhou, and E. J. Lavernia (2004), *Appl. Phys. Lett.*, 85:5049–5051.

66. S. Lee, M. Furukawa, Z. Horita, and T. G.Langdon (2003), *Mater. Sci. Eng. A*, 342:294.

Chapter 6

MECHANICAL BEHAVIOR AND DEFORMATION MECHANISM OF FCC METALS WITH NANOSCALE TWINS

L. Lu and K. Lu

Shenyang National Laboratory for Materials Sciences, Institute of Metal Research,
Chinese Academy of Sciences,
72 Wenhua Road, Shenyang 110016, PR China
E-mail: llu@imr.ac.cn, lu@imr.ac.cn

This chapter will present an overview of mechanical behaviors and deformation mechanisms in face-centered-cubic (fcc) polycrystalline metals containing nanoscale twins. The twin-thickness (λ) dependence on strength, ductility, work hardening, and strain-rate sensitivity of Cu samples will be included. For the ultrafine-grained (ufg) Cu with fixed grain size, the strengthening effect of the twin thickness is analogous to that of grain size and follows the Hall–Petch (H–P) relationship when λ decreases down to several tens of nanometers. At λ = 15 nm, a highest strength is observed, followed by a softening at smaller λ and a significant enhancement in both strain hardening and tensile ductility. The enhanced strain-rate sensitivity is also observed with a decreasing twin thickness. A H–P-type relationship fitting on the experiment data of activation volume as a function of the twin thickness suggests a transition of the rate-controlling mechanism from the intratwin- to twin-boundary (TB)-mediated processes with decreasing λ. The quantitative and mechanics-based models with the related transmission electron microscopy observations and molecular dynamics (MD) simulation are discussed with particular attention to the TB-mediated deformation processes. These findings provide insights into the possible routes for optimizing the strength and ductility of nanostructured metals by tailoring internal interfaces.

Mechanical Properties of Nanocrystalline Materials
Edited by James C. M. Li
Copyright © 2011 Pan Stanford Publishing Pte. Ltd.
www.panstanford.com

6.1 INTRODUCTION

The strength, ductility, and many other mechanical characteristics of metals and alloys are strongly dependent upon their micro- or nanoscale structures. Traditional methodologies for strengthening materials are based on the generation of, and interactions among, various internal defects to resist the dislocation motions. The defects generally include the atomic vacancies and interstitials (point defects), dislocations (line defects), grain and interphase boundaries, and stacking faults (SFs) that introduce crystallographic disregistry between adjacent regions of the atomic lattice (planar defects), and dispersed reinforcement particles (volume defects) of a different material than the surrounding matrix.[1] Several commonly used strengthening approaches for metallic metals and alloys, such as solid solution strengthening,[2] second-phase strengthening, strain-hardening strengthening (or Taylor strengthening),[3] and grain refinement strengthening, are summarized in Fig. 6.1a,b.[4] However, these approaches invariably suffer from the undesirable consequence that increase in strength facilitated by dislocation interactions with internal barriers also causes reduced ductility and increased brittleness.

For grain refinement, the high concentration of incoherent grain boundaries (GBs), which do not create close crystallographic registry between regions separated by the boundaries, provides barriers to transmission of dislocations from one grain to the next. A higher stress is needed to deform a polycrystalline metal with a smaller grain size d (more GBs), which usually follows the so-called traditional H–P relationship.[5-6] The H–P relation is derived on the basis of the strengthening mechanism[7] at internal interfaces. Proliferating to an extreme case, nanocrystalline (nc) materials, which are characterized by an extremely small grain size (typically finer than 100 nm) and high GB density (which is inversely proportional to the grain size, being 10^{6-8} m^2/m^3),[8-11] usually exhibit substantially high strength and hardness, typically 5–10 times higher than that of the coarse-grained (cg) counterparts.[10-12] These high-energy, incoherent GBs in nc metals are effective in obstructing dislocation motion (following the H–P relationship); however, their ability to accommodate plastic deformation is limited by reducing ductility.[12-14] The tensile elongation-to-failure of pure fcc nc metals is typically smaller than a few percent, and the uniform strain is even more limited.[13-16] The brittleness of nc metals, which significantly limits their technological applications, is believed to originate intrinsically from serious suppression of dislocation activities inside the grains and at the high density of GBs.

In contrast to GBs, coherent internal boundaries with low excess energies are seldom introduced as major strengthening agents for structural materials. It is well known that TB is a special kind of coherent GB. These planar defects form interfaces, one side of which contains arrangements of atoms that are mirror reflections of those on the other side divided by the twin composition plane (Fig. 6.1c).

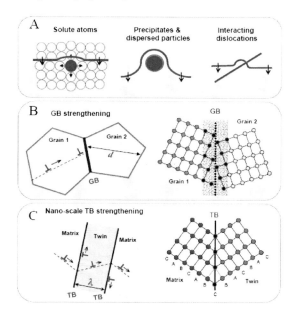

Figure 6.1 Schematic illustration of examples of structural modifications for strengthening metals and alloys. (a) Solid solution strengthening, second-phase strengthening, and strain-hardening strengthening; (b) GB strengthening, in which the dislocation (red ⊥ symbol) motion is blocked by GB (whose incoherent structure is schematically shown on the right) so that a dislocation pileup against GB is formed; and (c) nanoscale TB strengthening based on the dislocation–TB interactions from which mobile and/or sessile dislocations could be generated, either in the neighboring domains (twin or matrix) or at TBs. Higher strength and high ductility are achieved simultaneously with a smaller twin thickness (λ) in the nanometer scale.[4]

Compared with the traditional high-angle GBs, TBs usually exhibit much higher thermal and mechanical stability. The formation of annealing twins and deformation twins in fcc and body-centered cubic (bcc) metals have been well studied by Mahajan *et al.* since 1970s.[17-19] Early works had also investigated the possible interactions between the slip dislocations and coherent TBs from transmission electron microscopy (TEM) observations and crystallographic analysis.[20-22] In an annealed brass, it is even found that the strengthening of TBs is quantitatively identical to that of ordinary GBs.[23] However, the TB is rarely considered a rivalrous mechanism for strengthening in conventional metals. This is because strengthening from coherent TBs is relatively less pronounced than that from grain refinement when the TB spacing is of micrometer scale. Furthermore, generation of a sufficiently high density of stable coherent (or semicoherent) TBs for achieving even moderate strength enhancements is a technical challenge in material processing for a long time.

In the present chapter, we will aim at microstructure exploration, mechanical properties improvement, and the deformation mechanism involving the high density of nanoscale twins in fcc metals. The experimental observations, discussed herein, of the twin lamellar thickness (λ) or TB density dependence of strength, ductility, strain-rate sensitivity, and work hardening of nanotwinned (nt) materials reveal challenges to the understanding of its intrinsic deformation mechanism. Considering the high thermal stability and mechanical stability of the coherent TBs, the critical twin thickness and the maximum strength in nt Cu samples have also been explored experimentally.

6.2 SAMPLE PREPARATION AND MICROSTRUCTURE CHARACTERISTIC OF NANOSCALE TWIN LAMELLAE

Twin defects are not uncommon in nature and can be formed by various approaches such as plastic deformation (so-called deformation twins),[22, 24–27] phase transformation, thermal annealing (annealing twins),[18, 28–31] and other physical or chemical processes (growth twins)[32–33] in a large variety of metals and alloys. However, it is not easy to obtain a high twin density in fine-grained materials experimentally. Recently, high density of twins has been reported in electro- or vapor-deposited low-stacking-fault-energy alloys such as Ni–Co.[34]

Electrodeposition is an effective technique for introducing twin structures in fcc metals.[35] Lu *et al.*[35] found that high-purity copper foils with nanosized growth twin lamellae could be synthesized by means of the pulsed electrodeposition technique from an electrolyte of $CuSO_4$. TEM observations (Fig. 6.2) indicate that the as-deposited Cu consists of irregular-shaped grains with random orientations (see the selected electron diffraction pattern in Fig. 6.2a). The grain sizes are between 100 nm and 1 $\propto$m with an average value of about 400 nm (Fig. 6.2b). Each grain contains a high density of growth twins of {111}/[112] type. Measurements of the lamella thickness along the [110] orientation show a wide distribution from several nanometers to about 150 nm (Fig. 6.2c), due to the fact that in this orientation only $(\bar{1}11)$ and $(1\bar{1}1)$ twins are edge-on while (111) and $(11\bar{1})$ twins are inclined to the surface. The lamella thickness distribution shows a peak at about 15 nm in this case, corresponding approximately to the average lamella thickness (λ)

for the edge-on twins. For simplicity, the nt Cu sample is identified by its mean twin thickness hereafter (e.g., the sample with λ = 15 nm is referred to as nt-15). The length of twin lamellar geometry varies according to its grain diameter. The high-density-growth twins separate submicro-sized grains into nanometer-thick twin/matrix lamellar structures.

Close TEM observations and high-resolution TEM (HRTEM) images show that most TBs are perfectly coherent and atomically sharp (as seen in Fig. 6.3a,b). Lattice dislocations cannot be detected in most lamellae, but a few dislocations are observed in the thick lamellae. This agrees with the X-ray diffraction result that a negligible (atomic level) lattice strain is identified. X-ray diffraction patterns show an evident (110) texture in the sample, which is consistent with previous observation in electrodeposited Cu specimens with growth twins.[36]

By modifying the physical and chemical parameters of the pulsed electrodepositon process, the average twin thickness of the nt Cu specimens can be varied while keeping their grain size and texture constant. For comparison, an ufg Cu sample with a similar grain size but without twins was prepared by means of direct-current electrodeposition.

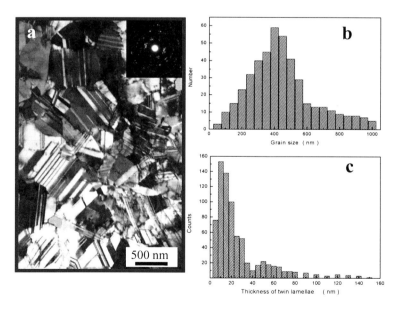

Figure 6.2 TEM observations of the typical microstructure in an as-deposited Cu sample. A bright-field TEM image (a) with SEAD (inset in [a]) shows roughly equiaxed submicron-sized grains with random orientations separated by high-angle GBs. The statistical distributions for grain size (b) and for thickness of the twin/matrix lamellae (c) were obtained from the many TEM images of the same sample.[35]

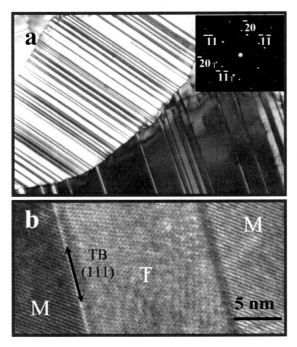

Figure 6.3 Close TEM (a) and HRTEM observations (b) of the TBs in as-deposited Cu samples. SEAD (inset of [a]) indicates that the twins in each grain are parallel to each other in {111} planes (a), and HRTEM images (b) show that the twins follow a sequence of MTMTM. . . .[35]

The thickness of the growth twins formed during the deposition depended on the nucleation rate. The average twin thickness can be easily reduced from 100 nm to 15 nm when the current density increases from 10 to 40 mA/cm^2. In order to produce nt Cu specimens with twin thickness $\lambda < 15$ nm, a high average current density and a high average deposition rate were applied to facilitate the twin nucleation. Systematically increasing the deposition average current density from 70 to 300 mA/cm^2, a series of nt Cu samples with λ of 10 nm, 8 nm, and 4 nm were synthesized, as shown in Fig. 6.4.

The series of nt Cu specimens with variable λ but similar grain size offers a possibility to systematically investigate the twin-thickness dependence on the mechanical properties of nt structures through the direct traditional mechanical tests. Thermal stability of the as-deposited Cu sample was studied by examining the twin lamellar structure at elevated temperatures. It was found that thickening of twin lamellae began when the as-deposited sample was isothermally annealed above 250°C (for a duration of 300 s).[35]

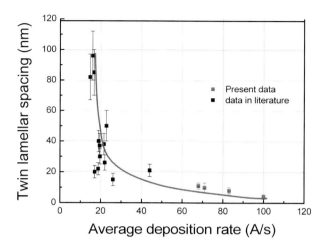

Figure 6.4 Variation of twin thickness as a function of the average deposition rate during pulsed electrodeposition.

The formation mechanism of multiple twins in the pulsed electrodeposition process is still an open question. From a thermodynamic point of view, formation of twins will decrease the total interfacial energy, as the excess energy for coherent TBs is much smaller than that for conventional high-angle GBs.[29,37] Twins prefer to nucleate at GBs or triple junctions to reduce the total GB energies by means of the twinning-induced orientation change. Although an extra TB is formed, the sum of the interfacial energies (including GBs and TBs) will be reduced by twinning. The nucleation and growth rate of twins is controlled by the deposition conditions. This may originate from the fact that at a higher deposition rate and/or a lower temperature, GBs and triple junctions with higher excess energies will be formed due to the limited relaxation, so the nucleation rate of twins will be higher to reduce the total interfacial energy. Apparently, formation of twins is also strongly dependent upon the nature of the TBs and the ratio of the TB energy to the GB energy ($\alpha = \gamma_{TB}/\gamma_{GB}$). High twin densities prefer to form in those metals and alloys with smaller α values (e.g., with lower SF energies as well as lower TB energy) under proper conditions. For example, in austenite stainless steel 330, which has a low SF energy (~ 20 mJ/m^2), growth twins with λ of ~ 4–5 nm can be formed in a DC sputter deposition.[38–40] For Cu with a SF energy of about 45 mJ/m^2, the twin lamellar spacing is larger (about a few tens of nanometers) compared with that in stainless steel 330 under the same sputter-deposition conditions.[33, 39]

6.3 MECHANICAL PROPERTIES

6.3.1 Strengthening Effect of Nanoscale Twins

The typical tensile true stress–strain curves for the as-deposited nt Cu samples with different twin lamellar thickness, in comparison with that for the annealed cg Cu sample (grain size > 100 ∝m) are shown in Fig. 6.6a.[41] Obviously, the as-deposited nt Cu exhibits much higher strength than that of the cg counterpart. With a decreasing twin thickness from 100 nm to 15 nm, the strength of the nt Cu sample increases gradually. For the nt-15, the tensile yield strength σ_y (at 0.2% offset) reaches as high as 900 MPa and the ultimate tensile strength (σ_{UTS}) is 1068 MPa, which is similar to, or even larger than, that reported for polycrystalline pure Cu with three-dimensional nanosized grains.[42-45]

The ultrahigh strength of nt Cu originating mainly from the strengthening effect of a high density of TBs due to a yield strength of about 200 MPa is expected for the pure Cu with an average grain size of 400 nm, which is apparently much lower than the measured strength from the present nt Cu samples. When the twin thickness is taken as a characteristic structure dimension, analogous to the grain size, the dependent strengthening effect can be represented analogously to the classical H–P relation[46-47]:

$$\sigma_y = \sigma_0 + K_T \cdot \lambda^{-l} \tag{6.1}$$

where K_T is a constant, λ is the mean twin thickness, and l is a coefficient ranging from 0.5 to 1. It is obvious that all σ_y for nt Cu with λ varying from 100 to 15 nm are in good agreement with the traditional H–P relationship for cg and nc Cu. Such an agreement suggests that the strengthening effect of TBs is analogous to that of GBs, even when λ decreases down to several tens of nanometer. This λ-dependent strengthening behavior is consistent with previous studies reported in the literature, in which TBs were found to act as barriers to dislocation motions during plastic deformation.[22-23,48]

6.3.2 The Strongest Twin Thickness

The H–P relationship suggests that the strength of polycrystalline materials increases with decreasing grain size. However, below a certain critical size (in the nanometer scale), the dominating deformation mechanism may change from lattice dislocation activities to other mechanisms such as GB-related processes and a softening behavior (rather than *strengthening*) is expected instead.[49-50] Although such a softening phenomenon has been demonstrated by atomistic simulations and a critical size of maximum strength has been predicted,[49-53] experimental verification of this prediction has never been achieved in high-quality metals.[16,54,55] One of the major barriers in exploring the strongest size in metals is the practical difficulties in obtaining stable nanostructures in pure metals with an extremely small size of structural domains (e.g., several nanometers). The driving force for growth of nanosized grains in pure metals, originating from the high excess energy of numerous GBs, becomes so large that grain growth may take place easily even at ambient temperature or below.[56-57]

Coherent TBs are much more stable against migration (a fundamental process of coarsening) than conventional GBs as the excess energy of coherent TBs is one order of magnitude lower than that of GBs[37]. Hence, nt structures are energetically more stable than the nanograined counterparts with the same chemical compositions. The stable nt structure with extremely small domain (twin) sizes may provide unique samples for exploring the softening behavior below the strongest size in metals.

Figures 6.5a–c show TEM plane-view observations of three as-deposited samples with a mean λ of 96, 15, and 4 nm, respectively.[41] TEM and transverse SEM observations of cross sections and plane views show the grains in nt Cu are roughly equiaxed in three dimensions. Statistic grain size measurements showed a similar distribution and a similar average size of 400–600 nm for all nt Cu samples.[58] Twins (see the SEDP in Fig. 6.6d) are formed in most grains. Note that in all samples the edge -on twins formed in different grains are aligned randomly around the foil normal (growth) direction.[35,59] For each sample, λ was statistically measured from a large number of grains to generate a distribution. Figure 6.5e illustrates the λ distributions for the sample with the finest twins, showing that the majority of measurements are smaller than 10 nm and the mean value is only 4 nm.

Figure 6.6 shows the uniaxial tensile true S-S curves for nt Cu samples of various λ values. First of all, a remarkable feature is the occurrence of the λ giving the highest strength. For $\lambda > 15$ nm (Fig. 6.6a), the S-S curves shift upward with decreasing λ, similar to the strengthening behavior reported previously in the nt Cu[35,58] and nc Cu samples.[10,60-62] However, with further decrease of λ down to extreme dimensions, that is, less than 10 nm, the S-S curves shift downward (Fig. 6.6b). As plotted in Fig. 6.7a, the measured yield strength σ_y (at 0.2% offset) shows a maximum value of 900 MPa at about $\lambda = 15$ nm.

The strength of the nt Cu samples has been considered controlled predominantly by the nanoscale twins via the mechanism of slip transfer across the TBs,[39,63] and increases with decreasing λ following an H–P-type relationship,[58] similar to that of GB strengthening in nc metals.[10] However, such a relationship breaks down when λ < 15 nm, although other structural parameters such as grain size and texture are unchanged. The grain sizes of the nt Cu samples are in the submicrometer regime, which is too large for GB sliding to occur at room temperature, as expected for nc materials with grain sizes below 20 nm.[49] Therefore the observed softening cannot be explained by the initiation of GB-mediated mechanisms such as GB sliding and grain rotation as proposed by MD simulations for nc materials.[49-50,53]

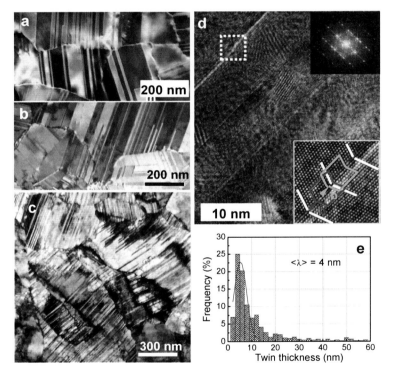

Figure 6.5 TEM images of as-deposited Cu samples with various λ values. (a) λ = 96 nm, (b) λ = 15 nm, (c) λ = 4 nm, (d) the same sample as (c) but a higher-resolution image with a corresponding ESDP (inset upper right) and an HRTEM image showing the presence of Shockley partials at the TB (inset lower right), and (e) a statistical distribution of the lamellar twin thicknesses determined from TEM and HRTEM images.[41]

HRTEM observations showed that in each sample TBs are coherent Σ3 interfaces associated with the presence of Shockley partial dislocations (as steps), as shown in Fig. 6.5d. These partial dislocations have their Burgers vector parallel to the twin plane and are an intrinsic structural feature of twin growth during electrodeposition. The distribution of the preexisting partial dislocations is inhomogenous, but the density per unit area of TBs is found to be rather constant among samples with different twin densities. This suggests an insignificant effect of the deposition parameters and the twin refinement on the nature of TBs. Therefore, as a consequence of decreasing λ, the density of such TB-associated partial dislocations per unit volume increases. It was also noticed that GBs in the nt Cu samples with $\lambda \leq 15$ nm are characterized by straight segments (facets) that are often associated with dislocation arrays,[64] while in samples with coarser twins, GBs are smoothly curved, similar to conventional GBs.

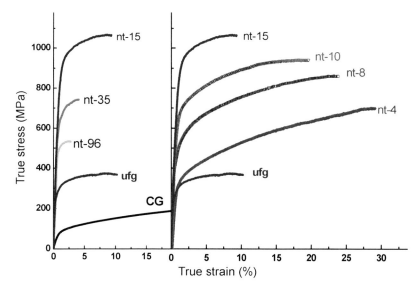

Figure 6.6 Uniaxial tensile true stress–strain curves for nt Cu samples with different twin thicknesses varying from 96 to 15 nm (a) and from 15 to 4 nm (b) tested at a strain-rate of 6×10^{-3} s^{-1}. For comparison, curves for a twin-free ufg Cu with a mean grain size of 500 nm and for a cg Cu with a mean grain size of 10 μm are included.[41]

In nt Cu samples with ultrathin λ, both the dislocation arrays associated with GBs and the preexisting partial dislocations along TBs could be potential dislocation sources, which are expected to affect the initiation of plastic deformation.[65] The preexisting partial dislocations can act as readily mobile dislocations, and their motion may contribute to the plastic yielding when

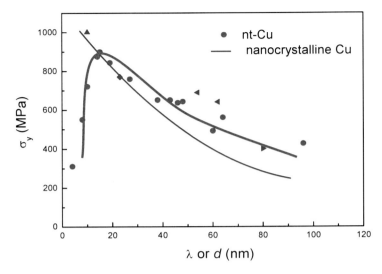

Figure 6.7 Variation of (a) yield strength as a function of mean λ for nt Cu samples. For comparison, the yield strength and n values for cg Cu, ufg Cu, and nc Cu samples reported in the literature are also included (▲from[10], ◀[60], ▶[61], ◆[62]). A maximum in the yield stress is seen for the nt Cu with λ = 15 nm, but this has not happened for the nc Cu, even when the grain size is as small as 10 nm.[41]

an external stress is applied to the sample. The plastic strains induced by the motion of preexisting partial dislocations can be simply estimated as $\varepsilon = \rho_0 b_s d/M$ (where ρ_0 is the initial dislocation density, b_s is the Burgers vector of Shockley partial dislocation, d is the grain size, and M is the Taylor factor). Calculations showed that for the samples with λ > –15 nm, the preexisting dislocations induce a negligibly small plastic strain (<0.05%). However, for the nt-4, a remarkable amount of plastic strain, as high as ~0.1%, can be induced even only by the motions of high density of preexisting dislocations at TBs (roughly 10^{14} m^{-2}), which could essentially control the macroscopic yielding of the sample. This analysis suggests that only when an extremely small λ is achieved, an unusual softening phenomenon will occur due to the yielding mechanism transition from slip transfer across TB (TB strengthening) to the activity of preexisting easy dislocation sources at TBs and GBs (softening).

6.3.3 Ductility and Work Hardening

Besides the ultrahigh strength achieved in nt Cu, another impressive feature is the substantially increased tensile ductility and strain hardening with decreasing the twin thickness or increasing the twin density. The nt-15, which has the highest strength, shows an elongation to failure more than 13%, as illustrated in Fig. 6.6. When λ < 15 nm, the uniform tensile elongation exceeds that of the ufg Cu sample, and it reaches a maximum value of 30% at λ = 4 nm (Fig. 6.6b). Note that such a high uniform elongation has never been reported in any other nanostructured copper with a similar strength. The tensile elongation of the nt Cu samples increases monotonically with decreasing λ or increasing the twin density, as summarized in Fig. 6.8.

To quantify the strain hardening, strain-hardening coefficients (*n*) were determined according to $\sigma = K_1 + K_2\varepsilon^n$, where K_1 represents the initial yield stress and K_2 the increment in strength with ε = 1.[47, 66] The *n* values determined for all the nt Cu samples increase monotonically with decreasing λ (Fig. 6.9), similar to the trend of uniform elongation versus λ. When λ < 15 nm, *n* exceeds the value for cg Cu (0.35)[47, 66] and finally reaches a maximum of 0.66 at λ = 4 nm. Note that the twin refinement induced increase in *n* is opposite to the general observation in ufg and nc materials, where *n* continuously decreases with decreasing *d*, as seen in Fig. 6.9.

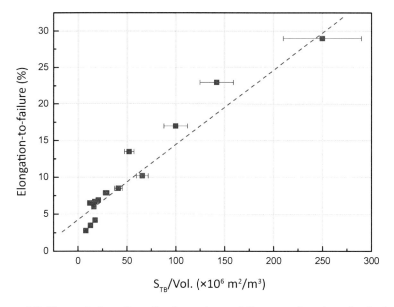

Figure 6.8 The variation of tensile elongation to failure as a function of twin density in the nt Cu.

It is speculated that high densities of interfacial dislocations could be accumulated near the regions of TBs and facilitate uniform plastic deformation.[58,67] Meanwhile, nt Cu has multimodel distributions of length

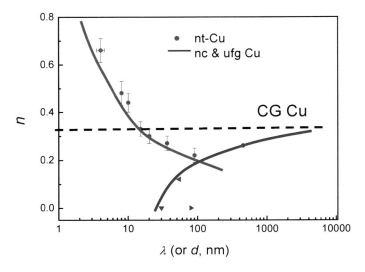

Figure 6.9 Variation of the strain-hardening coefficient, n, as a function of the mean twin thickness for the nt Cu samples.[41] The n value level of cg Cu[47,66] is marked by the dashed line.

scales, which is known to benefit ductility[68]. The twins subdivide the submicron-sized grains into nanometer-sized twin/matrix lamellar structures, of which the length scale parallel to the TBs (plastically soft direction) is on the order of submicrometers, whereas that in the direction perpendicular to TBs (plastically hard direction) is at the nanometer scale. In the former direction, dislocation glide/accumulation is relatively easier, while it is more difficult in the latter direction. These experimental result suggests that the nanostructuring strategy of using high-density nanoscale, low-energy CTBs to "replace" the more typical general high-angle GBs in nc metals can result in a combination of high strength and high ductility for pure metals.[35]

6.3.4 Strain-Rate Sensitivity

A well-known, effective experimental technique to probe the active deformation mechanism is to measure the sensitivity of flow stress to the rate of loading, because both the sensitivity index m and the associated activation volume v^* can vary by orders of magnitude for different rate-limiting processes.[69–70] In cg fcc metals, a typical rate-determining process, such as the intersections of forest dislocations, gives a large activation volume on the order of several hundred to a few thousand $\mathbf{b}^3$ [69,71]. At another extreme, GB sliding or GB-diffusion-mediated creep (Coble creep) gives a small activation volume that is as small as $1\mathbf{b}^3$. When the activation volume is between 1 and $100\mathbf{b}^3$, the rate process typically involves cross slip or nucleation of dislocation; the activation of either process requires a locally high stress at the GPa level.[70-73] Experimental studies suggest that for fcc metals such as Cu and Ni, grain refinement into the nc regime leads to an increase in m by up to an order of magnitude relative to metals with grain size in the micrometer range and a concomitant decrease in the activation volume v^* by two orders of magnitude.[74-76]

The strength and ductility of nanostructured metals are strongly influenced by the plastic strain hardening and strain-rate sensitivity. In order to develop an understanding of the trade-off between strength and ductility, we have performed systematic experiments to study the twin-thickness dependence of strain-rate sensitivity and activation volume. Before discussing the results, it is useful to establish a framework for the analysis. The strain-rate sensitivity of a ductile metal in uniaxial deformation is commonly written as[69, 74]:

$$m = \frac{\sqrt{3}kT}{V^*\sigma} = \frac{3\sqrt{3}kT}{V^*H_v} \tag{6.2}$$

where m is a nondimensional rate sensitivity index, k is the Boltzmann constant, T is the absolute temperature, σ is the flow stress, H_v is the hardness (which is usually assumed to be three times the flow stress), and V^* is the activation volume, which is the rate of decrease of the activation enthalpy with respect to flow stress at a fixed temperature:

$$V^* = \sqrt{3}kT\left(\frac{\partial \ln \varepsilon}{\partial \sigma}\right) \tag{6.3}$$

where ε is the strain-rate.

True stress–strain curves from the tensile strain-rate up-jump tests on nt Cu are shown in Fig. 6.10. When the strain-rate increases, the corresponding variation in plastic flow characteristics are observed to depend strongly on the twin density. With increasing strain-rate, the strength increases monotonically.[77] For example, the flow strength for nt-15 rises from about 900 to 1070 MPa within the strain-rate range of 1×10^{-5} to 1×10^{-2} s^{-1}, as seen in Fig. 6.10b. The increment of flow stress for nt-15 is close to 300 MPa within the range of strain-rates considered. However, a weak dependence of stress is found for nt-100 with increasing strain-rate. The increment of flow strength for nt-100 is less than 50 MPa (from 492 MPa to 540 MPa) within a strain-rate range of 1×10^{-5} to 6×10^{-2} s^{-1}, as seen in Fig. 6.10c. As generally expected for fcc mc metals, a very small increase in strength is observed with increasing strain-rate for ufg Cu. This is consistent with the trends reported for ufg Ni from instrumented nanoindentation tests at different strain-rates.[76] Figure 6.11 summarizes the experimental results on the variation of m as a function of λ for nt Cu[59,77–78] from a series of experiments, including constant loading rate tests in nanoindentation,[59] constant strain-rate tests in tension,[79] and tensile strain-rate jump tests.[77] When λ is reduced to about 15 nm, m can be nearly one order of magnitude higher than that for samples with micron-sized grains. In addition, the m values for nc Cu are also shown in Fig. 6.10, with d replacing λ as the characteristic structural length scale. It is seen that m increases slightly with reducing d from the micrometer to submicrometer scale, while an obvious "take-off" appears when d is further reduced below 100 nm or so. It is interesting to note that the dependence of rate sensitivity on λ is very close to that on d in the size range measured for nc Cu.[10]

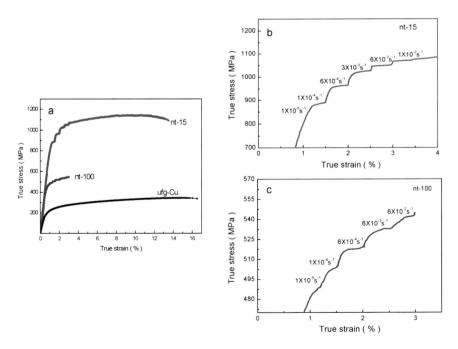

Figure 6.10 (a) Tensile stress–strain curves of the nt Cu specimens with different twin thicknesses. Magnified strain-rate jump tensile curves for nt-15 and nt-100 are shown in (b) and (c), respectively.[77]

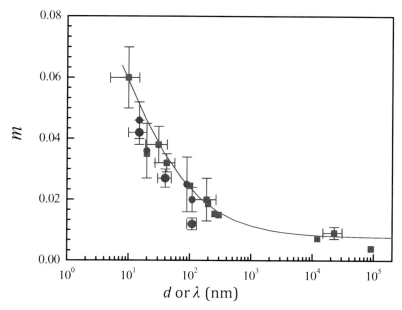

Figure 6.11 Effect of twin thickness (λ) on rate sensitivity for nt Cu (red symbols)[59,77,78]. For comparison, the rate sensitivities are also plotted as a function of grain size (d) for nc Cu (blue symbols).[10]

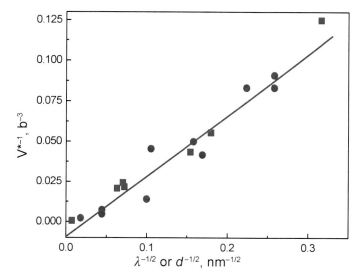

Figure 6.12 Effect of twin thickness (λ) on activation volume (measured in units of **b**[3]) for nt Cu (red symbols)[59,77,78]. The blue symbols are from nc Cu in the literature[10] for comparison.

Figure 6.12 shows the effects of λ and d on the activation volume, V^*, measured in units of $\mathbf{b}^3$, where $\mathbf{b}$ is the Burgers vector of a perfect dislocation. v^* decreases from 1,000$\mathbf{b}^3$ to about tens of $\mathbf{b}^3$ when d (or λ) decreases from the micrometer to the nanometer range. An H–P-type V^{*-1} dependency has been observed in a wide range of grain sizes from the micrometer,[80–83] submicrometer regime to nanoscale metals,[84] that is, according to Conrad,[70,71] Armstrong, and Rodriguez,[84] V^{*-1} varies linearly with $d^{-1/2}$ for fcc metals. This size dependence was rationalized by combining the dislocation pileup model and the thermal activation analysis of plastic flow rate.

The atomic processes are different when impinging dislocations react with GBs and TBs because of their structure differences.[4] However, the size dependence of activation volume in the nt system should be captured well by the H–P-type relation. This is because the essence of the dislocation pileup model is to bring out the length scale effects by invoking the size-mediated number of dislocations in the pileup, $n(d)$ or $n(\lambda)$. As shown by Eqs. 6.2 and 6.3 in Conrad,[85] $n(d)$ is a magnification factor that connects the applied stress and the local concentrated stress acting on the rate-determining processes at the GBs. Considering that the functional form of $n(d)$ and $n(\lambda)$ is independent of the nature of the GB and TB in the pileup model, we generalize the H–P -type relation of activation volume to include both GB and TB effects

$$\frac{1}{V} = \frac{1}{V_0} + \left[\frac{k_{GB}}{2m\tau_c V_{GB}^*}\right] d^{-1/2} + \left[\frac{k_{TB}}{2m\tau_c V_{TB}^*}\right] \lambda^{-1/2} \qquad (6.4)$$

where V is the activation volume of the specimen and V_0^* is the activation volume associated with the intragrain or intratwin dislocation mechanism, that is, intersection of lattice dislocations; V_{GB}^* and V_{TB}^* are the activation volumes associated with the GB- and TB-mediated mechanisms, respectively. In Eq. 6.4, m is the Taylor factor ($m \approx 3.1$). The constant k and the local shear resistance τ_{GB}^c or τ_{GB}^c are determined from the classic H–P model of dislocation pileup. Considering that the measured tensile yield strengths for nt Cu vary with twin thickness, λ, in the same manner as d for nc Cu, we approximate $V_{GB}^* = V_{TB}^* = 25b^3$ and $k_{TB} = k_{GB} = 5.0$ MPa/mm$^{-1/2}$ ($\tau_c = 450$ MPa, and the shear modulus $G = 30.5$ GPa).

The size dependence of activation volume shown in Fig. 6.12, along with its agreement with the dislocation pileup model of Eq. 6.4, suggests a size-dependent transition of the dominant rate-controlling mechanism from the intratwin to intertwin processes with decreasing twin lamellar thickness. At a large λ, the dislocation mechanism within the twin lamellae dominates, giving a characteristically large activation volume of about 1,000$\mathbf{b}^3$, which arises because of dislocation intersections.[70] In the limiting case of λ about 20 nm or below, the lattice dislocation tangles can hardly be stabilized inside the twin lamellae. Then the TB-mediated dislocation activities are expected to dominate. In between, both mechanisms are operative; the relative contribution of each mechanism is controlled by the twin size as indicated by Eq. 6.4.

6.4 THE DEFORMATION MECHANISM

6.4.1 TEM Characterization of Deformed Structures

As pointed out earlier, metals with an nt structure impart ultrahigh strength, hardness, and enhanced rate sensitivity, similar to those observed in nc metals, while still retaining an adequate strain-hardening capacity and considerable ductility, which cannot be achieved in nc metals. Developing an understanding of these mechanical properties would provide valuable insights into the deformation mechanism. Such an understanding would also offer helpful information for microstructure designing, controlling, and optimizing of engineering metals and alloys.

It is generally accepted that the interactions between dislocations and TBs play a dominant role in the plastic deformation process of the metals with an nt structure. Postmortem TEM observations of the nt metals have indicated that plenty of dislocations are identified in as-deformed nt samples[58,59]. In contrast to the clear and sharp coherent TBs in the as-deposited state, TBs in the deformed nt Cu, with copious dislocations and debris, are much strained and stepped (or even curved and disappeared), while the grain size does not change and grains remain almost equiaxed with high-angle GBs, as shown in Fig. 6.13. TEM images indicate that the distribution of dislocations is not uniform. Plenty of dislocations and dislocation tangles are identified inside the thick lamellae, similar to those in the deformed cg material. However, on the contrary, dislocations are hardly seen inside the thin lamellae. A great amount of interfacial dislocation debris locates along TBs with a thickness ranging from several to several tens of nanometers.

Lots of Shockley partial dislocations ($\frac{1}{6}[1\bar{2}1]$) were indeed found at TBs, as shown in Fig. 6.13b.[58] High density of Shockley dislocations (with a spacing of several nanometers) results in a deviation in the relative orientation between the neighboring lamellae from the standard $\Sigma 3$ TB. Noticeable displacement and movement of coherent TBs and formation of steps and jogs along TBs were detected. It is anticipated that most of the plastic strain of nt metals is carried by the dislocation piling up along TBs, which results in shear strain accumulation at TBs. The higher the TB density, the larger the capacity of dislocation accumulation at TBs. Consequently more twins result in an enhanced ductility.

The strengthening effect of coherent TBs is much like that of the conventional GBs. It is obvious that coherent TBs in the metals could serve as effective barriers to the motion of dislocations. The stress concentration acting upon the leading dislocation is proportional to the number of piled-up dislocations and the applied external shear stress. With the decrease of the twin lamellae thickness, fewer dislocations are expected to pileup inside the twins. In other words, higher external stress is necessary for the dislocations to cross the TBs. If the twin lamellae are too thin for a dislocation pileup, a single dislocation may penetrate the TBs. In such a scenario, extremely high stress is required for such dislocation–TB interactions. Rao *et al.*[86] estimated from MD simulation that the applied stress could be as high as 0.03μ to 0.04μ (with μ being the shear modulus), that is, the stress could be $\sim$1.4–1.9 GPa in Cu. Therefore, the interactions of dislocations and TBs may provide one of the principal mechanistic reasons for the high strength of twinned Cu. This supports the experimental finding of improved strength/hardness in the Cu specimen with thinner twin lamellae (i.e., higher twin density).[58]

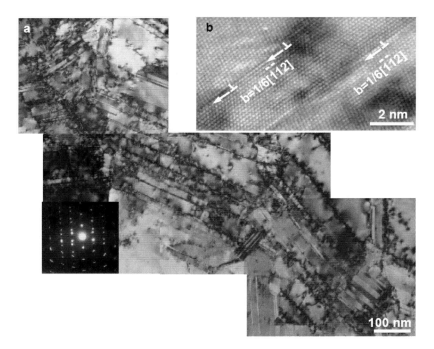

Figure 6.13 A TEM image of the microstructure close to the failure surface of nt-15 after tensile tests (a), containing lots of Shockley (partials) dislocations ($1/6[\overline{1}\overline{1}2]$) (b).[58]

In addition, TBs with a density of defects or the displaced TBs not only behave as obstacles to motion of dislocations but also serve as dislocation sources.[46,65] These processes appear to lead to enhanced strain-rate sensitivity and a thermal activation volume during plastic flow, which could be three orders of magnitude smaller than that found for cg metals. Such a significant dislocation accumulation accompanying the plastic deformation of nt metals is much different from that observed in nc metals, inside which dislocations cannot be easily trapped.[87]

Recently, Lu *et al.* reported the stress relaxation experiments for the nt Cu specimens.[78] The result indicated that the exhaustion rate of mobile dislocations reduces with decreasing twin thickness, which means that the higher the twin density, the larger the value of mobile dislocations. For nt-15, the value of mobile dislocations decreases to about 85% at the end of the first cycle, implying a large portion of mobile dislocations remaining in the specimen. However, it decreases to 78% in nt-35, and it is only 64% in ufg Cu. A large value of mobile dislocations can be attributed to either a lower rate of annihilation or a higher rate of nucleation of mobile dislocations or both.[78] In contrast, a sharp decrease of the mobile dislocation was measured in the relaxation process for nc Ni samples.[75,88,89] The high rate of mobile dislocation exhaustion in nc metals was related to the ready availability of sources near the random GBs where dislocations can nucleate readily. But these GB sources are not regenerative and can be used up quickly upon straining, causing a rapid decay of the exhaustion rate of mobile dislocations,[13] which is much different from that in TB case.[78]

To understand the extraordinary strain hardening, the deformation structures of the tensile-deformed samples were analyzed. In samples with coarse twins, tangles and networks of perfect dislocations were observed within the lattice between the TBs (Fig. 6.14a), and the dislocation density was estimated to be of the order of 10^{14}–10^{15} m^{-2}.[41] In contrast, high densities of SFs and Shockley partials associated with the TBs, which is consistent with the MD simulations,[63,86,90] were found to characterize the deformed structure of the nt-4 sample (Fig. 6.14b,c), indicating the interactions between dislocations and TBs. The density of partial dislocations in the deformed nt-4 sample was estimated to be $5 \cdot 10^{16}$ m^{-2} on the basis of the spacing between the neighboring partials and λ. Such a density is two orders of magnitude higher than that of the preexisting dislocations and the lattice dislocations stored in the coarse twins, suggesting that decreasing the twin thickness facilitates the dislocation–TB interactions and affords more room for storage of dislocations which sustain more pronounced strain hardening in the nt Cu.[72,79]

The above observations suggest that the strain-hardening behavior of nt Cu samples is governed by two competing processes: dislocation–dislocation interaction hardening in coarse twins and dislocation–TB interaction hardening in fine twins. With a refining of λ, the contribution from the latter mechanism gradually increases and eventually dominates the strain hardening, which provides an increasing effect on the strain hardening, as revealed from the continuous increase of n values (Fig. 6.9). However, the former hardening mechanism usually leads to an inverse trend, diminishing with size refinement.[47]

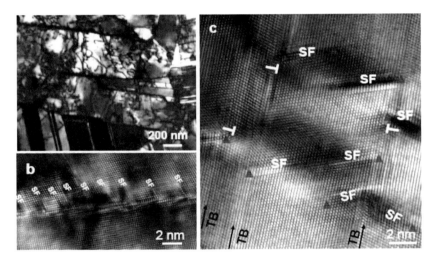

Figure 6.14 (a) A typical bright TEM image of the deformed nt-96 sample, showing the tangling of lattice dislocations. (b) An HRTEM image of the nt-4 tensile deformed to a plastic strain of 30%, showing a high density of SFs at the TB. (c) The arrangement of Shockley partials and SFs at TBs within the lamellae in nt-4. Solid triangles indicate the Shockley partial dislocations associated with SFs. The partials with their Burgers vector parallel to the TB plane are labeled "$\perp$".[41]

6.4.2 Computational Analysis of the Deformation Mechanism

Atomistic simulations have been performed to understand the TB-mediated dislocation mechanisms.[63,90] These atomically detailed studies are central to understanding the role of TBs in controlling strain hardening and ductility. As shown in Fig. 6.15a, there are generally two types of dislocation–TB interactions, the screw case and the nonscrew case. As a screw dislocation impinges on a TB, it may either (1) be absorbed into the TB with two Shockley partials propagating along the TB in the opposite directions or (2) be transmitted into the adjoining twin by cutting through the boundary without leaving any residual on the TB.[63] The nonscrew dislocation similarly involves absorption and transmission. However, the situation for transmission is much more complicated.[90] As shown in Fig. 6.15b, when a 60° nonscrew dislocation of type $\vec{b}_1$ impinges on the TB, it may be transferred directly through the boundary into the adjoining twin accompanied by emission of an additional partial dislocation along the TBs. In contrast, when a 60° dislocation of type $\vec{b}_2$ proceeds, a 30° leading Shockley partial can be released from the TB, and a long SF ribbon is left behind. The remaining partial is pinned down at the boundary, which can be identified as a Hirth lock of 1/3[001] type, as seen in Fig. 6.15c. Such a sessile dislocation in the TB opposes the subsequent transmission across the TB as well as the glide of nearby partials within the TB.[90]

Whereas MD simulations provide physical insights into the TB-mediated dislocation mechanisms, one may wonder whether a quantitative connection can be established between atomistic modeling and experimental measurement. In order to determine the deformation mechanisms relevant to laboratory experiments, Zhu *et al.*[72] have used a novel approach of reaction pathway calculations to analyze slip-transfer reactions between the lattice dislocations and TB. Specifically, using the free-end-nudged elastic band method, they determined the minimum energy paths for reactions of absorption, desorption, and slip transmission of a screw dislocation at the TB. On the basis of the three-dimensional saddle-point states, they further predicted yield stress, activation volume, and rate sensitivity, which are consistent with the experimental measurements. This agreement demonstrates that the slip-transfer reactions at TBs are the rate-controlling mechanisms when the twin lamellar thickness is around or below 20 nm. The simulation results by Zhu *et al.*[72] also suggested that the relatively high ductility of nt Cu is closely related to the hardening of TBs as they gradually lose coherency during plastic deformation. As singular interfaces, TBs are much more hardenable when they gradually lose coherency during deformation.

Along the same line, Asaro and Kulkarni[91] have recently performed the dislocation mechanics analysis of rate sensitivity and strength mediated by cross slip in nt fcc metals. They pointed out that the resolved shear stress is naturally proportional to the reciprocal square root of twin lamellar thickness, $\lambda^{-1/2}$, via a clear and quite specific mechanism for dislocation–TB interaction.[91] Their results are consistent with those of Zhu *et al.*[72]

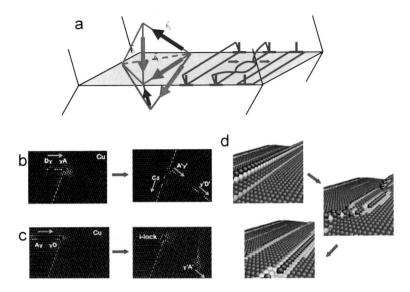

Figure 6.15 Atomistic modeling of the dislocation–TB interaction. (a) Thompson's tetrahedron showing the Burgers vectors of incoming and outgoing dislocations; also shown is a schematic of the reaction pathway of dislocation absorption into the TB. (b and c) MD simulations of slip transmission of the nonscrew 60° dislocation of type $\vec{b}_1$ and type $\vec{b}_2$, respectively[63,90]. (d) An atomic structure from the reaction pathway calculation showing the absorption of a screw dislocation by cross slip.[72]

A TB-affected zone (TBAZ) model was proposed by Dao *et al.*[79] on the basis of the TEM observations and atomistic modeling results. A two-dimensional[79] and most recently a three-dimensional[92] continuum crystal plasticity model were formulated. The predicted stress–strain behavior, strain-rate sensitivity, and failure initiation limit captured the experimentally observed trends very well in terms of the twin thickness, λ. These continuum models took into account the size and orientation dependent dislocation blocking and absorption at the TBs. An important underlying assumption of the TBAZ-based crystal plasticity model is that there are coexisting deformation mechanisms operating simultaneously, including TB-mediated dislocation activities and orientation-dependent lattice dislocation activities. Only by combining these different contributing factors, instead of exclusively considering the dominant deformation mechanism, a fairly good match between model predictions and experimental results can be achieved for the size (λ)-dependent trends of strength, rate sensitivity, and ductility. These findings suggest that the strength and ductility can be further optimized by exploring the microstructure design space involving both grain and twin sizes.

6.5 CONCLUDING REMARKS AND PERSPECTIVES

The unusual mechanical behaviors of the fcc metals with nanoscale twinned structures are investigated. For the fcc metals with high concentration of growth twins in the nanometer scale, an impressive strengthening has been achieved while keeping considerable ductility and work hardening. The twin-thickness (size) dependence of strength, ductility, work hardening, strain-rate sensitivity, and activation volume in nt Cu samples has been demonstrated. The result indicates that TB-mediated dislocation activities play an increasingly dominant role with decreasing twin thickness. The optimization of mechanical properties of metals through introducing high density of TBs provides a novel strengthening methodology for engineering internal coherent boundaries in the nanoscale. An understanding of the coherent nanoscale-twins-induced strengthening provides values insight into the optimization of the comprehensive mechanical properties.

While various attractive mechanical properties are induced, desirable physical properties may also be resulted by introducing nanoscale coherent TBs in metals. It is known that strengthening approaches of metals generally cause a pronounced decrease in electrical and thermal conductivity owing to an increasing amount of scattering barriers, so that a trade-off must be made between conductivity and mechanical strength. However, by introducing nanoscale coherent TBs in ufg pure Cu, electrical conductivity was found to be slightly changed (comparable to that of cg high-conductivity Cu) even though its mechanical strength is increased by about one order of magnitude.[35] The high conductivity originates from the numerous coherents possessing an extremely low electrical resistivity.

Electromigration is known as a main failure mode of integrated circuits, which is basically enhanced atomic diffusion under a high-density electric current. Less coordinated GBs are generally regarded as preferential electromigration paths. Recently, Chen *et al.* found that by generating nanoscale twins in Cu grains, atomic transport at GBs driven by an electric current is slowed down by one order of magnitude due to presence of the triple junctions of coherent TBs and GBs.[93] This result indicates that the electromigration-induced failure of Cu lines in integrated circuits would be significantly suppressed if numerous nanoscale TBs are formed in Cu grains.

There are still many challenges relevant to the nanoscale twins strengthening. Producing materials with nanoscale TBs in bulk form, which is essential to exploit their beneficial mechanical properties in structural applications, is one of the major challenges. Further, the syntheses techniques for preparing a high density of twin structures are only restricted within a few methods, such as electrodeposition, vapor deposition, and magnetron sputtering. Recently, dynamic plastic deformation at a high strain-rate and/or at cryogenic temperature has provided practical approaches for producing materials in bulk form with nanoscale deformation twins[25,26,94]. Another challenge is the applicability of these strategies to a broad variety of engineering materials. While nanoscale twins form in many metallic materials, especially those with low SF energy, some materials with high SF energy (such as Al and Ni) may not easily form twins except under extreme conditions. Despite these challenges, the major gains in mechanical and physical characteristics and damage tolerance arising from the engineering of coherent internal boundaries in the nanoscale offers significant new opportunities in materials research.

References

1. Hosford WF, *Mechanical Behavior of Materials* (Cambridge University Press, NY, 2005).
2. Hirth JP, Lothe J, *Theory of Dislocations* (John Wiley & Sons, NY, 2nd ed., 1982).
3. Widersich H, *J. Met.*, 15 (1963) 423.
4. Lu K, Lu L, Suresh S, *Science*, 324 (2009) 349.
5. Hall EO, *Proc. Phys. Soc. Lond.*, B64 (1951) 747.
6. Petch NJ, *J. Iron Steel Inst.*, 174 (1953) 25.
7. Li JCM, *Trans. Metall. Soc. AIME*, 227 (1963) 239.
8. Schuh CA, Nieh TG, Iwasaki H, *Acta Mater.*, 51 (2003) 431.
9. Li H, Ebrahimi F, *Acta Mater.*, 54 (2006) 2877.
10. Chen J, Lu L, Lu K, *Scripta Mater.*, 54 (2006) 1913.
11. Knapp JA, Follstaedt DM, *J. Mater. Res.*, 19 (2004) 218.
12. Kumar KS, Van Swygenhoven H, Suresh S, *Acta Mater.*, 51 (2003) 5743.
13. Koch CC, *Scripta Mater.*, 49 (2003) 657.
14. Ma E, *Scripta Mater.*, 49 (2003) 663.
15. Koch CC, Morris DG, Lu K, Inoue A, *MRS Bulletin*, 24 (1999) 54.
16. Meyers MA, Mishra A, Benson DJ, *Prog. Mater. Sci.*, 51 (2006) 427.
17. Mahajan S, Chin GY, *Acta Metall.*, 21 (1973) 1353.
18. Mahajan S, Pande CS, Imam MA, Rath BB, *Acta Mater.*, 45 (1997) 2633.
19. Mahajan S, *Acta Metall.*, 23 (1975) 671.
20. Remy L, *Acta Metall.*, 25 (1977) 711.
21. Remy L, *Metall. Mater. Trans. A*, 12 (1981) 387.

22. Christian JW, Mahajan S, *Prog. Mater. Sci.*, 39 (1995) 1.

23. Babyak JW, Rhines FN, *Trans. Metall. Soc. AIME*, 218 (1960) 21.

24. Zhao WS, Tao NR, Guo JY, Lu QH, Lu J, *Scripta Mater.*, 53 (2005) 745.

25. Li YS, Tao NR, Lu K, *Acta Mater.*, 56 (2008) 230.

26. Tao NR, Lu K, *Scripta Mater.*, 60 (2009) 1039.

27. Tao NR, Lu K, *J. Mater. Sci. Technol.*, 23 (2007) 771.

28. Meyers MA, McCowan C, in *International Symposium on Interface Migration and Control or Microstructure held in conjuction with ASM's Metals Congress and TMS/AIME Fall Meeting* (Detroit, MI, vol. 8408–037, pp. 1–25, 1984).

29. Fullman RL, Fisher JC, *J. Appl. Phys.*, 22 (1951) 1350.

30. Gleiter H, *Acta Metall.*, 17 (1969) 1421.

31. Baro G, Gleiter H, *Z. Metallkd.*, 63 (1972) 661.

32. Zhang X, Misra A, Wang H, Lima AL, Hundley MF, and Hoagland RG, *J. Appl. Phys.*, 97 (2005) 094302.

33. Zhang X, Wang H, Chen XH, Lu L, Lu K, Hoagland RG, Misra A, *Appl. Phys. Lett.*, 88 (2006) 173116.

34. Wu BYC, Ferreira PJ, Schuh CA, *Metall. Mater. Trans. A*, 36A (2005) 1927.

35. Lu L, Shen Y, Chen X, Qian L, Lu K, *Science*, 304 (2004) 422.

36. Rasmussen AA, Jensen JAD, Horsewell A, Somers MAJ, *Electrochim. Acta*, 47 (2001) 67.

37. Fullman RL, *J. Appl. Phys.*, 22 (1951) 448.

38. Zhang X, Misra A, Wang H, Swadener JG, Lima AL, Hundley MF, Hoagland RG, *Appl. Phys. Lett.*, 87 (2005) 233116.

39. Zhang X, Misra A, Wang H, Shen TD, Nastasi M, Mitchell TE, Hirth JP, Hoagland RG, Embury JD, *Acta Mater.*, 52 (2004) 995.

40. Zhang X, Misra A, Wang H, Nastasi M, Embury JD, Mitchell TE, Hoagland RG, Hirth JP, *Appl. Phys. Lett.*, 84 (2004) 1096.

41. Lu L, Chen X, Huang X, Lu K, *Science*, 323 (2009) 607.

42. Youssef KM, Scattergood RO, Murty KL, Horton JA, Koch CC, *Appl. Phys. Lett.*, 87 (2005) 091904.

43. Sanders PG, Eastman JA, Weertman JR, *Acta Mater.*, 45 (1997) 4019.

44. Cheng S, Ma E, Wang YM, Kecskes LJ, Youssef KM, Koch CC, Trociewitz UP, Han K, *Acta Mater.*, 53 (2005) 1521.

45. Wang YM, Wang K, Pan D, Lu K, Hemker KJ, Ma E, *Scripta Mater.*, 48 (2003) 1581.

46. Konopka K, Wyrzykowski JW, *J. Mater. Proc. Technol.*, 64 (1997) 223.

47. Meyers MA, Chawla KK, *Mechanical Behavior of Materials* (Prentice Hall, Upper Saddle River, NJ, 1999).

48. Mahajan S, Chin GY, *Acta Metall.*, 21 (1973) 173.

49. Schiotz J, Jacobsen KW, *Science*, 301 (2003) 1357.

50. Yip S, *Nature*, 391 (1998) 532.

51. Schiotz J, Di Tolla FD, Jacobsen KW, *Nature*, 391 (1998) 561.

52. Wolf D, Yamakov V, Phillpot SR, Mukherjee A, Gleiter H, *Acta Mater.*, 53 (2005) 1.

53. Argon AS, Yip S, *Philos. Mag. Lett.*, 86 (2006) 713

54. Sanders PG, Eastman JA, Weertman JR, *Acta Mater.*, 45 (1997) 4019.

55. Koch CC, Youssef KM, Scattergood RO, Murty KL, *Adv. Eng. Mater.*, 7 (2005) 787.

56. Gunther B, Kumpmann A, Kunze HD, *Scripta Metall. Mater.*, 27 (1992) 833.

57. Gertsman VY, Birringer R, *Scripta Metall. Mater.*, 30 (1994) 577.

58. Shen YF, Lu L, Lu QH, Jin ZH, Lu K, *Scripta Mater.*, 52 (2005) 989.

59. Lu L, Schwaiger R, Shan ZW, Dao M, Lu K, Suresh S, *Acta Mater.*, 53 (2005) 2169.

60. Cheng S, Ma E, Wang YM, Kecskes LJ, Youssef KM, Koch CC, Trociewitz UP, Han K, *Acta Mater.*, 53 (2005) 1521.

61. Champion Y, Langlois C, Guerin-Mailly S, Langlois P, Bonnentien J-L, Hytch MJ, *Science*, 300 (2003) 310.

62. Wang YM, Wang K, Pan D, Lu K, Hemker KJ, Ma E, *Scripta Mater.*, 48 (2003) 1581.

63. Jin ZH, Gumbsch P, Ma E, Albe K, Lu K, Hahn H, Gleiter H, *Scripta Mater.*, 54 (2006) 1163.

64. Chen XH, Lu L, Lu K, *J. Appl. Phys.*, 102 (2007) 083708.

65. Konopka K, Mizera J, Wyrzykowski JW, *J. Mater. Proc. Technol.*, 99 (2000) 255.

66. Misra A, Zhang X, Hammon D, Hoagland RG, *Acta Mater.*, 53 (2005) 221.

67. Dao M, Lu L, Shen YF, Suresh S, *Acta Mater.*, 54 (2005) 5421.

68. Ma E, Wang YM, Lu QH, Sui ML, Lu L, Lu K, *Appl. Phys. Lett.*, 85 (2004) 4932.

69. Conrad H, in *High-Strength Materials,* ed. Zackey VF (Wiley, NY, 1964).

70. Conrad H, *Mater. Sci. Eng. A*, 341 (2003) 216.

71. Conrad H, *Metall. Mater. Trans. A*, 35 (2004) 2681.

72. Zhu T, Li J, Samanta A, Kim HG, Suresh S, *Proc. Natl. Acad. Sci. USA*, 104 (2007) 3031.

73. Zhu T, Li J, Samanta A, Leach A, Gall K, *Phys. Rev. Lett.*, 100 (2008) 025502.

74. Wei Q, Cheng S, Ramesh KT, Ma E, *Mater. Sci. Eng. A*, 381 (2004) 71.

75. Dalla Torre F, Spatig P, Schaublin R, Victoria M, *Acta Mater.*, 53 (2005) 2337.

76. Schwaiger R, Moser B, Dao M, Chollacoop N, Suresh S, *Acta Mater.*, 51 (2003) 5159.

77. Shen YF, Lu L, Dao M, Suresh S, *Scripta Mater.*, 55 (2006) 319.

78. Lu L, Zhu T, Shen Y, Dao M, Lu K, Suresh S, *Acta Mater.*, 57 (2009) 5165.

79. Dao M, Lu L, Shen YF, Suresh S, *Acta Mater.*, 54 (2006) 5421.

80. Flynn PW, Mote J, Dorn JE, *Trans. Metall. Soc. AIME*, 221 (1961) 1148.

81. Conrad H, Armstrong R, Wiedersich H, Schoeck Q, *Phil. Mag.*, 6 (1961) 177

82. Yang Q, Ghosh AK, *Acta Mater.*, 54 (2006) 5159.

83. del Valle JA, Ruano OA, *Scripta Mater.*, 55 (2006) 775.

84. Armstrong RW, Rodriguez P, *Phil. Mag.*, 86 (2006) 5787.

85. Conrad H, *Nanotechnology*, 18 (2007) 325701.

86. Rao SI, Hazzledine PM, *Phil. Mag. A*, 80 (2000) 2011.

87. Budrovic Z, Van Swygenhoven H, Derlet PM, Van Petegem S, Schmitt B, *Science*, 304 (2004) 273.

88. Wang YM, Hamza AV, Ma E, *Appl. Phys. Lett.*, 86 (2005) 241917.

89. Wang YM, Hamza AV, Ma E, *Acta Mater.*, 54 (2006) 2715.

90. Jin ZH, Gumbsch P, Albe K, Ma E, Lu K, Gleiter H, Hahn H, *Acta Mater.*, 56 (2008) 1126.

91. Asaro RJ, Kulkarni Y, *Scripta Mater.*, 58 (2008) 389.

92. Jerusalem A, Dao M, Suresh S, Radovitzky R, *Acta Mater.*, 56 (2008) 4647.

93. Chen K-C, Wu W-W, Liao C-N, Chen L-J, Tu KN, *Science*, 321 (2008) 1066.

94. Zhang Y, Tao NR, Lu K, *Acta Mater.*, 56 (2008) 2429.

Chapter 7

GRAIN BOUNDARIES IN NANOCRYSTALLINE MATERIALS

James C. M. Li

Materials Science Program, Department of Mechanical Engineering, University of Rochester, Rochester, NY 14627, USA

E-mail: li@me.rochester.edu

Grain boundaries have more than five degrees of freedom if the material has been treated by heat, deformation, irradiation, etc. The five degrees of freedom are from the relative orientation of the two grains on the two sides of the grain-boundary as well as the orientation of the grain-boundary. These five degrees of freedom are statistically averaged over many grains in the material and hence are not really free to be adjusted unless the material has a texture. Two more freedoms are introduced here, namely, the impurity content and the porosity content. These two degrees of freedom can be adjusted externally. It is shown how these two extra parameters may influence the Hall–Petch relationship between strength and grain size.

7.1 INTRODUCTION

Grains are the most basic microstructure for polycrystalline materials, and hence the grain-size effect is the first to be investigated for any property. For mechanical properties Hall (1951) and Petch (1953) were the first ones to discover the inverse reciprocal grain size effect which was explained by dislocation pileups or grain-boundary sources of dislocations. For dislocation pileups, dislocations moving along the same slip plane were assumed to pile up against the grain-boundary. The number of dislocations in the pileup was proportional to the grain size and inversely proportional to the effective stress applied uniformly over the pileup. Yielding was assumed

Mechanical Properties of Nanocrystalline Materials
Edited by James C. M. Li

to take place when the stress concentration at the grain-boundary exceeds a critical value. This critical value was assumed independent of grain size. For the grain-boundary source mechanism, dislocations were assumed to be emitted from grain boundaries. The total length of dislocations emitted per unit area of grain-boundary was assumed constant independent of grain size. Hence in each case the grain-boundary structure was assumed constant. This assumption may not be valid when the grain size gets into the nanometer regime. Two parameters are discussed in this chapter, the impurity content and the porosity of the grain boundaries. It will be shown how these two parameters affect the Hall–Petch relationship.

In view of the Hall–Petch relationship, to increase strength is to reduce the grain size. Some early efforts were successful to make 1-micron grains. Gleiter (see Gleiter, 2000) was the one who made the initial efforts to reduce the grain size to the nanometer regime. His effort stimulated a lot of interest to make nanocrystalline materials and the accompanied studies of such materials, including the breakdown of the Hall–Petch relation in the nanometer grains. The reason of the breakdown is not clear, and this chapter will try to shed some light over this problem.

7.1.1 Dislocation Pileups

A simple pileup of dislocations is shown in Fig. 7.1.

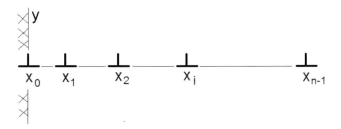

Figure 7.1 A simple pileup of dislocations.

Let the dislocations, all of Burgers vector $\mathbf{b}(b_x, b_y, b_z)$ and parallel to the z-axis, be numbered 0, 1, 2, ..., $n - 1$. The leading disloction (0) is locked in position on the z-axis by a grain-boundary. The rest of the dislocations (1, 2, ..., $n-1$) are in the $+x$ half of the xz plane. The external stress $\sigma = - (b_x \sigma_{xy} + b_y \sigma_{yy} + b_z \sigma_{yz})/b$ is such that it tends to push all the free dislocations toward the z-axis.

The equilibrium positions of the free dislocations were obtained by Eshelby, Frank, and Nabarro (1951). Their method is as follows. At equilibrium, the force on the ith dislocation is zero with $i = 1, 2, ..., n - 1$.

$$\frac{A}{x_i} + \sum_{\substack{j=1 \\ j \neq i}}^{n-1} \frac{A}{x_i - x_j} - \sigma b = 0 \tag{7.1}$$

where $A = G[b_x^2 + b_y^2 + (1 - v)b_z^2]/2\pi(1-v)$ for an isotropic medium, with G being the shear modulus of the material.

For simplicity, let the unit of distance be $A/2\sigma b$ so that Eq. 7.1 becomes

$$\frac{1}{x_i} + \sum_{\substack{j=1 \\ j \neq i}}^{n-1} \frac{1}{x_i - x_j} - \frac{1}{2} = 0, \, i = 1, 2, \dots n - 1 \tag{7.2}$$

This system of nonlinear algebraic equations was solved by Stieltjes (1885) by using the zeros of a polynomial:

$$f(x) = \prod_{j=1}^{n-1} (x - x_j) \tag{7.3}$$

Then $\dfrac{f'(x)}{f(x)} - \dfrac{1}{x - x_i} = \displaystyle\sum_{\substack{j=1 \\ j \neq i}}^{n-1} \dfrac{1}{x - x_j}$ $\tag{7.4}$

When $x \to x_i$ the right-hand side of Eq. 7.4 approaches $(1/2) - (1/x_i)$ according to Eq. 7.2 and the left-hand side approaches $f''(x_i)/2f'(x_i)$, Eq. 7.4 then becomes

$$f''(x_i) + \left(\frac{2}{x_i} - 1\right) f'(x_i) = 0 \tag{7.5}$$

Clearly, each x_i satisfies the following differential equation:

$$f''(x) + \left(\frac{2}{x} - 1\right) f'(x) + q(n, x) f(x) = 0 \tag{7.6}$$

If $q(n, x)$ can be so chosen that a polynomial solution exists, the roots of the polynomial will be x_i. For the present case, let $q(n, x) = (n-1)/x$. The solution of the differential equation

$$xf''(x) + (2 - x) f'(x) + (n - 1) f(x) = 0 \tag{7.7}$$

is the first derivative of the nth Laguerre polynomial (see Szego, 1948, p. 99):

$$L_n'(x) = -\sum_{k=0}^{n-1} \binom{n}{k+1} \frac{(-x)k}{k!} \tag{7.8}$$

Some examples are as follows:

$$L_2'(x) = x - 2 \qquad\qquad L_3'(x) = -(x^2/2) + 3x - 3$$
$$L_4'(x) = (x^3/6) - 2x^2 + 6x - 4$$
$$L_5'(x) = -(x^4/24) + 5(x^3/6) - 5x^2 + 10x - 5 \tag{7.9}$$

The coefficients of $L_n(x)$ for $n = 0$ through 12 are given in Abramowitz and Stegun (1965, p. 799). The roots of these polynomials were computed by Head (1959), by Chou, Garofalo, and Whitmore (1960), and by Mitchell, Hecker, and Smialek (1965). Now it is very simple to program on the computer using Eq. 7.2.

The distance x_{n-1} is called the length of the pileup, l, which may be equal to the grain size. An upper limit can be obtained from the asymptotic relation for large zeros of the Laguerre polynomial (see Abramowitz and Stegun, 1965, p. 787, or Szego, 1948, p. 127).

$$l < 4n \, (A/2\sigma b) = 2nA/\sigma b \tag{7.10}$$

Some better estimates are (Szego, 1948, p. 131; Bilby and Entwisle, 1956):

$$l \cong [4n - 3.7115 \, (4n)^{1/3}] \, (A/2\sigma b)$$

$$l \cong [4n \cos^2 \, (9\pi/16n)^{1/3}] \, (A/2\sigma b) \tag{7.11}$$

A comparison with exact values is shown in Fig. 7.2.

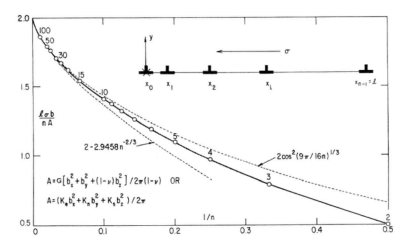

Figure 7.2 The length of a simple pileup of dislocations.

The force exerted on the locked dislocation divided by its Burgers vector is known as the *tip stress* or the stress concentration of the pileup. In this case it is

$$\sigma_{\text{tip}} = \sigma + \frac{A}{b} \sum_{i=1}^{n-1} \frac{1}{x_i} \left(\frac{2\sigma b}{A} \right) = n\sigma \tag{7.12}$$

The summation of x_i^{-1} can be obtained from Eq. 7.4 and is equal to $-f'(0)/f(0)$, which is seen to be $(n-1)/n$ from Eq. 7.7. Equation 7.12 can be obtained also from a virtual work argument of Cottrell (1953, p. 105). By replacing the locking agent by an external force, **F** per unit length of dislocation, and displacing the whole pileup by a small distance δx, the total amount of external work done should be zero because the configuration of the pileup is unchanged. Since the work done by the external stress is $-n\sigma\, b\delta x$ and that by the locking force is $F\delta x$, the tip stress is $F/b = n\sigma$, the same as Eq. 7.12. This is a special case of a more general theorem known as Moutier's theorem (see Guggenheim, 1949).

Yielding can be assumed to take place when the tip stress reaches a critical value σ_c. Then a combination of Eqs. 7.10 and 7.12 gives:

$$\sigma = \sigma_0 + \sqrt{\frac{2\sigma_c A}{b}}\; l^{-1/2} = \sigma_0 + kl^{-1/2} \tag{7.13}$$

where σ has been replaced by $\sigma - \sigma_0$ to allow for a friction stress σ_0, which could be the yield stress for single crystals, namely, without the effect of grain boundaries. Equation 7.13 is known as the Hall–Petch relation since they found it valid experimentally. Now it has a basis from the dislocation pileup model. But as Eq. 7.13 shows, k will be constant only if σ_c is constant or the grain-boundary structure remains the same for all grain sizes. Furthermore the factor 2 in Eq. 7.13 is really for large number of dislocations in the pileup, as shown in Fig. 7.2. It decreases with a decreasing number of dislocations or decreasing grain size. Hence the Hall–Petch relation is not expected to hold all the way to the nanometer-grain-size region.

7.1.2 Dislocation Densities

Another way of understanding the Hall–Petch relation when there are no pileups is to consider grain boundaries as sources of dislocations (Li, 1963). Direct observations of dislocation emission from grain-boundary ledges have been reported by Murr's group (Murr, 1974; Venkatesh and Murr, 1976–1978; Murr and Hecker, 1979; Murr and Wang, 1982; Esquivel and Murr, 2005). A recent molecular dynamics (MD) simulation confirmed that dislocations can be emitted from grain-boundary ledges easier than from planar regions of the grain-boundary (Capolungo *et al.*, 2007). If the grain-boundary structure remains the same for all grain sizes, their ability of emitting dislocations could be the same too. Let m be the total length of dislocations emitted into the two neighboring grains per unit area of the grain-boundary. The total density of dislocations inside the grain at the time of yielding is

$$\rho = \frac{\pi l^2 m/2}{\pi l^3/6} = \frac{3m}{l} \tag{7.14}$$

A spherical grain of diameter l is assumed in the calculation. By using the well-known relation between the flow stress and dislocation density, the yield stress is given by

$$\sigma = \sigma_0 + \alpha Gb \sqrt{\rho} = \sigma_0 + \alpha Gb \sqrt{3ml}^{-1/2} = \sigma_0 + kl^{-1/2} \qquad (7.15)$$

where α is of the order of 0.5 or $2/2\pi$ $(l - v)$. It is seen that the Hall–Petch relation can result without dislocation pileups. Here the Hall–Petch slope is a constant only if m is a constant or the grain-boundary structure remains the same for all grain sizes, the same requirement for the dislocation pileup model.

That the two models of yielding can give similar results is shown in Fig. 7.3. A low-angle-tilt boundary of misfit angle θ is used to block the pileup in the first model. The tip stress required to push the pileup through is $\sigma_c = 0.45 G\theta/2$ $(l - v)$. Substituting this into Eq. 7.13 gives the Hall–Petch slope shown in Fig. 7.3 for the pileup model. For the dislocation density model, the dislocations in the subboundary are moved out into the subgrain, which forms a forest of dislocations with a dislocation density corresponding to $m = 2\theta/3\pi b$ in Eq. 7.15. This dislocation density gives a Hall–Petch slope, as given in Fig. 7.3 for the dislocation source model. For details, see Li (1963). Hence the two models could give similar Hall–Petch slopes.

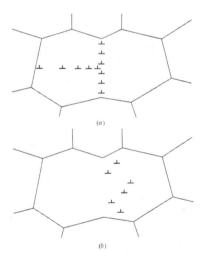

Figure 7.3 A comparison between the pileup model and the dislocation source model of yielding (Li, 1963).

Hall–Petch slope k:
$$\sigma = \sigma_0 + k/\sqrt{1}$$

Pileup mechanism:

$k = 1.7G \sqrt{b\theta} / 2\pi (1 - v)$

Grain-boundary source mechanism:

$k = 1.6G \sqrt{b\theta} / 2\pi (1 - v)$

Can the grain-boundary structure remain constant when the grain size decreases into the nanometer region? The answer is mostly likely no due to severe mechanical and heat treatment needed to make the nanocrystalline materials. At least two additional parameters must be considered, the impurity content and the porosity. The purpose of this chapter is to discuss these two parameters.

7.2 IMPURITY CONTENT

If the total impurity content of the material remains the same when the grain size decreases, it is possible that the grain-boundary becomes purer at smaller grain sizes. Zhang and Weertman (2004, 2005) observed grain growth in nanocrystalline Cu even at liquid nitrogen temperatures. This was explained by Li (2006), who proposed that the nanograin boundaries may be purer than micrograin boundaries. Malow and Koch (1997) studied grain growth in nanocrystalline iron prepared by mechanical attrition and found two regions with different growth exponents and activation energies. For grain sizes smaller than 25 nm, the activation energy was 125 kJ/mole, and for those larger than 25 nm, it was 248 kJ/mole. They thought the difference was due to impurity content. More recently, Novikov (2008) suggested that precipitate particles are needed to stabilize the nanograins. It will be shown later that there is a transition grain size between impure and pure grain boundaries.

Suppose that N_β is the total concentration of an impurity β per unit volume of the material and a fraction of that, $N_{\beta m}$, is dissolved per unit volume of the grain interior and the rest, $N_{\beta b}$, is segregated in the grain boundaries per unit area of the boundary. Then a materials balance gives

$$N_{\beta m} V + N_\beta b A = N_\beta V \tag{7.16}$$

where V is the volume of grains and A is the area of grain boundaries. Following Gibbs, the grain boundaries have areas but occupy no volumes, so the volume of all the grains is the same as the volume of the material. By assuming a spherical grain, $A/V = 3/l$, where l is the grain size. Equation 7.16 becomes

$$N_{\beta m} + \frac{3N_{\beta b}}{l} = N_\beta \tag{7.17}$$

Now if $N_{\beta b}^0$ is the saturation amount of impurity β per unit area of grain-boundary and $x = N_{\beta b}/N_{\beta b}^0$ is the extent of saturation or the fraction of possible sites occupied by impurity atoms, a quantity between 0 and 1, the equilibrium between the impurity in the matrix and that in the grain boundaries is assumed to obey the Langmuir isotherm:

$$\frac{x}{1-x} = KN_{\beta m} = K\left(N_\beta - \frac{3xN_{\beta b}^0}{l}\right) \tag{7.18}$$

where K is the equilibrium constant, which is a function of temperature. Equation 7.18 gives a relation between the grain size l and x:

$$\frac{l}{3KN_{\beta b}^0} = \frac{x}{KN_\beta - \dfrac{x}{1-x}} \tag{7.19}$$

Since l must be positive, x has a maximum value which is not unity and is given by

$$x_m = KN_\beta/(1 + KN_\beta) \tag{7.20}$$

A plot of l versus x reveals a sudden transition at some grain size below which the grain boundaries are purer. If a critical grain size l_c is defined at $x = 0.5$

$$\frac{l_c}{3\,KN_{\beta b}^0} = \frac{0.5}{KN_\beta - 1} = \frac{0.5\,(1 - x_m)}{2x_m - 1} \tag{7.21}$$

All the grain boundaries for grain sizes below l_c are purer than those above l_c, confirming the suspicion of Malow and Koch (1997). It is seen from Eq. 7.21 that the critical grain size increases with decreasing total impurity content.

The relation between the impurity content of the grain-boundary and the Hall–Petch slope was shown before (Li, 1963). The idea is that the impurity in the grain-boundary stabilizes the ledges, which serve as dislocation sources. Hansen (2005) supported this idea by comparing the observed ledge density with that predicted from the Hall–Petch slope. The relation between the yield stress and the impurity content x of the grain-boundary is

$$\sigma = \sigma_0 + A\sqrt{\frac{x}{l/3KN_{\beta b}^0}} \tag{7.22}$$

which gives the Hall–Petch relation if x is a constant. For micro grain sizes, all the grain boundaries are saturated with impurities or $x = x_m$ so that the Hall–Petch slope is

$$k = A\sqrt{3x_m KN_{\beta b}^0} \tag{7.23}$$

Now since x varies with l, as shown in Eq. 7.19, the yield stress–grain size relation is shown in Fig. 7.4 on the basis of Eq. 7.22. When x or l approaches zero, the yield stress approaches a limit, as shown. This limit is given by

$$\sigma = \sigma_0 + A \sqrt{KN_\beta} \tag{7.24}$$

It is seen that this limit depends on the total impurity content. Figure 7.4 reproduces the trend of many sets of experimental data, such as shown in Figs. 15, 19, and 46 of Meyers *et al.* (2006) for Cu, Fe, Ni, and Ti; Fig. 4 of Koch and Narayan (2001) for Zn; and Fig. 3 of El-Sherik, *et al.* (1992), which showed the Hall–Petch breakdown at 25 nm for Ni. Equation 7.24 suggests that impurities can reduce the grain size below which the Hall–Petch relation breaks down. This is confirmed by Schuh *et al.* (2003), who found that alloying of W in Ni can suppress the Hall–Petch breakdown from 14 nm for pure Ni (see also Nieh and Wang [2005]) to about 8 nm for the alloy. This is consistent also with the recent finding of Chen *et al.* (2005), who made a nanocrystalline (40 nm grain size) 316L austenitic stainless steel (loaded with impurities) with a tensile yield strength of 1.45 GPa and that still obeys the Hall–Petch relation extrapolated from the microcrystalline regime. The effect of solute segregation on the strength of nanocrystalline alloys was discussed recently by Shen *et al.* (2007).

7.3 POROSITY CONTENT

The foregoing is for grain boundaries with the lowest-energy structure consistent with the constraints of the neighboring two grains and equilibrated with impurities. As pointed out by many authors, the grain boundaries in the nanocrystalline materials may not have the lowest-energy structure (Gryaznov, *et al.* 1993; Masumura *et al.* 1998; Horita *et al.* 1998; Wu and Yu, 2000; Huang *et al.* 2001; Valiev, *et al.* 2002; Hasnaoui *et al.* 2002; Liao, *et al.* 2003; Ma, 2003; Armstrong *et al.* 2004; Valiev, *et al.* 2006; Wu and Zhu, 2006; Armstrong, Chapter 3 of this book). However, so far there has not been a suitable parameter characterizing such nonequilibrium structure. The number and kind of extra dislocations have been suggested (Li, 2006). Another possible parameter is the free volume or porosity. While many parameters may be needed, we will start with just one.

The term "free volume" was borrowed from the literature dealing with amorphous polymers, glasses, or amorphous metals. It is defined as the excess volume over and above the equilibrium volume extrapolated from below-the-glass transition temperature (see Ward, 1985). Since the grain-boundary is not really an amorphous region and has no glass transition temperature, free volume is not an appropriate term to describe the porosity or the vacant spaces in the grain-boundary. In fact porosity is not a bad term for a grain-boundary. The equilibrium porosity was discussed before on the basis of the dislocation-core model (Li, 1961).

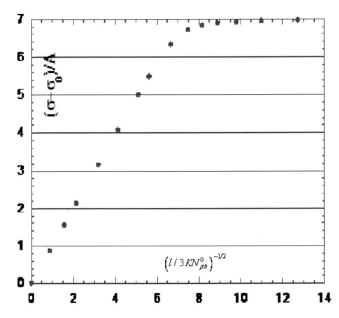

Figure 7.4 The yield stress–grain size relationship as affected by impurity content in the grain boundaries (Li, 2007).

Grain-boundary porosities could result from many processes such as mutual diffusion along grain boundaries, a kind of grain-boundary Kirkendall effect (see Rabkin *et al.*, 2000). These large porosities could be seen, but most others may not be that obvious. Another possibility is grain-boundary melting, which expands the volume, and upon resolidification the extra space becomes porous (see Campbell, 2009). Adsorbed dislocations are a source of porosity due to their cores. Carlson and Ferreira (2007) suggested that nanograin boundaries adsorb dislocations more easily than large grain boundaries. If so, nanograin boundaries will have more porosity.

An average grain-boundary porosity can be formally defined as

$$\phi = \frac{V - V_c}{A} = \frac{d}{3}\left(1 - \frac{\rho}{\rho_c}\right) \tag{7.25}$$

where V is the volume of the material excluding any voids or excess vacancies inside the grains, V_c is the volume for a single crystal of the same mass, and

A is the total grain-boundary area. The second expression follows if A/V is taken as $3/l$, with l being the grain size, ρ the density of the material without any voids inside the grains, and ρ_c the density for the single crystal. This parameter can be determined experimentally and calculated in any computer model, so the experiments can be compared with simulations for the same porosity.

When the porosity deviates from the equilibrium porosity, the property of the grain-boundary is expected to differ accordingly. For example, Hasnaoui *et al.* (2002) found in their computer simulation that annealed grain boundaries are stronger and deformed grain boundaries are weaker revealing the effect of grain-boundary structure (or porosity) on the mechanical properties of the grain boundaries. Porous grain boundaries may facilitate and enhance grain-boundary diffusion. Chokshi *et al.* (1989) suggested Coble creep as a deformation mechanism for nanocrystalline materials. However, since the porosity may change with grain size, the usual cubic law for grain size dependence should not be expected. Porous grain boundaries also may facilitate grain-boundary sliding. Van Swygenhoven *et al.* (1999, 2001) showed by MD simulation that grain-boundary sliding was the primary mechanism for the deformation of nanocrystalline materials. Chinh *et al.* (2006) showed grain-boundary sliding in ultrafine-grained Al at low temperatures. Grain-boundary sliding was regarded as a softening mechanism by Zhao *et al.* (2009) for nanograins. Their model suggested that the stress associated with a dislocation mechanism decreases with increasing grain size, which leads to the Hall–Petch relation. The stress associated with grain-boundary sliding increases with increasing grain size, which leads to the inverse Hall–Petch relation. See Conrad and Narayan for some more discussion (Chapter 1 of this book) on grain-boundary sliding.

Porous grain boundaries and the accompanied effect of grain-boundary diffusion and sliding also can cause grain rotation. In fact Jia *et al.* (2000, 2001) suggested that nanograins rotate during deformation based on their observations. Porous, nonequilibrium, and purer grain boundaries may enhance mobility (see Li, 2006; Mishra *et al.*, 2007).

When the grain-boundary has some extra porosity, its internal stress distribution must be altered and so must its equilibrium impurity content. It is possible that impurities are less segregated onto a more porous grain-boundary. So the equilibrium segregation x calculated from Eq. 7.19 must be modified if the porosity deviates from the equilibrium porosity. Since the segregation should decrease with increasing porosity due to the diminishing internal stresses, a simple modification is to subtract from the equilibrium segregation a quantity which is proportional to the excess porosity:

$$x - B(\phi - \phi_0) \tag{7.26}$$

where B is a constant and ϕ_0 is the equilibrium porosity. To illustrate, Eq. 7.22 could be modified if the grain-boundary has excess porosity such that $B(\phi - \phi_0) = 0.1$ or 0.2, which can be assumed a constant independent of grain size:

$$\sigma = \sigma_0 + A \sqrt{\frac{x - 0.1}{d/3 \, KN_{\beta b}^0}} \tag{7.27}$$

Equation 7.27 is plotted in Fig. 7.5 for both 0.1 and 0.2. It is seen that now the Hall–Petch plot has a maximum without invoking another mechanism at small grain sizes. The position of the maximum would depend on the quantity 0.1 or 0.2 assumed for the excess porosity and its effect on impurity segregation. The softening is simply due to a purer grain-boundary for smaller grain sizes. Figure 7.5 reproduces the trend for many sets of data—see Figs. 15–17 and 49 in Meyers *et al.* (2006), Fig. 2 in Schuh, *et al.* (2002) for Ni, Fig. 42 in Lu (1996) for nanograined materials crystallized from amorphous solids, Fig. 14 in Kumar *et al.* (2003) for Ni and Cu, Fig. 3 in Jang and Atzmon (2003) for Fe, Figs. 1 and 5 in Giga *et al.* (2006) for Ni–W alloys in a tensile test, and Figs. 1 and 2 in Masumura *et al.* (1998). See also a recent review by Pande and Cooper (2009).

The assumption of the same porosity for all grain sizes can be removed by allowing the quantity 0.1 or 0.2 to vary with grain size such as the one used by Aifantis and Konstantinidis (2009), who regard the triple junctions as vacant. However, without a direct determination of porosity as a function of grain size, any assumed function would be just a way to fit the data.

Obviously the present model suggests that the dislocation mechanism is still operating even in the nanograin range. An evidence is shown by Cai *et al.* (2001), who found a large amount of internal stress in cold-rolled (1,400% rolling extension) electrolytically deposited nc Cu (30 nm grain size, which did not change much after rolling). In their creep experiments, about 90% of the applied stress was internal stress. If the mechanism of cold rolling was Coble creep, no such amount of internal stress should be developed. Lu *et al.* (2001) also showed an increase of grain-boundary energy upon cold rolling of nc Cu, indicating the accumulation of extra dislocations in the boundary. Hansen (2004) showed that dislocation mechanisms are operating for all grain sizes. Youssef *et al.* (2005) observed dislocation activity in 23 nm grains in Cu during *in situ* transmission electron microscopy (TEM) deformation. Shan *et al.* (2007) observed dislocations in nanocrystlline Ni with grain sizes as small as 5 and 10 nm. Wu *et al.* (2009) found strong strain hardening in 20 nm Ni by rolling due to dislocation accumulation inside the grains. Rodriguez and Armstrong (2006), Armstrong (Chapter 3 of this book), and Malygin (2007a and 2007b) successfully analyzed the deformation of nanocrystalline materials using dislocation dynamics. Jannalagadda *et al.* (2010) measured the strain-rate dependence of flow stress of a 25 nm Pt film and concluded that it is due to dislocation plasticity with only some effects of grain-boundary deformation.

If the grain-boundary porosity is the same for all grain sizes, the Hall–Petch slope for micro grains is

$$k = A \sqrt{3 \, (x_m - 0.1) \, KN_{\beta b}^0} \qquad\qquad (7.28)$$

which decreases with increasing porosity of the grain boundaries. There are indications of this effect in Meyers *et al.* (2006), Van Vliet *et al.* (2003), and Zao *et al.* (2003). This model is different from the Coble creep model of Masumura *et al.* (1998), in which the initial Hall–Petch slope does not change when the extent of Coble creep changes.

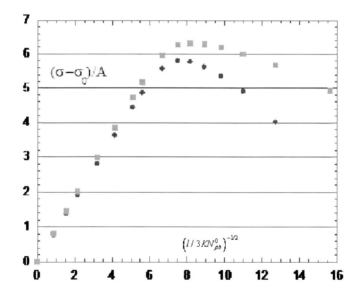

Figure 7.5 The yield stress–grain size relationship as affected by grain-boundary porosity and impurity redistribution. The upper curve is for Eq. 7.27 with 0.1 subtracted from *x*, and the lower curve is the same with 0.2 subtracted (Li, 2007).

7.4 CONCLUDING REMARKS

It is seen that impurity content and grain-boundary porosity should be considered when the grain size gets down to the nanometer region. They could explain all the recent findings about the grain-size dependence of yield stress of nanocrystalline materials without invoking any new mechanisms. By varying these parameters it offers hope of synthesizing ultra strong and ductile materials by increasing the total impurity content and also the porosity of grain boundaries. Some successes have already been shown by Wang *et al.* (2003), Youssef *et al.*, (2004–2006), Cheng *et al.* (2005), Nieh and Wang (2005), Xu *et al.* (2008), and Wu *et al.* (2009). Xu *et al.* (2008) thought that the strength was derived from solid solutions and the ductility originated from nonequilibrium grain boundaries and the bimodal grain structure. We expect more successes are forthcoming.

References

M. Abramowitz and I. A. Stegun, *Handbook of Mathematical Functions* (Dover, NY, 1965).

K. Aifantis and A. A. Konstantinidis, "Hall–Petch Revisited at the Nanoscale," *Mat. Sci. Eng.*, **B163** (2009), 139–144.

R. W. Armstrong, "Strength and Strain-rate Sensitivity of Nanopolycrystals," Chapter 3 of this book.

R. W. Armstrong, H. Conrad, and F. R. N. Nabarro, "Meso-to-Nano-scopic Polycrystal/ Composite Strengthening," *Mat. Res. Soc. Proc.*, **791** (2004), 69/77.

B. A. Bilby and A. R. Entwisle, "Dislocation Arrays and Rows of Etch Pits," *Acta Met.*, **4** (1956), 257–261.

B. Cai, Q. P. Kong, P. Cui, L. Lu, and K. Lu, "Creep Behavior of Cold-Rolled Nanocrystalline Pure Copper," *Scripta Mater.*, **45** (2001), 1407–1413.

J. Campbell, "Incipient Grain-boundary Melting," *Mater. Sci. Technol.*, **25** (2009), 125–126.

L. Capolungo, D. E. Spearot, M. Cherkaoui, D. L. McDowell, J. Qui, and K. I. Jacob, "Dislocation Emission from Bicrystal Interfaces and Grain-boundary Ledges: Relationship to Nanocrystalline Deformation," *J. Mech. Phys. Sol.*, **55** (2007), 2300–2327.

C. E. Carlton and P. J. Ferreira, "What Is Behind the Inverse Hall–Petch Effect in Nanocrystalline Materials," *Acta Mater.*, **55** (2007), 3749–3756.

X. H. Chen, J. Lu, L. Lu, and K. Lu, "Tensile Properties of a Nanocrystalline 316L Austenitic Stainless Steel," *Scripta Mater.*, **52** (2005), 1039–1044.

S. Cheng, E. Ma, Y. M. Wang, L. J. Kecskes, K. M. Youssef, C. C. Koch, U. P. Trociewitz, and K. Han, "Tensile Properties of in situ Consolidated Nanocrystalline Cu," *Acta Mater.*, **53** (2005), 1521–1533.

N. Q. Chinh, P. Szommer, T. Csanadi, and T. G. Langdon, "Flow Processes at Low Temperatures in Ultrafine-Grained Aluminum," *Mat. Sci. Eng.*, **A434** (2006), 326–334.

H. Chokshi, A. Rosen, J. Karch, and H. Gleiter, "On the Validity of the Hall Petch Relationship in Nanocrystalline Materials," *Scripta Metall.*, **23** (1989), 1679–1684.

Y. T. Chou, F. Garofalo, and R. W. Whitmore, "Interactions between Glide Dislocations in a Double Pileup in Alpha Iron," *Acta Mater.*, **8** (1960), 480–488.

Y. T. Chou, "Dislocation Pileups against a Locked Dislocation of a Different Burgers Vector," *J. Appl. Phys.*, **38** (1967), 2080–2085.

V. N. Chuvil'deev, V. I. Kopylov, and W. Zeiger, "A Theory of Non-Equilibrium Grain Boundaries and Its Applications to Nano- and Micro-Crystalline Materials Processed by ECAP," *Ann. Chim. Sci. Mat.*, **27** (2002), 55–64.

H. Conrad and J. Narayan, "Mechanisms Governing the Plastic Deformation of Nanocrystalline Materials including Grain Size Softening," Chapter 1 of this book.

H. Cottrell, *Dislocations and Plastic Flow in Crystals* (Clarendon, Oxford, 1953).

M. El-Sherik, U. Erb, G. Palumbo, and K. T. Aust, "Deviations from Hall–Petch Behavior in As-Prepared Nanocrystalline Nickel," *Scripta Mater.*, **27** (1992), 1185–1188.

J. D. Eshelby, F. C. Frank, and F. R. N. Nabarro, "The Equilibrium of Linear Arrays of Dislocations," *Phil. Mag.*, **42** (1951), 351–364.

E. V. Esquivel and L. E. Murr, "Grain-boundary Contributions to Deformation and Solid-State Flow in Severe Plastic Deformation," *Mat. Sci. Eng.*, **A409** (2005), 13–23.

A. Giga, Y. Kimoto, Y. Takigawa, and K. Higashi, "Demonstration of an Inverse Hall–Petch Relationship in Electrodeposited Nanocrystalline Ni–W Alloys through Tensile Testing," *Scripta Mater.*, **55** (2006), 143–146.

H. Gleiter, "Nanostructured Materials: Basic Concepts and Microstructure," *Acta Mater.*, **48** (2000), 1–29.

V. G. Gryaznov, M. Yu. Gutkin, A. E. Romanov, and L. I. Trusov, "On the Yield Stress of Nanocrystals," *J. Mat. Sci.*, **28** (1993), 4359–4365.

E. A. Guggenheim, *Thermodynamics* (Interscience, NY, 1949), p. 78.

E.O. Hall, "The Deformation and Aging of Mild Steel," *Proc. Phys. Soc. Lond.*, **B64** (1951), 747.

N. Hansen, "Hall–Petch Relation and Boundary Strengthening," *Scripta Mater.*, **51** (2004), 801–806.

N. Hansen, "Boundary Strengthening in Undeformed and Deformed Polycrystals," *Mat. Sci. Eng.*, **A409** (2005), 39–45.

A. Hasnaoui, H. Van Swygenhoven, and P. M. Derlet, "On Non-equilibrium Grain Boundaries and Their Effect on Thermal and Mechanical Behavior: A Molecular Dynamics Computer Simulation," *Acta Mater.*, **50** (2002), 3927–3939.

A. K. Head, "The Positions of Dislocations in Arrays," *Phil. Mag.*, **4** (1959), 295–302.

Z. Horita, D. J. Smith, M. Nemoto, R. Z. Valiev, and T. G. Langdon, "Observation of Grain-boundary Structure in Submicrometer-Grained Cu and Ni using High-resolution Electron Microscopy," *J. Mat. Res.*, **13** (1998), 446–450.

J. Huang, Y. T. Zhu, H. Jiang, and T. C. Lowe, "Microstructures and Dislocation Configurations in Nanostructured Cu Processed by Repetitive Corrugation and Straightening," *Acta Mater.*, **49** (2001), 1497–1505.

D. Jang, and M. Atzmon, "Grain-Size Dependence of Plastic Deformation in Nanocrystalline Fe," *J. Appl. Phys.*, **93** (2003), 9282–9286.

D. Jia, K. T. Ramesh, and E. Ma, "Failure Mode and Dynamic Behavior of Nanophase Iron under Compression," *Scripta Mater.*, **42** (2000), 73–78.

D. Jia, Y. M. Wang, K. T. Ramesh, E. Ma, Y. T. Zhu, and R. Z. Valiev, "Deformation Behavior and Plastic Instabilities of Ultrafine-Grained Titanium," *Appl. Phys. Lett.*, **79** (2001), 611–613.

K. N. Jonnalagadda, I. Chasiotis, S. Yagnamurthy, J. Lambros, J. Pulskamp, R. Polcawich, and M. Dubey, "Experimental Investigation of Strain-rate Dependence of Nanocrystalline Pt Film," *Exp. Mech.*, **50** (2010), 25–35.

K. S. Kumar, H. Van Swygenhoven, and S. Suresh, "Mechanical Behavior of Nanocrystalline Metals and Alloys," *Acta Mater.*, **51** (2003), 5743–5774.

J. C. M. Li, "High-Angle Tilt Boundary—A Dislocation Core Model," *J. Appl. Phys.*, **32** (1961), 525–541.

J. C. M. Li, "Petch Relation and Grain-boundary Sources," *Trans. AIME*, **227** (1963), 239–247.

J. C. M. Li, "Mechanical Grain Growth in Nanocrystalline Copper," *Phys. Rev. Lett.*, **96** (2006), 215506.

J. C. M. Li, "Grain-boundary Impurity and Porosity Effects on the Yield Strength of Nanocrystalline Materials," *Appl. Phys. Lett.*, **90** (2007), 041912.

X. Z. Liao, J. Y. Huang, Y. T. Zhu, F. Zhou, and E. J. Lavernia, "Nanostructures and Deformation Mechanisms in a Cryogenically Ball-Milled Al-Mg Alloy," *Phil. Mag.*, **83** (2003), 3065–3075.

K. Lu,, "Nanocrystalline Metals Crystallized from Amorphous Solids: Nanocrystallization, Structure and Properties," *Mat. Sci. Eng.*, **R16** (1996), 161–221.

L. Lu, M. L. Sui, and K. Lu, "Cold Rolling of Bulk Nanocrystalline Copper," *Acta Mater.*, **49** (2001), 4127–4134.

E. Ma, "Instabilities and Ductility of Nanocrystalline and Ultrafine-Grained Metals," *Scripta Mater.*, **49** (2003), 663–668.

T. R. Malow and C. C. Koch, "Grain Growth in Nanocrystalline Iron Prepared by Mehanical Attrition." *Acta Mater.*, **45** (1997), 2177–2186.

G. A. Malygin, "Plasticity and Strength of Micro- and Nanocrystalline Materials," *Phys. Sol. Stat.*, **49** (2007a), 1013–1033.

G. A. Malygin, "Analysis of the Strain-Rate Sensitivity of Flow Stresses in Nanocrystalline FCC and BCC Metals," *Phys. Sol. Stat.*, **49** (2007b), 2266–2273.

R. A. Masumura, P. M. Hazzledine, and C. S. Pande, "Yield Stress of Fine Gained Materials," *Acta Mater.*, **46** (1998), 4527–4534.

M. A. Meyers, A. Mishra, and D. J. Benson, "Mechanical Properties of Nanocrystalline Materials," *Prog. Mater. Sci.*, **51** (2006a), 427–556.

M. A. Meyers, A. Mishra, and D. Benson, "The Deformation Physics of Nanocrystalline Metals: Experiments, Analysis and Computations," *J. Metals*, **58** (2006b), 41–48.

A. Mishra, V. Richard, F. Gregori, R. J. Asaro, and M. A. Meyers, "Microstructural Evolution in Copper Processed by Severe Plastic Deformation," *Mat. Sci. Eng.*, **A410–A411** (2005), 290–298.

Mishra, B. K. Kad, F. Gregori, and M. A. Meyers, "Microstructural Evolution in Copper Subjected to Severe Plastic Deformation: Experiments and Analysis," *Acta Mater.*, **55** (2007), 13–28.

T. E. Mitchell, S. S. Hecker, and R. L. Smialek, "Dislocation Pileups in Anisotropic Crystals," *Phys. Stat. Sol.*, **11** (1965), 585–594.

L. E. Murr, "Yielding and Grain-Boundary Ledges: Some Comments on the Hall–Petch Relation," *Appl. Phys. Lett.*, **24** (1974), 533.

L. E. Murr and S. S. Hecker, "Quantitative Evidence for Dislocation Emission from Grain Boundaries," *Scripta Metall.* **13** (1979), 167–171.

L. E. Murr and S. H. Wang, "Comparison of Microstructural Evolution Associated with the Stress-Strain Diagrams for Nickel and 304 Stainless-Steel–An Electron-Microscope Study of Micro-Yielding and Plastic-Flow," *Res. Mech.*, **4** (1982), 237–274.

T. G. Nieh and J. G. Wang, "Hall–Petch Relation in Nanocrystalline Ni and Be-B Alloys," *Intermetallics*, **13** (2005), 377–385.

V. Y. Novikov, "MiIcrostructure stabilization in Bulk Nanocrystalline Materials: Analytical Approach and Numerical Modeling," *Mater. Lett.*, **62** (2008), 3748–3750.

S. Pande and K. P. Cooper, "Nanomechanics of Hall–Petch Relationship in Nanocrystalline Materials," *Prog. Mater. Sci.*, **54** (2009), 689–706.

N. J. Petch, "The Cleavage Strength of Polycrystals," *J. Iron Steel Inst.*, **174** (1953), 25–28.

E. Rabkin, L. Klinger, T. Izumova, and V. N. Semenov, "Diffusion-Induced Grain-boundary Porosity in NiAl," *Scripta Mater.*, **42** (2000), 1031–1037.

P. Rodriguez and R. W. Armstrong, "Strength and Strain-rate Sensitivity for hcp and fcc Nanopolycrystal Metals," *Bull. Mater. Sci.*, **29** (2006), 717–720.

A. Schuh, T. G. Nieh, and T. Yamasaki, "Hall–Petch Breakdown Manifested in Abrasive Wear Resistance of Nanocrystalline Nickel," *Scripta Mater.*, **46** (2002), 735–740.

A. Schuh, T. G. Nieh, and H. Imasaki, "The Effect of Solid Solution W Additions on the Mechanical Properties of Nanocrystalline Ni," *Acta Mater.*, **51** (2003), 431–443.

Z. W. Shan, J. M. K. Wiezorek, E. A. Stach, D. M. Follstaedt, J. A. Knapp, and S. X. Mao, "Dislocation Dynamics in Nanocrystalline Nickel," *Phy. Rev. Lett.*, **98** (2007), 095502.

T. D. Shen, R. B. Schwarz, S. Feng, J. G. Swadener, J. Y. Huang, M. Tang, Jianzhong Zhang, S. C. Vogel, and Yusheng Zhao, "Effect of Solute Segregation on the Strength of Nanocrystalline Alloys: Inverse Hall–Petch Relation," *Acta Mater.*, **55** (2007), 5007–5013.

T. J. Stieltjes, "Sur Certains Polynomes qui Verifient une Equation Differentielle Lineaire du Second Odre et sur la Theorie des Functions de Lame," *Acta Math*, **6** (1885), 321–326.

N. Stroh, "A Theoretical Calculation of Stored Energy in a Work Hardened Material," *Proc. Roy. Soc.*, **A218** (1953), 391–400.

A. N. Stroh, "The Formation of Cracks as a Result of Plastic Flow," Proc. Roy. Soc., **A223** (1954), 404–414.

G. Szego, "Orthogonal Polynomials," *Am. Math. Soc. Colloquium* Publications, **23** (1948).

R. Z. Valiev, I. V. Alexandrov, Y. T. Zhu, and T. C. Lowe, "Paradox of Strength and Ductility in Metals Processed by Severe Plastic Deformation," *J. Mat. Res.*, **17** (2002), 5–8.

R. Z. Valiev, Y. Estrin, Z. Horita, T. G. Langdon, M. J. Zehetbauer, and Y. T. Zhu, "Producing Bulk Ultrafine-Grained Materials by Severe Plastic Deformation," *J. Metals*, **58** (2006), 33–39.

H. Van Swygenhoven, M. Spaczer, and A. Caro, "Microscopic Description of Plasticity in Computer Generated Metallic Nanophase Samples: A Comparison between Cu and Ni," *Acta Mater.*, **47** (1999), 3117–3126.

H. Van Swygenhoven and P. M. Derlet, "Grain-Boundary Sliding in Nanocrystalline fcc Metals," *Phys. Rev. B*, **64** (2001), 224105.

K. J. Van Vliet, S. Tsikata, and S. Suresh, "Model Experiments for Direct Visualization of Grain-boundary Deformation in Nanocrystalline Metals," *Appl. Phys. Lett.*, **83** (2003), 1441–1443.

S. Venkatesh and L. E. Murr, "Variations in Grain-boundary Ledge Structure with Thermomechanical Treatment in High-Purity Aluminum," *Scripta Metall.*, **10** (1976), 477–480.

S. Venkatesh and L. E. Murr, in reprint from *35th Annual Proceedings of Electron Microscopy Society of America* (Ed. G. W. Bailey, 1977), p. 2840.

E. S. Venkatesh and L. E. Murr, "The Influence of Grain-boundary Ledge Density on the Flow Stress in Nickel," *Mater. Sci. Eng.*, **33** (1978), 69–80.

Y. M. Wang, K. Wang, D. Pan, K. Lu, K. J. Hemker, and E. Ma, "Microsample Tensile Testing of Nanocrystalline Copper," *Scripta Mater.*, **48** (2003), 1581–1586.

I. M. Ward, *Mechanical Properties of Solid Materials* (Wiley, 1985), p. 150.

M. S. Wu, and Y. Yu, "Analysis of Cracks Nucleated by Dislocation Pileups against Nonequilibrium Grain Boundaries," *Mech. Mater.*, **32** (2000), 511–529.

X. L. Wu and Y. T. Zhu, "Partial Dislocation Mediated Processes in Nanocrystalline Ni with Nonequilibrium Grain Boundaries," *Appl. Phys. Lett.*, **89** (2006), 031922.

X. L. Wu, Y. T. Zhu, Y. G. Wei, and Q. Wei, "Strong Strain Hardening in Nanocrystalline Nickel," *Phys. Rev. Lett.*, **103** (2009), 205504.

W. Xu, X. Wu, D. Sadedin, G. Wellwood, and K. Xia, "Ultrafine- Grained Titanium of High Interstitial Content with a Good Combination of Strength and Ductility," *Appl. Phys. Lett.*, **92** (2008), 011924.

K. M. Youssef, R. O. Scattergood, K. L. Murty, and C. C. Koch, "Ultratough Nanocrystalline Copper with a Narrow Grain Size Distribution," *Appl. Phys. Lett.*, **85** (2004), 929–931.

K. M. Youssef, R. O. Scattergood, K. L. Murty, J. A. Horton, and C. C. Koch, "Ultrahigh

Strength and High Ductility of Bulk Nanocrystalline Copper," *Appl. Phys. Lett.*, **87** (2005), 091904.

K. M. Youssef, R. O. Scattergood, K. L. Murty, and C. C. Koch, "Nanocrystalline Al-Mg Alloy with Ultrahigh Strength and Good Ductility," *Scripta Mater.*, **54** (2006), 251–256.

K. Zhang, J. R. Weertman, and J. A. Eastman, "The Influence of Time, Temperature and Grain Size on Indentation Creep in High Purity Nanocrystalline and Ultrafine Grain Copper," *Appl. Phys. Lett.*, **85** (2004), 5197–5199.

K. Zhang, J. R. Weertman, and J. A. Eastman, "Rapid Stress-Driven Grain Coarsening in Nanocrystalline Cu at Ambient and Cryogenic Temperatures," *Appl. Phys. Lett.*, **87** (2005), 061921.

M. Zhao, J. C. Li, and Q. Jiang, "Hall–Petch Relationship in Nanometer Size Range," *J. Alloys Compd.*, **361** (2003), 160–164.

J. Zhou, N. Xu, R. Zhu, Z. Zhang, T. He, and L. Cheng, "A Polycrystalline Mechanical Model for Bulk Nanocrystalline Materials Using the Energy Approach," *J. Mater. Process. Tech.*, **209** (2009), 5407–5416.

Chapter 8

DYNAMIC MECHANICAL BEHAVIOR OF ULTRAFINE AND NANOCRYSTALLINE REFRACTORY METALS

Qiuming Wei[a] and Laszlo J. Kecskes[b]

[a] *Department of Mechanical Engineering and Engineering Science,*
  University of North Carolina at Charlotte, 9201 University City Blvd.,
  Charlotte, NC 28223, USA
  E-mail: qwei@uncc.edu
[b] *Weapons and Materials Research Directorate,*
  AMSRD-ARL-WM-MB, U. S. Army Research Laboratory,
  Aberdeen Proving Ground, Aberdeen, MD 21005-5069, USA
  E-mail: kecskes@arl.army.mil

Mechanical behavior under extreme conditions such as high rate of loading (or dynamic strain-rate $\dot{\varepsilon} > 10^2$ s^{-1}) of ultrafine-grained (UFG) (grain size, $d < 500$ nm but > 100 nm) and nanocrystalline (NC) ($d < 100$ nm) metals are of paramount importance for a number of applications. In this chapter, we present an overview of studies of the dynamic mechanical behavior of some refractory body-centered cubic (BCC) metals with UFG/NC microstructures which were produced by either bottom-up or top-down methods. The dynamic compressive behavior was evaluated by means of the Kolsky bar (or Split-Hopkinson Pressure Bar [SHPB]) technique. Experimental results indicate that the majority of the UFG/NC BCC refractory metals are prone to localized plastic flow under dynamic compression, with UFG-Ta being an exception. This tendency is rationalized as a consequence of the combined effect of reduced strain hardening and strain-rate sensitivity (SRS), increased strength, and enhanced adiabatic heating in such metals.

Mechanical Properties of Nanocrystalline Materials
Edited by James C. M. Li
Copyright © 2011 Pan Stanford Publishing Pte. Ltd.
www.panstanford.com

8.1 INTRODUCTION

Improving the strength of structural materials has long been the "holy grail" for many applications. Toward this end, a number of avenues are at the disposal of materials engineers [1]. For example, metals and alloys can be made strong by second-phase precipitation, by alloying, by plastic deformation (work hardening), by reduction of the grain-size, and so forth. Among these strengthening mechanisms, grain size reduction has stood out to be the only one that does not necessarily lead to a concurrent decrease in toughness. What is more, recent investigations have brought forth great hope that high strength accompanied by good ductility in "microstructure-engineered" materials is no longer an elusive dream [2–10]. Furthermore, metals and alloys with UFG and NC microstructures exhibit a number of interesting properties [11–13]. Among the attractive properties, the most well known is the Hall–Petch [14, 15] effect that relates the yield strength of metals and alloys to their grain sizes, wherein the yield strength, σ_y, is proportional to the inverse square root of d, such that

$$\sigma_y = \sigma_0 + k_{H-P}\, d^{-1/2}, \tag{8.1}$$

In Eq. 8.1, σ_0 corresponds to the yield strength of a single crystal (without grain-boundary [GB] effects) and k_{H-P} is the Hall–Petch coefficient. Recent investigations have reported the breakdown of Eq. 8.1 [16, 17], which is manifested in a leveling-off of the effect of d, when d is very small (e.g., $d < 50$ nm); this corresponds to a decrease in yield strength with decreased d, especially at the lower bound of the NC regime. Such effects have been referred to as the inverse Hall–Petch relation and have been substantiated via experimental observations and molecular dynamics simulations [17–19]. However, early reports of the inverse Hall–Petch effect have been proven to be a result of processing artifacts such as poor interparticle bonding, porosity, contamination, impurities, etc., common in NC metals produced by most bottom-up or two-step processing routes [17].

The techniques used to produce UFG/NC materials have been reviewed by a number of experts [20–27]. Generally speaking, fabrication routes to produce UFG/NC metals can be classified into two distinct categories [28], namely, the top-down and the bottom-up methods. In the former, one starts with a coarse-grained (CG) bulk metal and gradually refines d into the UFG/NC regime. Within this category, severe plastic deformation (SPD) with its many variants is the best-known example. It can even be said that SPD is the only or exclusive top-down method. In the latter category, one starts with production of NC powders, followed by compaction of the powders via hot pressing, hot isostatic pressing (HIPing), etc. Accordingly, many bottom-up techniques are also two step by nature. Exceptions include electrodeposition (ED) or vapor deposition in a vacuum chamber.

A survey of literature leads to at least two observations. The first is that the majority of published investigations, both experimental and theoretical, have mostly concentrated on metals with face-centered cubic (FCC) lattice structures. In this regard, Cu and Ni have received the most attention due to the ease in making NC structures of these metals [11]. In contrast, much fewer publications can be found on processing and properties of UFG/NC metals with BCC lattice structures. Only a few reports are available on processing and testing of UFG/NC Fe before 2004 [29–32]. The primary reason is believed to be the difficulty in producing fully dense BCC metals free of processing artifacts. This is because most representative BCC metals are refractory metals, which by nature are highly susceptible to interstitial impurities, translating into having relatively high ductile-to-brittle-transition temperatures (DBTT). As a consequence, neither bottom-up nor top-down techniques can be easily employed to produce UFG/NC microstructures of such metals. The second observation is that studies of mechanical properties of UFG/NC metals have primarily focused on quasi-static behavior, viz., the loading rate is quite low. Behavior of such metals under extreme conditions, such as at high temperature, high-rate (dynamic) loading, or the combined effect of both, is not well investigated. It is known that the dynamic mechanical properties of materials are important in many applications such as high-speed machining, vehicular crash and impact, detonation of explosives, fragmentation and fracture, etc. [33]. Consequently, more research efforts are needed for a better understanding of the dynamic behavior of UFG/NC materials.

Recent work has uncovered that, in addition to the well-known Hall–Petch relation (or its inverse), metals with UFG/NC structures exhibit some other unique behavior [34, 35] associated with the strain-rate dependence of their mechanical properties. For example, FCC metals have been found to show increased SRS when d is refined (particularly into the UFG/NC regime) [3, 4, 7, 36, 37], whereas BCC metals show an opposing trend [34]. Such behavior will certainly impart a strong influence on the constitutive response of UFG/NC metals [34].

In the past few years, the authors of this chapter have investigated the dynamic mechanical properties of a number of UFG/NC BCC metals, including Fe, Ta, W, and V. Both bottom-up and top-down methods have been employed to produce the UFG/NC BCC metals. The microstructural characteristics, such as grain morphology and size, structure of GBs, etc., have been determined via transmission electron microscopy (TEM) and X-ray diffraction (XRD) analysis.

The intent of this chapter is to provide a relatively comprehensive overview of what has been accomplished so far in understanding the dynamic mechanical behavior of a few UFG/NC metals with BCC structure, most of them being refractory metals (a commonly accepted definition of a refractory metal is that the metal's melting point should be no less than that of Cr, that is, 1,907°C or 2,180 K [38]). We have included Fe because Fe is also a typical BCC metal and its melting point is not far away from that of Cr. Occasionally the quasi-static stress–strain curves are given as a reference for strain-hardening information only. Uniaxial dynamic mechanical properties, including plastic deformation and fracture (at strain rates $\sim 10^3 \, \text{s}^{-1}$), will be the focus of this chapter. Generally speaking, the Kolsky bar (or SHPB) technique has been used to perform dynamic loading experiments. The experimental results will be analyzed in connection with a mechanics-based model.

8.2 MATERIALS PROCESSING AND MECHANICAL TESTING

The dynamic mechanical behavior of the following BCC metals has been investigated: Fe, Ta, W, and V. Among them, Ta and V belong to group VB, W to group VIB, and Fe to group VII, respectively, in the periodic table. Experimental results of the mechanical properties of UFG/NC Fe, made by the bottom-up method, have been reported by several groups [29–32, 39], including its quasi-static and dynamic mechanical behavior. Under quasi-static compression, consolidated UFG/NC Fe deforms via localized shear banding [31, 32, 39], and the shear banding process is believed to be nonadiabatic due to the slow loading rate and, as seen by TEM [32], to the observed strong remaining texture within the shear band (SB). In this chapter, we will focus on SPD-processed Fe with UFG microstructure. We also used SPD to obtain UFG/NC Ta and W. We used equal-channel angular extrusion (ECAE) to obtain the UFG microstructure and high-pressure torsion (HPT) to obtain the NC microstructure, respectively. Detailed information on ECAE and HPT can be found in Valiev *et al.* [24] and Valiev and Langdon [25]. All SPD processing was performed at room temperature for Fe and Ta. Four passes of route-E (4E) ECAE was imposed on Fe. Further grain refinement and strengthening of Fe were conducted by low-temperature rolling. For Ta, the maximum number of ECAE passes was 16, using route B_c ($16B_c$). For the ECAE of W, instead of a 90° die angle, a 120° die angle was used to mitigate the poor workability of CGW. ECAE was performed between 1,000°C and 1,200°C (the recrystallization temperature of moderately deformed W is $\sim 1,250$°C [40]). To further refine the W grain size, the ECAE-processed W was confined (to avoid cracking) and rolled at successively lower temperatures; the lowest rolling temperature was 600°C. Details of the procedure to make UFG W can be found in Wei *et al.* [41]. For the HPT of W, electropolished W disks of initial dimensions of 9 mm diameter and 2 mm thickness were upset by 30% strain at 600°C under a protective Ar gas cover. Five turns of HPT were then performed on the W disks using Bridgman anvils preheated to 500°C under a pressure of 4 GPa. Details of the HPT procedure for NC W can be found in Wei *et al.* [42].

Microstructures of the processed metals were examined using TEM and XRD. Tungsten TEM specimens were prepared by mechanical thinning and dimpling, followed by dual ion-milling for electron transparency. In contrast, a double-jet electropolishing unit was used to prepare TEM specimens of Fe, while the Ta TEM specimens were prepared by a chemical immersion method [43]. We used a Philips EM-420 microscope for conventional microstructural analysis and used a Philips CM-300 for high-resolution TEM. X-ray line-broadening analysis was used to measure the average grain size (or more appropriately, the coherent domain size) of the materials [44, 45].

Quasi-static compression tests at various strain rates were performed on an MTS hydro-servo system with strain rates controlled by adjusting the cross-head speed. The main purpose of the quasi-static compression experiments was to evaluate the isothermal strain-hardening behavior of the materials. Uniaxial dynamic compression (strain-rate $\sim 10^3 \, \text{s}^{-1}$) was conducted using a Kolsky bar system. The technique is based on well-articulated one-dimensional stress wave theory [46]. In this test, the specimen is sandwiched between two elastic bars (called the incident and transmitter [input and output] bars, respectively, both made of high-strength maraging steel). Strain gages are cemented on the elastic bars to measure (i) the incident stress wave pulse generated by an impacting projectile bar, (ii) the reflected pulse from the incident bar/specimen interface, and (iii) the pulse transmitted through the specimen to the transmitter bar. Detailed description of the Kolsky bar technique can be found elsewhere [47]. Dynamic behavior of small specimens (with a dimension less than 1.0 mm) were evaluated using a desktop Kolsky bar (DTKB) system [48]. Figure 8.1 displays a schematic of the DTKB system.

In some cases, particularly in the testing of UFG/NC W, since the specimens might be stronger than the elastic bars, in order to protect the elastic bars from plastic deformation, the specimen is first sandwiched between two tungsten carbide (WC) platens shrink-fitted into Ti-6Al-4V collars (the outside diameter of the collar is the same as the diameter of the elastic bars). The mechanical impedance of the WC platens is carefully matched to that of the maraging steel bars. Standard precautions were practiced during specimen preparations. Specifically, prior to dynamic loading, side faces of the dynamic specimens were polished to a mirror finish either for high-speed photography or for postmortem examinations with optical microscopy or scanning electron microscopy (SEM). It should be noted here that the as-received microstructures of the dynamic specimens and the quasi-static specimens are the same for a specific metal.

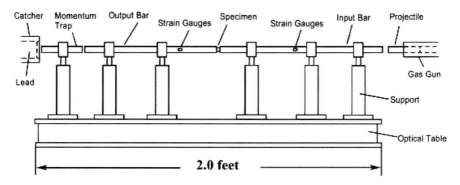

Figure 8.1 Schematic of the DTKB system. The bars and their support structure are seated on a precision optical table (~600 mm in length). The bars are made of high-strength maraging steel. The diameter of the bars is ~3.18 mm. The specimen is sandwiched between the input and output bars. Due to the unusual high strength of the HPT W, we used impedance-matched tungsten carbide (WC) platens, shrink-fitted into Ti–6Al–4V collars, to protect the input and output bars. Thus the specimen is first sandwiched between the WC platens and then mounted between the two bars.

8.3 DYNAMIC MECHANICAL BEHAVIOR OF UFG/NC BCC METALS

We first present the microstructures of some UFG/NC BCC metals prior to any mechanical testing, followed by a brief summary of the quasi-static mechanical properties. In turn, we will then focus on the dynamic behavior of the UFG/NC BCC metals.

Figure 8.2a displays a bright-field TEM micrograph of Fe processed by ECAE (4E). It shows that the grain size (d) has been refined into the UFG regime ($d < 500$ nm). Selected area diffraction (SAD) (see inset in Fig. 8.2a) exhibits discontinuous rings and concentrated reflections, suggesting the presence of small-angle GBs. Dislocation debris [4] can be identified within some grains. Further low-temperature rolling of the ECAE-processed Fe leads to smaller average grain size, but the grain interiors show dislocation densities higher than the as-ECAE Fe [49]. It has been reported that the yield stress and average grain size of Fe saturates rapidly after the first ECAE pass [50]. However, further SPD will render more low-angle GBs to evolve into high-angle ones [24], thus leading to a smaller effective grain size.

Figure 8.2b displays a bright-field TEM micrograph of Ta after 16 passes of ECAE using the B_c route. The SAD pattern (not shown) consists of almost continuous rings, indicating that a large number of the GBs are the high-angle type. Figure 8.2c provides the grain size distribution of this material. It shows that a significant number of grains have a size below 250 nm. Work by Wei *et al.* [43] on Ta after 4E ECAE passes revealed an average apparent grain size close to that of the 16B_c Ta. However, in the 4E Ta most of the grains appeared elongated and had strong texture [43].

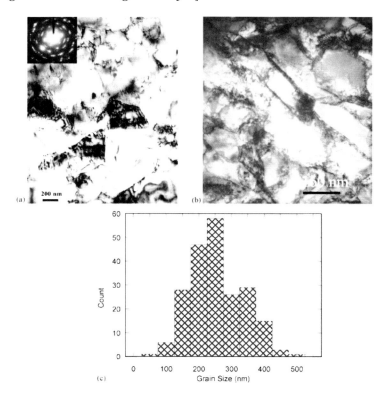

Figure 8.2 (a) TEM image and SAD pattern (inset in [a]) of ECAE-processed Fe (via 4E); (b) TEM image of ECAE-processed Ta (via 16B_c); and (c) grain size distribution of Ta after 16B_c (obtained from TEM images). All evidence shows that the microstructure is in the UFG regime. Images are taken from Ref. 51.

On the basis of TEM observations, the grain size of ECAE W (processed at 1,000°C followed by low-temperature rolling) is within the UFG regime [41]. Figure 8.3a shows a bright-field TEM image of W processed by HPT at 500°C; Fig. 8.3b is the accompanying dark-field image. These images suggest that the HPT W has elongated grains, and the grains are heavily dislocated. However, the SAD pattern presented in Fig. 8.3c shows nearly continuous rings, implying that the GBs are mostly of high angle and the global microstructure is texture free. Grain breakup within the large elongated grains is also observed. Quantitative measurement shows that the grain size of HPT W is around 170 nm, counting the elongated grains [42]. The profuse presence of breakups apparently suggests that the effective grain size might be much smaller. A high density of dislocations, some of them of edge type, were revealed by high-resolution TEM [42].

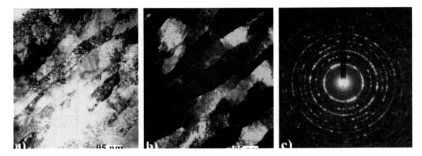

Figure 8.3 Bright-field (a), dark-field (b) images, and SAD pattern (c) of HPT W. Note the high density of defects in the grains and the breakup of the elongated grains. The SAD pattern shows nearly continuous rings, with no obvious intensity concentration along the rings, implying large-angle GBs. The HPT nominal equivalent strain in the specimen is around 70. Images are taken from Ref. 42.

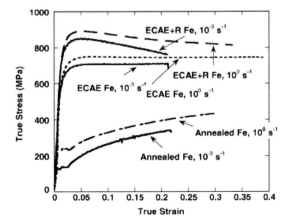

Figure 8.4 Quasi-static stress–strain curves under compressive loading for Fe in the annealed, ECAE-processed, and ECAE+R conditions. Note the significant strain hardening in the annealed Fe, which is absent in the SPD-processed Fe. Also note the apparently reduced rate dependence in the SPD Fe. The strain rates are given in the figure. See text for the explanation of the slight flow softening in the ECAE+R specimens. This plot is taken from Ref. 49.

Figure 8.4 displays representative quasi-static stress–strain curves of Fe in the annealed condition and after various grain-refining processing steps. In this plot, R stands for "rolling."

Apparently, ECAE has elevated the strength of Fe significantly. However, it is obvious that after SPD, the material exhibits an elastic–perfectly plastic behavior. After further low-temperature rolling, the specimen even shows slight flow softening when tested at room temperature, presumably due to dynamic recovery [49].

Figure 8.5 shows the dynamic stress–strain curves of Fe in the annealed condition as well as after various SPD-processing routes. Similar to the quasi-static curves, annealed Fe exhibits strain hardening but only to a strain of ~0.2. The curve then flattens due to the adiabatic heating within the specimen, typical of dynamic stress–strain curves of metals [52]. Again, the ECAE Fe exhibits elastic–perfectly plastic behavior or slight flow softening. Further rolling of the ECAE material renders strong dynamic flow softening, manifested by the stress drop in the stress–strain curves.

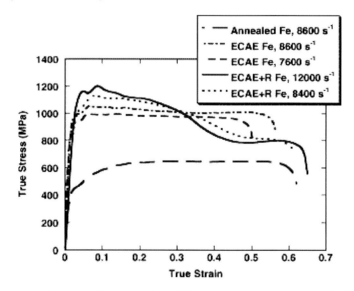

Figure 8.5 Dynamic stress–strain curves from Kolsky bar compression tests for Fe in the annealed, ECAE, and ECAE+R conditions. Note the strain hardening in the annealed Fe to a strain of 0.2. The strain hardening is absent in the SPD-processed samples. On the contrary, strong flow softening is observed in these samples, in particular in the ECAE+R state. The strain rates are given in the figure. This plot is taken from Ref. 49.

Figure 8.6 displays the postdynamic loading SEM image of the ECAE Fe [49]. It is interesting to note the localized shearing marks, indicative that the specimen was not deforming in a uniform manner. Instead, plastic instability had kicked in, whereby SBs formed, which account for the slight flow softening seen in the dynamic stress–strain curves (see Fig. 8.5).

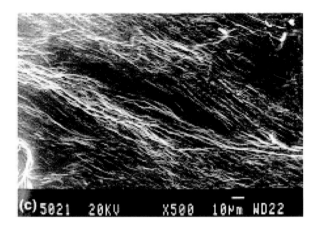

Figure 8.6 Postdynamic loading SEM image of ECAE Fe showing evidence (the lines of bright contrast) of localized shearing, or diffuse SBs, that explain the slight flow softening seen in the dynamic stress–strain curve (Fig. 8.5). This image is taken from Ref. 49.

Further cold rolling of the ECAE material has resulted in strong, dynamic flow softening. This is due to the severe plastic instability under dynamic loading, as illustrated by the postdynamic loading SEM image (see Fig. 8.7).

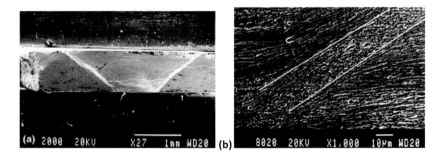

Figure 8.7 (a) Postdynamic loading SEM image of ECAE+R Fe showing strong evidence of SBs and (b) SEM image of an SB of (a) after polishing and chemical etching with a width of about 15 μm. The boundaries of the adiabatic SB have been marked by a pair of white lines for ease of identification. Flow lines can be clearly observed running in opposite directions from the band center. Images are taken from Ref. 49.

Figure 8.7a shows strong evidence of conjugate SBs that develop under dynamic compression in the further-rolled ECAE Fe. Close examinations of the SBs [49] indicate that severe localized flow accompanied by great adiabatic temperature rise had taken place within the SBs. If the specimen is polished and etched via standard metallographic procedure, the detailed microstructure of the SB can be revealed, as displayed in Fig. 8.7b, where two white lines are drawn to mark the width of the SB. Flow lines (canonical lines according to Wright [33]) along the two sides, outside the bounded region, are observed running in opposite directions, typical of adiabatic SBs observed in metals [33]. We envision that within the SB, very high temperatures have been reached during the dynamic loading process, causing dynamic recrystallization at least.

Figure 8.8 shows the TEM image and the accompanied SAD pattern of consolidated fully dense NC V [53]. The grain size is ~100 nm.

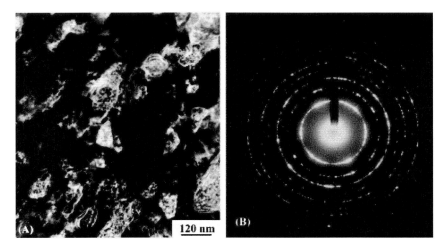

Figure 8.8 (a) TEM image and (b) accompanied SAD pattern of consolidated NC V from Ref. 53. The measured average grain size is ~100 nm.

Figure 8.9 presents the compressive quasi-static and dynamic stress–strain curves of the same consolidated NC V. It is observed that the strength of such NC V is very high. However, even under uniaxial compression, not much plastic deformation can be observed, indicating that the consolidated NC V is quite brittle. In fact, after dynamic loading, the specimen was broken into several pieces, though not pulverized. It is worth noting that under dynamic compression, though brittle, the consolidated NC V fails in a manner similar to that of bulk metallic glasses [54], as revealed by high-speed photography (see Fig. 8.10), viz., this NC V fails by nucleation and growth of SBs under dynamic compression.

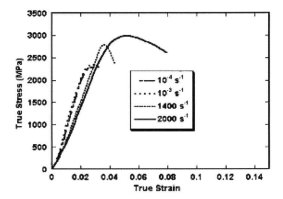

Figure 8.9 Compressive quasi-static and dynamic stress–strain curves of consolidated NC V. The strain rates given in the plot are taken from Ref. 53.

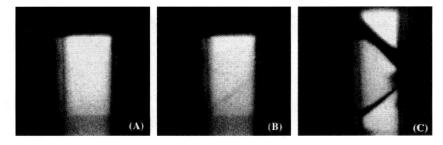

Figure 8.10 High-speed photographs taken during a dynamic compression experiment of consolidated NC V, showing (B) initiation of an SB and final failure of the specimen into three pieces via the shear-banding process, reminiscent of the behavior of bulk metallic glasses. Images are taken from Ref. 53.

The quasi-static and dynamic behavior of polycrystalline W have been reported by a number of investigators [55, 56]. Generally speaking, due to its strong susceptibility to intergranular failure from the weakening effect of interstitial impurities segregated along the GBs, polycrystalline W exhibits brittle fracture at room temperature [40]. It has also been noticed that there is strong compression/tension asymmetry in polycrystalline W. As such, it can be regarded as a ceramic-like material. However, rolling and other warm working can significantly enhance the tensile ductility of W [40, 57].

Figure 8.11 displays the quasi-static and dynamic compressive stress–strain curves of W after various SPD treatments [41]. Similar to Fe, ECAE (or ECAP) has almost exhausted the strain-hardening capacity of W, and the quasi-static stress–strain curves indicate elastic–perfectly plastic behavior. ECAE W also exhibits flow softening under dynamic loading. But further (warm) rolling has resulted in severe dynamic flow softening.

Postdynamic loading metallography of the UFG W reveals strong evidence for adiabatic shear banding. Figure 8.12 displays a series of postmortem optical and SEM images of UFG W. "Figure 8.12a shows of the side faces". The loading direction is vertical. Instead of cracks commonly observed in dynamically loaded polycrystalline W, obvious SBs are present. High-speed photography (shown later) indicates that the stress collapse in the stress–strain plot (Fig. 8.11) corresponds to the initiation and subsequent rapid growth of SBs. Figure 8.12b is an SEM micrograph, indicating that SBs have been developed on two conjugated faces (the bright contrast in the micrograph indicated by arrows corresponds to SBs). Figure 8.12c is an enlarged image of two SBs corresponding to the left branch of (a). Shear localization is evident, with the band oriented at an angle of ~40° with respect to the loading direction. At high magnifications (d–g), microstructural details in the SB are observed, that is, very large localized shear flow. In Fig. 8.12g and later in Fig. 8.13, one can see a crack as the result of the localized adiabatic shearing.

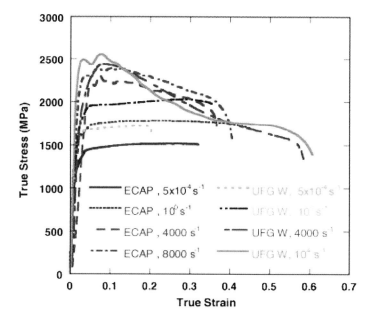

Figure 8.11 True stress–strain curves of the ECAP W and the ECAP + CR (UFG) W, under uniaxial quasi-static, and dynamic (Kolsky bar) compressive loading. All the samples are rectangular with square loading faces. The quasi-static tests used strain rates $\sim 10^{-4}$–10^{0} s^{-1} (MTS hydro-servo system). For the dynamic testing, the strain-rate was determined by the projectile mass and the pressure, as well as the prescribed strain level. The strain-rate of each curve is given in the plot. The plot is taken from Ref. 41.

More details of the microstructure of the SBs in UFG W can be revealed by polishing its surface followed by chemical etching. Figure 8.13 provides an example. Extensive plastic shear has been developed, presumably leading to subsequent cracking. The SB is ~40 μm wide. A simple estimate can be obtained for the maximum shear strain experienced at the center of the adiabatic shear band (ASB) using the slope of the flow lines. As such, a shear strain of the order of ~3.0 can be estimated. This information can be used to estimate the temperature rise within the SB using the Taylor–Quinney formula (given later in this chapter). Unlike the SBs in SPD Fe (described earlier in this chapter), these SBs do not show a well-defined central region that etches differently. This suggests that the temperature rise might not be sufficient to cause extensive recrystallization in the center of the SB. The flow lines suggest the canonical structure predicted for the ASB of metals [33], where the flow lines bend down through the boundary into the band and then curve away on the other side forming an antisymmetric pattern.

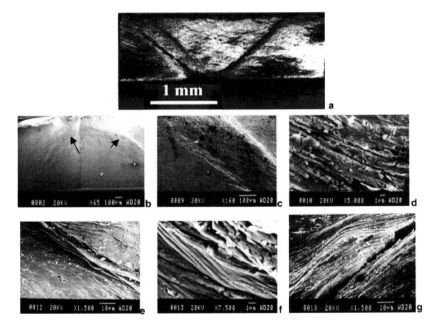

Figure 8.12 (a) Postmortem optical micrograph of ASB in dynamically loaded UFG W, (b) SEM micrograph showing ASBs on two conjugated faces of the sample after dynamic loading, and (c) enlarged image of two SBs corresponding to the left branch of (a). Shear localization is evident in the images. Image in (d) shows the details of the microstructure in the SB. Notice the voids and large, localized plastic deformation. Shown in images (e) and (f) are structures as a result from the very large localized shear flow. In (g), one can see a crack as the result of the localized adiabatic shearing. Images are taken from Ref. 41.

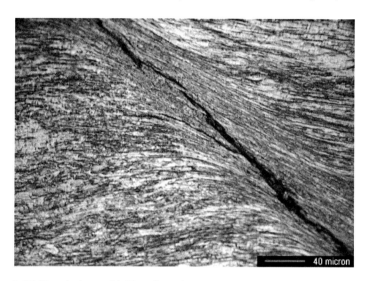

Figure 8.13 Morphology of ASB and crack as revealed by chemical etching. Flow lines are clearly observed. The SB width is ~40 μm. The image is taken from Ref. 41.

The initiation and evolution of ASBs in UFG W can be examined by using high-speed photography synchronized with the Kolsky bar system. Figure 8.14 displays seven frames from one such experiment and the corresponding stress–time/–strain curves. Squares on the curves give the times or strains at which each frame was taken. The first frame, (c), was taken before plastic deformation. The bright areas to the left and right of the specimen are reflections of the specimen in the polished ends of the bars. Compression occurs along the horizontal axis, with the incident wave arriving from the left. Frame (d) is taken at ~20 μs after frame (c) (strain ~0.1). The dark contrast at the upper-left corner indicates initiation of the SB. Time difference between subsequent consecutive frames is 4 μs. In frame (e) (strain ~0.15), the SB is obvious as is the overall compression of the sample; frame (f) shows the initiation of a second SB from the upper-right corner. Later frames (g–i) show the evolution of the SBs, and by frame (i) well-defined ASBs can be identified [41].

It turns out that further grain-size reduction using HPT enhances the tendency of W to adiabatic flow localization under dynamic compression. In the dynamic stress–strain curves, such behavior manifests itself as precipitous stress collapse, as shown in Fig. 8.15 [42]. In this plot, the dynamic stress–strain curves were measured using a DTKB system due to the small size of the specimens. Compared with UFG W (Fig. 8.11), the NC W (produced by HPT) shows much higher dynamic yield strength but also much stronger flow softening and much earlier stress collapse.

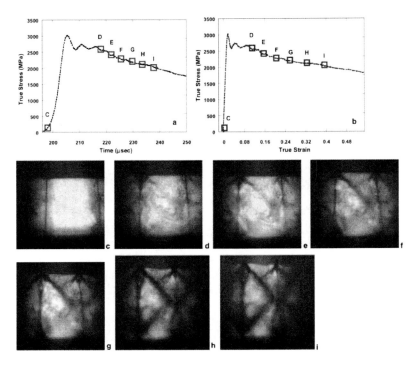

Figure 8.14 (a) Stress–time and (b) stress–strain curves and (c–i) high-speed photographs recorded in a dynamic test of UFG W. Squares (a and b) indicate the time or strain level at which each frame was captured. Letters above the squares correspond to frames (c–i). The loading wave came from the left horizontally. The final rolling temperature for this sample is 700°C. Images are taken from Ref. 41.

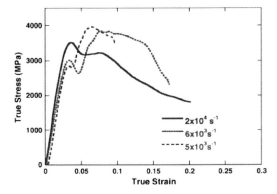

Figure 8.15 Uniaxial dynamic stress–strain curves of HPT W from DTKB testing. Note the sharp stress collapse after dynamic yielding. Such stress collapse occurs much earlier than in UFG W (compare Fig. 8.11). This plot is taken from Ref. 42.

Again, we rely on the postmortem morphology of the specimen for detailed information relevant to the dynamic stress–strain behaviors. Figure 8.16 displays the optical images of two dynamically loaded HPT W specimens [42]. Apparent evidence for adiabatic shear banding is observed running at ~45° with respect to the loading direction (the loading wave comes horizontally). In Fig. 8.16b, a couple of SBs have formed, intersecting each other at roughly 45°. However, compared with UFG W, the SBs are much narrower, and no obvious severe plastic flow can be observed from the optical images. In Fig. 8.16a, a dominant SB runs through the specimen accompanied by subsequent cracking. In Fig. 8.16b, no obvious cracking of the specimen can be observed.

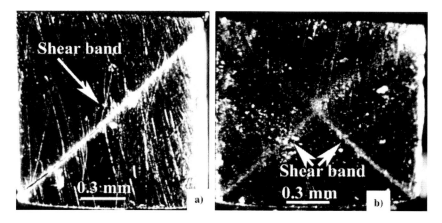

Figure 8.16 Postmortem optical images of dynamically compressed HPT W. Localized shearing resulted in the formation of a major SB and subsequent cracking of the dynamically loaded HPT W (a), and formation of two intersecting SBs (b). The SBs are at an angle of about 45° with respect to the loading direction—horizontal in both (a) and (b). The loading direction is along the HPT disk plane normal. Images are taken from Ref. 42.

To further examine the SB, SEM was used to reveal the details of the structure and evolution of the SB of Fig. 8.16. Figure 8.17 is an SEM micrograph of the SB of Fig. 8.16.

Here we observe an SB and the subsequent crack that formed. The severe localized shear flow is indicated by the bending of the preexisting scratches near the crack. (Those scratches were introduced during sample preparation.) The width of the ASB of HPT W is only about 5 μm, much smaller than that of UFG W. (Figure 8.13 displays the microstructure of an ASB of a UFG W. The SB width is ~40 μm.) This means that smaller grain size leads to increased tendency to ASB formation in W produced by top-down methods.

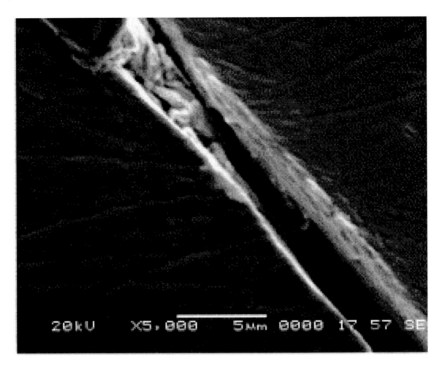

Figure 8.17 SEM image of the SB (and a crack) of the specimen that was shown in Fig. 8.16a. See text for detailed descriptions. The image is taken from Ref. 42.

It is also interesting to note that both UFG and NC W produced by SPD show much improved ductility under both dynamic and quasi-static loading [41, 42]. Reasons for this observation may be threefold. First, SPD created new GBs, which are nonequilibrium and highly energetic, and according to Valiev *et al.*, such GBs can improve ductility of UFG/NC metals [10, 58]. Second, the newly created GBs can share the preexisting interstitial impurities, which in turn entails cleaner GBs in toto and thus strengthens the preexisting GBs of W. Third, SPD generates a high density of dislocations, as shown in Fig. 8.3. Such dislocations can shield the crack tip stress field and thus increase the material's resistance to crack propagation [59, 60].

The last BCC refractory metal that has been investigated in the context of the UFG/NC microstructure and dynamic mechanical behavior is Ta. Experiments show that the yield and flow stresses of Ta never saturate even after 16 ECAE passes and continue to increase with an increased number of passes. The quasi-static stress–strain curves exhibit slight strain hardening in all stress–strain curves. This is exemplified in Fig. 8.18, wherein the stress–strain curve of Ta, even after 16 ECAE passes (route B_c), still shows moderate strain hardening. The grain-size distribution of this sample has been given in Fig. 8.2c. The yield strength of CG, annealed Ta is ~160 MPa [43]. Therefore, ECAE processing has greatly increased the yield strength of Ta, and it has also made the metal exhibit elastic-nearly perfect plastic behavior. However, it is suspected that 24 passes of ECAE can still increase the strength of Ta further. The strong strain-hardening capability of Ta needs further investigation. It is worth noting that among the BCC family, Ta stands out to be the most ductile metal, with a ductile to brittle transition temperature (DBTT) around 10 K [38]. We presume that this strong and lasting strain-hardening capability will also distinguish UFG/NC Ta from other UFG/NC BCC metals in terms of its dynamic deformation and failure mechanism.

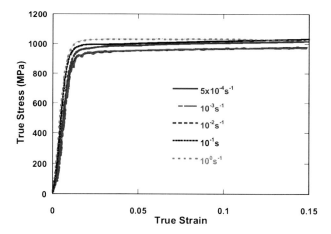

Figure 8.18 Quasi-static compression stress–strain curves of UFG Ta processed by ECAE with 16 passes of route B_c. The strain rates are given in the plot.

Figure 8.19 displays the dynamic stress–strain curves of UFG Ta processed by ECAE with 16 passes of route B_c. Note that in this plot, the stress drop is due to unloading at the prescribed strain of the experiment.

Figure 8.19 suggests that UFG Ta processed by ECAE with 16 passes of route B_c is very different from UFG/NC W or Fe presented earlier in this chapter in terms of its stress–strain behavior. While a slight dynamic flow softening is indeed observed in Fig. 8.19, the flow softening proceeds with strain in a homogeneous manner. That is, the precipitous stress drop seen in UFG/NC Fe and W is absent in Fig. 8.19. This observation is in line with image analysis, as neither *in situ* high-speed photography nor postmortem examination of the dynamically tested specimens reveals evidence for dynamic plastic instability such as ABS. In other words, the dynamic flow softening is merely a consequence of the adiabatic temperature rise during dynamic loading, and the specimen deforms in a homogeneous fashion. It should be pointed out that even after further rolling, UFG/NC Ta is still strongly resistant to ASB formation under dynamic compression.

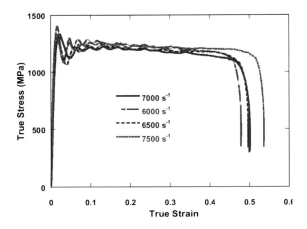

Figure 8.19 Dynamic compressive stress–strain curves of UFG Ta processed by ECAE with 16 passes of route B_c. The strain rates are given in the plot.

Therefore, on the basis of the experimental results, we presume that the major reason for the usually strong resistance of UFG/NC Ta to adiabatic shear localization is the strong and persistent strain-hardening capability of Ta that balances the thermal softening under dynamic compression.

8.4 MECHANISTIC UNDERSTANDING OF THE DYNAMIC MECHANICAL BEHAVIOR OF UFG/NC BCC METALS

In the preceding section, we have presented some representative experimental results on the dynamic mechanical behavior of some BCC metals with UFG/NC microstructures. We observed that UFG/NC Fe and W produced by SPD exhibit dynamic flow softening. This softening is accompanied, or is a consequence of, a transition in the dynamic deformation mode from uniform, stable plastic deformation to unstable, adiabatic shear localization. Adiabatic shear localization is a form of plastic instability where after a certain amount of uniform strain in the metal, ASBs appear and become the predominant deformation mechanism. The plastic strain-rate within the ASBs is very large (usually at least two orders of magnitude larger than the global strain-rate [33]) and is always accompanied by an adiabatic temperature rise. As a result, voids and cracks may start to form within this region [52]. Accordingly, subsequent cracking along the ASBs becomes the predominant dynamic failure mechanism in those UFG/NC BCC metals.

Under uniaxial dynamic compression, the heat generated within the specimen usually does not have sufficient time to be dissipated, and an adiabatic temperature rise within the specimen is anticipated [61]. The prevailing deformation mode of a material under such a loading condition is therefore a result of the competition among several factors. Among these are the strain and strain-rate hardening capability of the material that play the stabilizing roles, whereas thermal and geometrical softening tend to induce plastic instability [61].

The temperature rise during dynamic loading is usually estimated on the basis of an adiabatic assumption. The following equation is often used to compute the adiabatic temperature rise within a specimen in a dynamic loading experiment:

$$T = \frac{\beta}{\rho \cdot C_{\mathrm{p}}} \int_0^{\varepsilon_{\mathrm{f}}} \sigma d\varepsilon \tag{8.2}$$

In Eq. 8.2, β is the Taylor–Quinney coefficient that characterizes the portion of plastic work converted into heat. It is assumed to be 0.9 as is usually the case [61], though experiments have shown that it varies with a number of factors for different metals [62]. The physical meanings of other parameters in Eq. 8.2 are as follows: ρ the density of the material; C_p, the specific heat; σ, the flow stress, and ε_p the prescribed strain for the calculation. Table 8.1 lists the relevant physical properties of W, Fe, and Ta, including the density and the specific heat to be used to calculate the respective adiabatic temperature rise in each during dynamic loading.

Table 8.1 Densities and specific heats of Fe, W, and Ta.

Material	Density (g/cm³)	Specific Heat (Joule/g-K)
Iron (Fe)	7.90	0.440
Tungsten (W)	19.25	0.134
Tantalum (Ta)	16.65	0.140

As UFG/NC metals usually exhibit an exhausted strain-hardening capacity and, consequently, their stress–strain curves show elastic–nearly perfect plastic behavior even under quasi-static loading, we can safely assume that they are elastic–perfectly plastic under isothermal dynamic loading with a strength σ_y. Therefore, the flow stress will be a constant, independent of strain, and Eq. 8.2 can be rewritten as follows:

$$T = \frac{\beta}{\rho \cdot C_{\mathrm{p}}} \cdot \sigma_y \varepsilon_{\mathrm{f}} \tag{8.3}$$

We can now estimate the global adiabatic temperature rise within a specimen at a given plastic strain. For example, for a plastic strain of ~0.2, the overall temperature rise in ECAE W under dynamic loading would be ~140 K only, using a flow stress of ~2.0 GPa (see Fig. 8.11), while for UFG W, it would be ~210 K, using a flow stress of ~3.0 GPa (see Fig. 8.11, but keep in mind the stress–strain curves have been smoothed and the peak stress from the original data is ~3.0 GPa) [41]. In the case of HPT-processed NC W, at the same strain, the overall temperature rise is ~240 K. Comparing the densities and specific heats of W and Ta, it appears that UFG Ta will have a much smaller adiabatic temperature rise than UFG/NC W since the dynamic strength of UFG Ta is roughly one-third that of UFG/NC W. For UFG Fe, a similar estimation came to a value of ~100 K at a prescribed strain of 0.4.

Plastic instability has long been a concern for many reasons. Under tension, a specimen loses its stable, uniform deformation by the emergence and growth of necking [63, 64]. Under compression and torsion, plastic instability manifests itself in the form of shear banding [61]. In either case, strain hardening, strain-rate hardening, geometric factors (stress concentration, geometric constraints, etc.), mode of mechanical loading, thermal environment, etc., play important roles. Considering these factors, Wright [33] derived the following susceptibility parameter for ASB formation in viscoplastic materials such as metals:

$$\frac{\chi_{ASB}}{a/m} = \min\left\{1, \frac{1}{(n/m) + \sqrt{n/m}}\right\} \tag{8.4}$$

In Eq. 8.4, χ_{ASB} is the susceptibility to adiabatic shear banding; a is a dimensionless thermal softening parameter defined as $a = (\partial\sigma/\partial T)/\rho c$ (where σ is the flow stress, T the absolute temperature, ρ the density, and c the specific heat of the metal); n is the strain-hardening exponent as defined in the Hollomon equation ($\sigma = k\varepsilon^n$, σ and ε are true stress and true strain in the plastic portion of the stress–strain curve, respectively); and m is the strain-rate sensitivity (SRS) as defined by the power-law constitutive equation— $(\Sigma/\sigma_0) = (\dot{\varepsilon}/\dot{\varepsilon}_0)^{mn}$, $\dot{\varepsilon}$ is the imposed strain-rate, $\dot{\varepsilon}_0$ is a normalizing strain-rate, and σ_0 is a normalizing stress. For UFG/NC BCC metals, where the strain-hardening exponent (n) is usually vanishingly small (see the various stress–strain plots in this chapter), Eq. 8.4 can be rewritten as

$$\chi_{ASB} = \frac{a}{m} = \frac{\lambda\sigma_s}{\rho cm} \tag{8.5}$$

In Eq. 8.5, $\lambda = -(\lambda/\hat{\sigma}_0)(\partial\sigma/\partial T)$ is the thermal softening parameter, evaluated under isothermal conditions, and σ_s is the yield strength of the metal. If the global temperature rise in the specimen at relatively small dynamic strains is not sufficient to cause recovery, recrystallization and, especially, grain growth and the thermal softening parameter will only be weakly dependent on the microstructure. As such, the temperature dependence of the yield stress will be through that of the elastic modulus as it appears in the critical stress to turn on the Frank–Read sources. The temperature rise may also facilitate plasticity mechanisms that can be thermally activated, such as the Peierls–Nabarro model, forest cutting, and cross slip of dislocations, but these effects would remain equivalent for any microstructure. Thus, we envision that the influence of the microstructures prior to dynamic loading is through their strong effect on the yield strength of the material in consideration. In turn, this will impart a strong effect on the general susceptibility to adiabatic shear banding, as Eq. 8.5 suggests. Another important point is that the initial microstructure will affect the dynamic strain hardening in the material, as well as microstructure evolution within ASBs [65, 66]. We therefore assume that the thermal softening parameter in Eq. 8.5 is only weakly dependent on the microstructure, and ρ and c do not change with grain size; then Eq. 8.5 suggests that the combination of elevated yield strength and reduced SRS results in a greater susceptibility to ASB formation. It has been recognized that UFG/NC metals in general have much higher yield strengths than their CG counterparts. Therefore, they should exhibit increased susceptibility to ASB. However, the change of SRS, if any, with grain size may also be critical. Extensive experimental results have shown that the SRS of FCC metals increases with reduction in grain size [7, 11, 12, 35, 67]. For example, close to an order of magnitude increase of SRS has been reported for UFG/NC Ni, Al, and Cu [67]. The enhanced SRS of UFG/NC FCC metals have been considered the underlying mechanism to improve the tensile ductility of such metals [7]. As Lin *et al.*[64,68] have noticed that even for metals and alloys that have no strain hardening, respectable SRS alone can render considerable tensile ductility since increased SRS helps suppress the growth of plastic instability.

However, for BCC metals, available experimental results consistently show that their SRS is reduced when the grain size is refined into the UFG/NC regime [31, 35, 67]. Figure 8.20 displays a collection of SRS data as a function of grain size for BCC metals. Therefore, from Eq. 8.5, we understand why certain UFG/NC BCC metals exhibit ASBs under dynamic compression.

Another way to evaluate the propensity of a metal for ASB is to estimate the critical shear strain for localization. In the following we will present analysis on UFG W. Again, there are several models based on different assumptions about the constitutive laws of the material. The problem becomes very simple for a material without strain hardening such as is approximately the case in UFG W. Assuming an exponential dependence of the flow stress on temperature and no strain hardening, the constitutive law can be written as

$$\tau = \tau_0 \cdot \exp(-\alpha \cdot T) \cdot \dot{\gamma}^m \qquad (8.6)$$

where τ_0 is a constant, α a positive constant, T the absolute temperature, and m the SRS. The critical shear strain for shear localization can be derived as [69]

$$\gamma_A^c = \frac{\rho C}{\alpha \beta \tau_A^0} \left\{ \left[1 - g^{1/m} \exp(-\alpha(1-m)(T_B^0 - T_A^0)/m) \right]^{-m/(1-m)} - 1 \right\} \qquad (8.7)$$

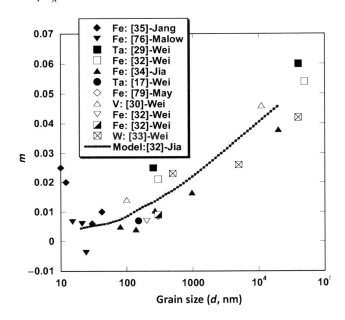

Figure 8.20 SRS of BCC metals. The grain size was refined by different methods, including ball milling followed by consolidation, SPD, ECAE plus low-temperature rolling, etc. The name of the metal is given in the inset, followed by the reference and its first author. The dashed curve is the calculation results based on the model presented in Ref. 31. The plot has been taken from Ref. 34, wherein the references could also be found.

In Eq. 8.7, τ_A^0 is the isothermal shear flow stress of the material. The factor g defines the geometric defect (taken to be unity for these compression tests), and m is the SRS according to its common definition; $(T_B^0 - T_A^0)$ describes the temperature difference between the center of the ASB and the matrix (taken to be ~1,000 K; this can be easily estimated using Eq. 8.3 given previously in this chapter). The experimental results by Argon and Maloof [70] on W at different temperatures can indeed be described very well by Eq. 8.6 with α being 0.003957. Using the above parameters for UFG W the critical shear strain is calculated to be about 0.1. Therefore, the normal critical strain for

ASB will be smaller than 0.1. The high-speed photographs and the associated stress–strain plot of Fig. 8.14 do show that SBs have already occurred at a strain level of 0.1. Keeping in mind the limited spatial resolution of the high-speed camera, it is then reasonable to believe that ASB might have initiated below this strain level.

Careful examination of the quasi-static true stress–true strain curves of the UFG W (see Fig. 8.11) indicates that the plastic region is not absolutely flat: a slight strain hardening is still present. Fitting the plastic part with a power law gives a strain-hardening exponent of ~0.023 (more than one order of magnitude smaller than that of CG W). Assuming that the constitutive behavior of UFG W can be described by the Litonski relation as [33]

$$\tau = A\gamma^n(1+B\dot{\gamma})^m(1-cT)$$
(8.8)

Here τ is the shear stress, γ the plastic shear strain, n the strain hardening exponent (~0.023 for UFG W of this work), $\dot{\gamma}$ the plastic shear strain-rate, m the SRS (usually the term $B\dot{\gamma}$ is much larger than unity, and Eq. 8.8 can be reduced to the power-law strain-rate hardening), T the temperature, and A, B, and c are materials constants. On the basis of Eq. 8.8 and stability analyses, Burns and Trucano [71] derived the following critical shear strain for a strain-hardenable metal:

$$\gamma_c = \left\{\frac{n\rho C}{Ac}\frac{1}{(1+B\dot{\gamma})^m}\right\}^{\frac{1}{n+1}}$$
(8.9)

Experimental results [72] on W ribbons (heavily deformed at low temperature) that have room-temperature flow stress close to the UFG W of this chapter show that their flow stresses vary as a function of temperature in the manner described by Eq. 8.8, with c being ~0.7 × 10^{-3}/K.

The values of other parameters to be used in Eq. 8.9 are as follows: A = 1,430 MPa, ρ = 19.25 g/cm^3 (density of W), C = 0.134 J/gK (specific heat of W), and B = 10^4 s [71]. An estimate of the critical shear strain from Eq. 8.9 is ~0.046, which corresponds to a normal strain of 0.026. We can see that even though work hardening is accounted for, because of its small contribution to stabilizing the plastic deformation (n is very small compared with conventional CG W), the estimated critical shear strain is not far from what is estimated on the basis of the perfectly plastic assumption.

We envisage that due to the concurrent effects of much elevated flow strength and ductility, and the diminished role of the stabilizing mechanisms of strain hardening and strain-rate hardening, BCC metals with UFG/NC microstructures will have low resistance to softening (thermal or geometric) mechanisms. As a consequence, plastic flow will be more likely to localize.

In this chapter, plastic flow in the form of ASB, and the subsequent cracking, is indeed observed upon uniaxial high-rate compressive loading of the high-strength, ductile UFG/NC W (Figs. 12, 13, 16, and 17). Note that the SPD and grain refinement have improved the ductility at the same time as obtaining higher strength.

In the past, flow localization has often been induced by intentionally generating localized shear stresses in the sample through special design of sample geometry. For example, adiabatic shear banding has been observed in CG W samples with hat-shaped or truncated cone geometry loaded under dynamic compression [55, 73, 74]. In contrast, we have reported the first observation of adiabatic shear localization in commercial purity UFG W using conventional Kolsky bar compression tests [75].

Finally, we would like to point out that HPT W shows much narrower ASBs than UFG W, with the former being only ~5 ∝m compared with ~40 ∝m of the latter. Various formulas have been derived for the estimation of ASB width (or thickness). For example, Dodd and Bai arrived at the following equation [76]:

$$\delta_{ASB} \cong 2\left(\frac{\lambda T}{\tau\dot{\gamma}}\right)^{1/2} \tag{8.10}$$

In Eq. 8.10, λ is the thermal conductivity, T the absolute temperature, τ the stress, and $\dot{\gamma}$ the strain-rate inside the SB. For a given material, the physical properties should be the same. Therefore the only factors that affect the SB width include the strength and strain-rate. It is noted that the strength of HPT W is roughly 1.4 times that of UFG W, and the strain-rate achieved by DTKB used to test the small HPT W specimens is about one order of magnitude higher than that achieved in the conventional Kolsky bar technique.

The above analysis should also apply to the adiabatic shear banding process in UFG/NC Fe.

As we have observed, UFG Ta does not exhibit ASB under uniaxial dynamic compression. We attribute this to the strong and lasting strain-hardening capability of Ta, as the quasi-static stress–strain curves of Ta after 16 ECAE passes still shows moderate strain hardening. The need for a better understanding of the microscopic mechanism of such strain hardening in Ta begs further effort to understand the behavior of this material. Recently, fully dense NC Ta has been obtained by HPT (grain size $d \sim 40$ nm). Evaluation of its dynamic behavior is underway. Preliminary nanoindentation on this NC Ta at different loa ding rates suggests increased SRS, which needs to be confirmed by other experiments [77]. It will be of interest to investigate the dynamic mechanical properties of such truly NC Ta.

8.5 CONCLUDING REMARKS

In this chapter, we have presented and described available experimental results primarily on the high strain-rate mechanical behaviors of BCC metals (mostly refractory, except for Fe) with UFG or NC microstructures. Such UFG/NC metals have been produced either by top-down methods, such as ECAE/ECAP, HPT, and a combination of ECAE/ECAP and rolling, or by bottom-up methods such as consolidation of nanometer-sized powders with powders made by mechanical attrition. The metals selected for investigation included Fe, V, Ta, and W.

Experimental results show that some BCC UFG/NC metals share certain common characteristics upon uniaxial dynamic compressive loading via the Kolsky bar technique (strain-rate $\sim 10^3\,s^{-1}$). These include significant dynamic flow softening in the true stress–true strain curves. In the case of W and Fe, sharp load drops have been observed. Such phenomena are associated with a change of the dynamic plasticity from a uniform, stable mode to an unstable, localized shearing mechanism. In other words, in such UFG/NC metals, adiabatic shear localization becomes the predominant dynamic plastic deformation mode. A mechanistic analysis shows that, in addition to the exhausted or nearly exhausted strain hardening of such metals, much increased strength and reduced strain-rate hardening also contribute to increasing the propensity to adiabatic shear banding.

However, Ta is found to stand out as an exception to these observations. UFG Ta shows little dynamic flow softening and persistent strain hardening under quasi-static loading. As a consequence, its dynamic plastic deformation proceeds uniformly over the specimen. Thus, even though further elaborations are needed to provide a clear understanding of the dynamic behaviors of UFG Ta, we presume at this point that the primary reason for the stable dynamic behavior of UFG Ta may be grounded in its strong and lasting strain-hardening capability and very high ductility and its relatively low strength.

Acknowledgments

This chapter is primarily based on the authors' previous work performed for the past several years in which many other colleagues have made contributions. The authors are particularly indebted to Professors Ma and Ramesh (the Johns Hopkins University) for many illuminating discussions. They are also grateful to Dr. Wright (U.S. Army Research Laboratory) for his generous help in the understanding of the physics and mechanics of adiabatic shear banding, which is the running theme of his monograph [33]. Dr. Magness (U.S. Army Research Laboratory) has kindly offered assistance to Qiuming Wei with his expert knowledge about penetrator performance and the associated materials issues. Discussions with Messrs. Dowding and Cho (U.S. Army Research Laboratory) are also appreciated. Some experimental results presented in this chapter were obtained at JHU-CAMCS through support by the U.S. Army Research Laboratory under the ARMAC-RTP Cooperative Agreement #DAAD19-01-2-0003. Many former colleagues participated in the experimental work described in this chapter, and the authors would like to extend their gratitude to Drs. Tonia Jiao, Brian Schuster, Haitao Zhang, Yulong Li, and Suveen Mathaudhu. Qiuming Wei would like to thank Professor F. H. Zhou for his essential help with the mathematics of mechanics models of adiabatic shear banding during his tenure at JHU.

Many of the SPD materials have been supplied by Professor Valiev's group (Ufa, Russia) or by Professor Hartwig's group (Texas A&M University at College Station, Texas).

Finally, the authors are thankful to Professor James Li, editor of this volume, for inviting them to write this chapter.

References

1. T. H. Courtney, *Mechanical Behavior of Materials*, Waveland Press, **2000**.

2. L. Lu, X. Chen, X. Huang, K. Lu, *Science*, **2009**, *323*, 607.

3. L. Lu, S. X. Li, K. Lu, *Scripta Mater.*, **2001**, *45*, 1163.

4. L. Lu, R. Schwaiger, Z. W. Shan, M. Dao, K. Lu, S. Suresh, *Acta Mater.*, **2005**, *53*, 2169.

5. L. Lu, Y. F. Shen, X. H. Chen, L. H. Qian, K. Lu, *Science*, **2004**, *304*, 422.

6. Y. M. Wang, M. W. Chen, F. H. Zhou, E. Ma, *Nature*, **2002**, *419*, 912.

7. Y. M. Wang, E. Ma, *Acta Mater.*, **2004**, *52*, 1699.

8. Y. H. Zhao, J. F. Bingert, Y. T. Zhu, X. Z. Liao, R. Z. Valiev, Z. Horita, T. G. Langdon, Y. Z. Zhou, E. J. Lavernia, *Appl. Phys. Lett.*, **2008**, *92*, 081903.

9. Y. H. Zhao, X. Z. Liao, S. Cheng, E. Ma, Y. T. Zhu, *Adv. Mater.*, **2006**, *18*, 2280.

10. R. Z. Valiev, I. V. Alexandrov, Y. T. Zhu, T. C. Lowe, *J. Mater. Res.*, **2002**, *17*, 5.

11. M. A. Meyers, A. Mishra, D. J. Benson, *Prog. Mater. Sci.*, **2006**, *51*, 427.

12. K. S. Kumar, H. Van Swygenhoven, S. Suresh, *Acta Mater.*, **2003**, *51*, 5743.

13. C. C. Koch, K. M. Youssef, R. O. Scattergood, K. L. Murty, *Adv. Eng. Mater.*, **2005**, *7*, 787.

14. E. O. Hall, *Proc. Phys. Soc. B, London*, **1951**, *64*, 747.

15. N. J. Petch, *J. Iron Steel Inst.*, **1953**, *174*, 25.

16. C. A. Schuh, T. G. Nieh, T. Yamasaki, *Scripta Mater.,* **2002**, *46*, 735.

17. C. C. Koch, J. Narayan, The inverse Hall–Petch effect-Fact or artifact? presented at *Structure and Mechanical Properties of Nanophase Materials-Theory and Computer Simulations versus Experiments*, Boston, MA, **2001**.

18. J. Schiotz, K. W. Jacobsen, *Science*, **2003**, *301*, 1357.

19. Z. L. Pan, Y. L. Li, Q. Wei, *Acta Mater.,* **2008**, *56*, 3470.

20. H. Gleiter, *Acta Mater.,* **2000**, *48*, 1.

21. C. C. Koch, I. A. Ovid'ko, S. Seal, S. Veprek, *Structural Nanocrystalline Materials-Fundamentals and Applications*, Cambridge University Press, Cambridge, **2007**.

22. R. Z. Valiev, *J. Mater. Sci.,* **2007**, *42*, 1483.

23. R. Z. Valiev, Y. Estrin, Z. Horita, T. G. Langdon, M. Zehetbauer, Y. T. Zhu, *J. Metals,* **2006**, *58*, 33.

24. R. Z. Valiev, R. K. Islamgaliev, I. V. Alexandrov, *Prog. Mater. Sci.*, **2000**, *45*, 103.

25. R. Z. Valiev, T. G. Langdon, *Prog. Mater. Sci.,* **2006**, *51*, 881.

26. A. P. Zhilyaev, T. G. Langdon, *Prog. Mater. Sci.,* **2008**, *53*, 893.

27. C. Suryanarayan, *Int. Mater. Rev.,* **1995**, *40*, 41.

28. C. C. Koch, *Nanostructured Materials: Processing, Properties and Potential Applications*, Noyes Publications, Norwich, **2002**.

29. T. R. Malow, C. C. Koch, *Acta Mater.,* **1998**, *46*, 6459.

30. W. W. Milligan, in *Comprehensive Structural Integrity*, Eds. I. Milne, R. O. Ritchie, B. Karihaloo, Elsevier-Pergamon, Amsterdam-Boston, vol. 8, **2003**, 529.

31. D. Jia, K. T. Ramesh, E. Ma, *Acta Mater.,* **2003**, *51*, 3495.

32. Q. Wei, D. Jia, K. T. Ramesh, E. Ma, *Appl. Phys. Lett.,* **2002**, *81*, 1240.

33. T. W. Wright, *The Physics and Mathematics of Adiabatic Shear Bands*, Cambridge Press, **2002**.

34. Q. Wei, *J. Mater. Sci.,* **2007**, *42*, 1709.

35. Q. Wei, S. Cheng, K. T. Ramesh, E. Ma, *Mater. Sci. Eng. A,* **2004**, *381*, 71.

36. Y. M. Wang, E. Ma, *Mater. Sci. Eng. A,* **2004**, *375–377*, 46.

37. S. Cheng, E. Ma, Y. M. Wang, L. J. Kecskes, K. M. Youssef, C. C. Koch, U. P. Trociewitz, K. Han, *Acta Mater.,* **2005**, *53*, 1521.

38. T. E. Tietz, J. W. Wilson, *Behavior and Properties of Refractory Metals*, Stanford University Press, Stanford, CA, **1965**.

39. S. Cheng, W. W. Milligan, X. L. Wang, H. Choo, P. K. Liaw, *Mater. Sci. Eng. A,* **2008**, *493*, 226.

40. E. Lassner, W.-D. Schubert, *Tungsten-Properties, Chemistry, Technology of the Element, Alloys and Chemical Compounds.*, Kluwer-Academic/Plenum Publishers, **1998**.

41. Q. Wei, T. Jiao, K. T. Ramesh, E. Ma, L. J. Kecskes, L. Magness, R. J. Dowding, V. U. Kazykhanov, R. Z. Valiev, *Acta Mater.,* **2006**, *54*, 77.

42. Q. Wei, H. Zhang, B. E. Schuster, K. T. Ramesh, R. Z. Valiev, L. J. Kecskes, R. J. Dowding, *Acta Mater.,* **2006**, *54*, 4079.

43. Q. Wei, T. Jiao, S. N. Mathaudhu, E. Ma, K. T. Hartwig, K. T. Ramesh, *Mater. Sci. Eng. A,* **2003**, *358*, 266.

44. H. P. Klug, L. E. Alexandor, *X-ray Diffraction Procedure*, John Wiley & Sons, **1974**.

45. B. E. Warren, *X-ray Diffraction*, Dover Publications, NY, **1990**.

46. H. Kolsky, *Stress Waves in Solids*, Dover Publications, NY, **1963**.

47. P. S. Follansbee, in *ASM Metals Handbook*, American Society of Metals, vol. 8, **1985**, 190.

48. D. Jia, K. T. Ramesh, *Exp. Mech.,* **2004**, *44*, 445.

49. Q. Wei, L. J. Kecskes, T. Jiao, K. T. Hartwig, K. T. Ramesh, E. Ma, *Acta Mater.,* **2004**, *52*, 1859.

50. B. Q. Han, E. J. Lavernia, F. A. Mohammed, *Metall. Mater. Tran. A*, **2003**, *34*, 71.

51. Q. Wei, B. E. Schuster, S. N. Mathaudhu, K. T. Hartwig, L. J. Kecskes, R. J. Dowding, K. T. Ramesh, *Mater. Sci. Eng. A,* **2008**, *493*, 58.

52. M. A. Meyers, *Dynamic Behavior of Materials*, John Wiley & Sons, NY, **1994**.

53. Q. Wei, T. Jiao, K. T. Ramesh, E. Ma, *Scripta Mater.,* **2004**, *50*, 359.

54. C. A. Schuh, T. C. Hufnagel, U. Ramamurty, *Acta Mater.,* **2007**, *55*, 4067.

55. T. Dummer, J. C. Lasalvia, G. Ravichandran, M. A. Meyers, *Acta Mater.,* **1998**, *46*, 6267.

56. A. M. Lennon, K. T. Ramesh, *Mater. Sci. Eng. A,* **2000**, *276*, 9.

57. Q. Wei, L. J. Kecskes, *Mater. Sci. Eng. A,* **2008**, *491*, 62.

58. R. Z. Valiev, V. Y. Gertsman, R. Kaibyshev, *Physica Status Solidi A,* **1986**, *97*, 11.

59. P. Gumbsch, *J. Nucl. Mater.,* **2003**, *323*, 304.

60. J. R. Rice, R. Thomson, *Philos. Mag.,* **1974**, *29*, 73.

61. Y. Bai, B. Dodd, *Adiabatic Shear Localization*, Pergamon Press, Oxford, **1992**.

62. P. Rosakis, A. J. Rosakis, G. Ravichandran, J. Hodowany, *J. Mech. Phys. Sol.,* **2000**, *48*, 581.

63. G. E. Dieter, *Mechanical Metallurgy*, McGraw-Hill, NY, **1986**.

64. E. W. Hart, *Acta Metall.,* **1967**, *15*, 351.

65. M. A. Meyers, Y. B. Xu, Q. Xue, M. T. Perez-Prado, T. R. McNelley, *Acta Mater.,* **2003**, *51*, 1307.

66. Q. Xue, E. K. Cerreta, G. T. Gray III, *Acta Mater.,* **2007**, *55*, 691.

67. Q. Wei, *J. Mater. Sci.,* **2006**, DOI 10.1007/s10853-006-0700-9, in press.

68. I. H. Lin, J. P. Hirth, E. W. Hart, *Acta Metall.,* **1981**, *29*, 819.

69. A. Molinari, R. J. Clifton, *J. Appl. Mech.,* **1987**, *54*, 806.

70. A. S. Argon, S. R. Maloof, *Acta Metall.,* **1966**, *14*, 1449.

71. T. J. Burns, T. G. Trucano, *Mech. Mater.,* **1982**, *1*, 313.

72. O. Boser, *J. Less Common Metals,* **1971**, *23*, 427.

73. Z. G. Wei, J. L. Yu, J. R. Li, Y. C. Li, S. S. Hu, *Int. J. Impact Eng.,* **2001**, *26*, 843.

74. J. R. Li, J. L. Yu, Z. G. Wei, *Int. J. Impact Eng.,* **2003**, *28*, 303.

75. Q. Wei, K. T. Ramesh, E. Ma, L. J. Kesckes, R. J. Dowding, V. U. Kazykhanov, R. Z. Valiev, *Appl. Phys. Lett.,* **2005**, *86*, 101907.

76. B. Dodd, Y. Bai, *Mater. Sci. Technol.,* **1985**, *1*, 38.

77. Q. Wei , Z.L. Pan, X.L. Wu, B.E. Schuster, L.J. Kecskes and R.Z. Valiev, Acta Materialia 2011;DOI: 10.1016/j.actamat.2010.12.042.

Chapter 9

FRACTURE BEHAVIOR OF NANOCRYSTALLINE CERAMICS

Ilya A. Ovid'ko and Alexander G. Sheinerman

Institute of Problems of Mechanical Engineering, Russian Academy of Sciences, Bolshoj 61, Vas.Ostrov, St.Petersburg 199178, Russia

E-mail: ovidko@gmail.com

An overview of experimental data and theoretical models of fracture processes in single-phase and composite nanocrystalline ceramics is presented. The key experimentally detected facts in this area are discussed. Special attention is paid to the theoretical models describing toughness enhancement in nanocrystalline ceramics. Also, we consider theoretical models of generation of cracks at grain boundaries (GBs) and their triple junctions in deformed nanocrystalline ceramics.

9.1 INTRODUCTION

The outstanding mechanical properties of single-phase and composite nanocrystalline ceramics (often called nanoceramics) represent the subject of intensive research efforts motivated by a wide range of their applications.[1-6] Commonly, nanoceramics have such technologically attractive characteristics as superior strength, superior hardness, and good fatigue resistance.[1-6] At the same time, in most cases, superstrong nanoceramics at ambient temperatures show both low fracture toughness and poor ductility/machinability,[3,7,8] which are undesired from an applied viewpoint. In particular, low fracture toughness of nanocrystalline ceramics, as with their conventional microcrystalline ceramics, is treated as the key factor limiting their practical utility.[1]

Mechanical Properties of Nanocrystalline Materials
Edited by James C. M. Li

However, recently, certain progress has been reached in enhancement of fracture toughness of ceramic nanocomposites at comparatively low homologous temperatures (for details, see, e.g., reviews,[1-5] book,[6] and original papers[9-17]). Also, several research groups reported on enhanced ductility or even superplasticity of nanoceramics at comparatively low homologous temperatures.[18-22] This experimental data serves as a basis for the technologically motivated hopes to develop new, superstrong nanocrystalline ceramics with good fracture toughness and machinability. To do so, of crucial interest are the specific structural features and generic phenomena responsible for optimization of mechanical characteristics (strength, ductility/machinability, and fracture toughness) of nanoceramics. In particular, it is very important to understand the fundamental fracture mechanisms operating in these materials and reveal the sensitivity of fracture processes to their structural and material parameters. The main aim of this chapter is to review experimental studies and theoretical models of fracture of nanocrystalline ceramics.

9.2 SPECIFIC STRUCTURAL FEATURES OF NANOCRYSTALLINE CERAMICS

The fracture processes in nanoceramics strongly depend on their structural features and phase content. This section briefly describes the specific structural features of single-phase and composite nanoceramics. Also, we discuss the peculiarities of plastic deformation processes affecting fracture of nanoceramics.

First, let us consider the structural features of single-phase nanocrystalline ceramics, compositionally homogeneous solids consisting of nanoscale grains (nanocrystallites) divided by GBs (Fig. 9.1). Their grains are characterized by the grain size $d < 100$ nm and have a crystalline structure. In most cases, nanocrystalline ceramics consist of approximately equiaxed grains with a narrow grain size distribution (Fig. 9.1a). At the same time, there are other examples of grain geometry in nanocrystalline ceramics.[20] Besides, in recent years, nanocrystalline ceramics with a bimodal structure have been produced during superplastic forming.[20] The bimodal ceramic structure consists of both nanoscopic and microscopic rodlike grains (Fig. 9.1b). (Formation of such rodlike grains during superplastic deformation can effectively occur either by the Li mechanism of grain rotation resulting in coalescence of neighboring grains[23] or through stress-driven collective GB migration.[24])

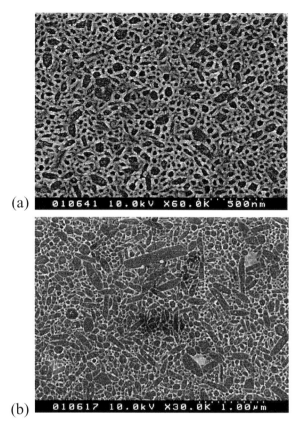

Figure 9.1 Scanning electron microscopy photographs of Si_3N_4 nanoceramics. (a) The microstructure of as-received nanoceramics and (b) nanoceramics after deformation at 1,500°C with a strain of 0.45 under an initial strain rate of $3 \cdot 10^{-5}$ s^{-1}. Reprinted from Xu *et al.*[20]; copyright 2006, with permission from Elsevier.

The crystal lattices of grains are misoriented relative to each other. Neighboring grains are divided by GBs, planes, or faceted layers that carry a geometric mismatch between adjacent misoriented crystalline grains. Typical thickness of conventional (nonamorphous) GBs is around 1 nm. Amorphous GBs have a thickness ranging from 1 to several nanometers. The atomic structure and properties of GBs are different from the structure and properties of grains. In particular, the arrangement of atoms in GBs is disordered compared with that in grain interiors. GBs join at triple junctions that are tubular regions with diameters around 1–2 nm when they adjoin conventional GBs. Triple junctions of conventional GBs are recognized as line defects, whose structure and properties are commonly different from those of GBs that they adjoin. Triple junctions of amorphous GBs are commonly amorphous and have a typical diameter ranging from 1 to several nanometers. With the nanoscale range of grain-sizes, both GBs and their triple junctions occupy large volume fractions in nanocrystalline ceramic materials and thereby strongly influence the mechanical properties of these materials.

In general, one can distinguish the following specific structural features of single-phase nanocrystalline ceramics, differentiating them from conventional coarse-grained polycrystalline ceramic materials:

(1) Grains have nanoscopic sizes. The grain size d does not exceed 100 nm.

(2) The volume fractions occupied by GBs and their triple junctions are large in nanocrystalline ceramics compared with those in coarse-grained polycrystals.

(3) GBs are short. The GB length does not exceed 100 nm.

Now let us turn to a discussion of the specific structural features of composite nanocrystalline ceramics, compositionally inhomogeneous solids containing nanoscale grains (nanocrystallites) of at least one component phase. Typical composite nanocrystalline bulk structures are schematically presented in Fig. 9.2.

These structures are as follows: a nanocrystalline composite consisting of approximately equiaxed nanoscale grains of different phases (Fig. 9.2a); a nanocrystalline composite consisting of grains of one phase, divided by GBs with a different chemical composition (second phase) (Fig. 9.2b); a nanocrystalline composite consisting of large grains of one phase, embedded into a nanocrystalline matrix of the second phase (Fig. 9.2c); a nanocrystalline composite consisting of large grains of one phase, with nanoscale particles of the second phase located at GBs between large grains (Fig. 9.2d); a nanocrystalline composite consisting of nanocrystallites of one phase, embedded into the polycrystalline matrix of the second phase (Fig. 9.2e); and a nanocrystalline composite consisting of grains of two phases, divided by GBs with a different chemical composition (third phase) (Fig. 9.2f). These and other typical classes of nanocomposite solids are discussed in detail by Niihara *et al.*[25] and Kuntz *et al.*[1]

It is difficult to identify the generic structural features of ceramic nanocomposites because of their variety (Fig. 9.2). In most cases, however, GBs occupy large volume fractions in nanocomposites and thereby strongly affect the mechanical and other properties of these materials. GBs in nanocomposites often serve as interphase boundaries characterized by geometric or, in other words, phase mismatch between different crystalline lattices of different phases matched at these boundaries. In this situation, GBs carry both misorientation and phase mismatches between crystallites matched at these boundaries.

(a) (d)

(b) (e)

(c) (f)

Figure 9.2 Typical composite nanocrystalline bulk structures. (a) Nanocrystalline composite materials consisting of tentatively equiaxed nanoscale grains of different phases; (b) nanocrystalline materials consisting of grains of one phase, divided by GBs with a different chemical composition (second phase); (c) nanocrystalline composite materials consisting of large grains of one phase, embedded into a nanocrystalline matrix of the second phase; (d) nanocrystalline composite materials consisting of large grains of one phase, with nanoscale particles of the second phase located at GBs between large grains; (e) nanocrystalline composite materials consisting of nanocrystallites of one phase, embedded into the polycrystalline matrix of the second phase; and (f) nanocrystalline materials consisting of grains of two phases, divided by GBs with a different chemical composition (third phase).

Note that nanoceramics are commonly fabricated at highly nonequilibrium conditions. Therefore, flaws (voids, contaminations) typically exist in nanoceramics. Such fabrication-produced flaws tend to be located at GBs and their triple junctions.

The special case is represented by amorphous GBs. Nanocrystalline ceramics, as with their microcrystalline counterparts, often contain amorphous GBs whose chemical composition can either coincide with or be different from that of nanocrystallites. The width of amorphous GBs ranges rather widely (from around 1 to several nanometers) depending on the chemical composition and fabrication conditions. Amorphous triple junctions in such nanoceramics are significantly extended compared with the triple junctions of conventional (nonamorphous) GBs in nanocrystalline ceramics. Nevertheless, in the nanoceramics with amorphous GBs, the crystalline phase commonly occupies the largest part of the volume fraction.

The specific structural features exert significant effects on fracture processes in nanoceramics. In next section, we discuss key experimental facts concerning the specific features of such processes in nanoceramics.

9.3 KEY EXPERIMENTAL FACTS ON FRACTURE PROCESSES IN NANOCRYSTALLINE CERAMICS

Fracture behavior in nanoceramics has the experimentally detected specific features which are briefly as follows:

(i) Nanoceramics exhibit improvement in strength compared with their conventional microcrystalline counterparts (see, e.g., pioneering papers[26,27] and reviews[1-5]). High values of the strength of nanoceramics are treated to be related to the combined effects of several factors, such as reduction in flaw sizes (close to nanoscale grain-sizes in nanocrystalline materials) and the composite structure homogenization (leading to the lowering of the thermal residual stresses that arise due to anisotropy and thermal expansion mismatches between different microstructural phases); for details, see reviews.[1-5]

(ii) Cracks in nanocrystalline ceramics with narrow grain size distributions tend to propagate along GBs. This statement is supported by the studies of fracture surfaces of such nanoceramics. For example, the fracture of TiN nanoceramics with a narrow grain size distribution (with the mean grain size = 18 nm) has an intergranular character.[28] At the same time, for TiN nanoceramics containing both nanoscale (with the size around 80 nm) and microscale (with the size about $2 \propto$m) grains, cracks propagate along the boundaries of smaller grains but penetrate the interior regions of larger grains.[3,28] Also, intragranular fracture processes dominate in composite nanoceramics having the microcrystalline structures with second-phase nanoparticles dispersed at GBs or in grain interiors (Fig. 9.2d).[3] Besides, Pei *et al.*[15] studied TiC/a-C:H nanocomposite coatings that consist of 2–5 nm TiC nanocrystallites embedded in an amorphous hydrocarbon (a-C:H) matrix. They observed that microcracks induced by nanoindentation or sliding wear propagate along the column boundaries, while the coatings without a columnar microstructure exhibit substantial toughness.

(iii) The mechanical properties and fracture mode of nanoceramics in ceramic nanocomposites are dependent on the strength of GBs. Therefore, the mode of crack propagation depends on the strength of different boundaries. For example, in Al_2O_3–SiC ceramic nanocomposites, the interfacial fracture energy between Al_2O_3 and SiC is twice that of the alumina GB fracture energy.[29] Besides, in such nanocomposites, the compressive stress generated by the SiC located within the matrix grain may reinforce the Al_2O_3/Al_2O_3 GBs.[17] Therefore, the fracture toughness can be greatly improved, and the strong boundary may force the crack to deflect into the Al_2O_3 grain, thus forming the transgranular fracture mode, as was observed by Liu *et al.*[17]

(iv) As with their microcrystalline counterparts, nanoceramics can be toughened by means of the implantation of second-phase inclusions, which hinder crack propagation. These comprise, in particular, inclusions of a ductile (metallic) phase or hard fibers. Following Kuntz *et al.*,[1] the ductile phase can lead to toughening of the composite either through ductile yielding in the region of a propagating crack or by ductile bridging ligaments in the crack wake. In the first case, crack propagation is hindered due to the plastic deformation of the ductile phase or blunting of the crack tip at a ductile particle. In the second case, crack propagation is retarded by ductile bridging ligaments in the crack wake. This occurs when the crack tip propagates past a ductile-phase grain that then bridges the crack wake and must be pulled to failure or debonds from the surrounding matrix. A similar situation takes place in the case of ceramic reinforced by hard second-phase fibers. When a propagating crack approaches a fiber, it can stretch, debond, or fracture. In the latter two cases, the fiber is pulled out of the ceramic matrix. The deformation and fracture of the fiber require additional energy expenses for crack advance and thus increase fracture toughness. As an alternative, the crack approaching the fiber can change the direction of propagation and continue its advance along the fiber–matrix interface. This leads to crack deflection from its favorable orientation with respect to the direction(s) of the applied load and also results in increased fracture toughness.[1]

Note that, as with the case of a ductile inclusion, the fracture of a fiber in a ceramic nanocomposite as a result of crack propagation can be preceded by the plastic deformation of the fiber. A theoretical analysis of such a mechanism of fiber fracture and pullout for the case of nanotube-reinforced ceramic nanocomposites, realized through the formation of circular dislocation loops around the nanotube, has recently been suggested.[30] It has been shown[30] that the generation and slip of circular prismatic dislocation loops along the fiber–matrix interface can be energetically favorable and barrier-less under high but still real local shear stress acting in the interface near the crack surface.

Among various kinds of fibers, one can distinguish carbon nanotubes that have recently drawn much attention as reinforcing fibers due to their high strength (see, e.g., reviews[31,32] and original papers[9,16,33-45]). For example, Siegel *et al.*[16] observed a 24% increase in fracture toughness for a 10 vol% multiwalled-carbon-nanotube (MWNT)–alumina composite compared with that for an unfilled nanophase alumina (from 3.4 to 4.2 MPa $m^{1/2}$). Siegel *et al.*[16] observed that damage delocalization occurs as a result of well-dispersed nanotubes in the matrix.

Besides MWNTs, single-walled carbon nanotubes (SWNTs) have been used to reinforce nanocrystalline ceramics.[9,37-45] SWNTs are preferred for making composites because the inner layers of MWNTs do not contribute considerably to carrying load, and therefore, for a given volume fraction of tubes, the stiffness of nanoceramic composites reinforced by MWNTs is lower than that of composites reinforced by SWNTs.[46] Moreover, the use of SWNTs as reinforcing elements in the Al_2O_3 matrix has resulted in a dramatic increase in fracture toughness.[9,39] In particular, following Zhan *et al.*,[9] the fracture toughness of the 5.7 vol% SWNTs–Al_2O_3 nanocomposite (8.1 MPa $m^{1/2}$) is more than twice that of pure alumina (3.3 MPa $m^{1/2}$), and there is almost no decrease in hardness. Moreover, for a 10 vol% SWNT–Al_2O_3 nanocomposite, the toughness appeared to be as high as 9.7 MPa $m^{1/2}$, which is nearly three times that of pure Al_2O_3 sintered under identical conditions. The higher fracture toughness of SWNT-reinforced alumina nanoceramics compared with their MWNT-reinforced counterparts has been attributed primarily to the differences in the ability to transfer load from the matrix to the nanotubes. (The internal shells of MWNTs are unable to bond to the alumina matrix, and therefore, tensile loads are carried entirely by the external shell.[9]) Also, the high fracture toughness of SWNTs–Al_2O_3 nanocomposites can be associated with a unique entangling network structure of nanotubes, which leads to crack deflection along the continuous interface between carbon nanotubes and nanocrystalline alumina matrix grains.[9]

At the same time, there is some controversy about the toughening in SWNTs–ceramics composites. For example, following Padture,[32] the fracture toughness may depend on the test used to measure it. Most researchers use the simple indentation test to measure the toughness, which may not be valid for these composites (see Padture[32] and references therein). Other toughness tests (not based on indentation) show no toughening,[37] significant toughening[41] (twofold increase over unreinforced ceramic),[41] or unclear results.[45]

Recently, Padture[32] has assumed that a high increase in fracture toughness observed in some experiments with SWNTs–ceramics composites is due to the action-specific toughening mechanism, namely, uncoiling and stretching of SWNTs. This mechanism is based on two peculiarities of SWNTs–ceramics composites. The first peculiarity is that the ends of SWNTs are located at GBs, which serve as anchors for SWNTs. The second specific feature of such composites lies in the fact that SWNTs, which are highly flexible, are commonly rolled, forming complicated tangles. As a result, when an advancing crack approaches a conventional rigid bridging fiber, the latter stretches or detaches from the matrix. At the same time, if a crack approaches SWNTs, the latter can uncoil (or unfurl) in the crack wake. With further propagation of the crack, the uncoiled SWNTs stretch, and eventually the SWNT bridges far behind the crack tip pullout or detach from the GBs and/or fail.[32]

Along with the toughening by means of the implantation of ductile phase or hard fibers, nanoceramics can be dramatically toughened through a phase transformation.[1,18,47–50] This concerns ceramics containing zirconia. Transformation-induced toughening is based on the transformation of zirconia from the tetragonal to the monoclinic phase. This transformation occurs under an applied stress concentrated near a crack tip and should be stabilized by a dopant (yttria or yttria-based composite). The transformation creates inelastic strain that locally relieves the stress field near the crack tip and thus significantly increases fracture toughness. The effect of transformation in zirconia on fracture toughness strongly depends on the zirconia grain size.[48,51] Besides the use as a major phase of ceramics, nanocrystalline zirconia is exploited as a toughening agent in alumina-based nanoceramics.[51]

Also, Choi and Awaji[52] suggested a model describing the role of misfit dislocations in enhancement of fracture toughness in composite nanoceramics having the microcrystalline structures with second-phase nanoparticles dispersed at GBs or in grain interiors (Fig. 9.2d). Within their approach, misfit dislocations are formed at interphase boundaries between nanoparticles and the matrix in such composite nanoceramics, and the dislocations are yielded during the cooling process after sintering. The dislocations contribute to the lowering of thermal residual stresses and serve as nuclei for nanocracks in the frontal process zone ahead of a growing crack tip. These effects hamper crack growth.[52]

In parallel with the mechanisms discussed earlier, the toughening of nanoceramics can be realized by means of the mechanisms associated with plastic deformation in the vicinities of crack tips. These toughening mechanisms will be examined in the next section.

9.4 TOUGHENING MECHANISMS ASSOCIATED WITH PLASTIC DEFORMATION IN NANOCRYSTALLINE CERAMICS

Along with the toughening mechanisms related to either the composite structure of nanoceramics or phase transformations, fracture toughness in nanoceramics can be enhanced due to the mechanisms associated with plastic deformation. In general, plastic deformation in nanoceramics has its specific features due to the specific structural features of nanoceramics, differentiating them from conventional microcrystalline ceramics. It is well known that lattice dislocation slip—the dominant mode of plastic flow in metals—is suppressed in ceramics at ambient temperatures. The suppression is dictated by a high Peierls barrier for lattice dislocation slip. (A high Peierls barrier in ceramics is attributed to narrow cores of dislocations as well as to ionic and covalent interatomic bonds.[53]) Nevertheless, lattice dislocation slip effectively operates in conventional microcrystalline ceramics at high homologous temperatures (exceeding temperature $\approx 0.5T_m$, where T_m is the melting temperature). At the same time, lattice dislocation slip is hampered or even suppressed in nanocrystalline ceramics, even at high homologous temperatures.

In nanocrystalline ceramic materials, as well as in nanocrystalline metals, the role of lattice dislocation slip is diminished due to nanoscale and interface effects. At the same time, in these materials alternative deformation modes mediated by grain and interphase boundaries come into play.[2,54–65] In particular, such alternative deformation modes represent GB sliding, GB diffusional creep (Coble creep), triple-junction diffusional creep, and rotational deformation mode in nanocrystalline materials.[2,56–65]

For instance, the first stage of plastic deformation in nanoceramics predominantly occurs in the interfacial phase.[55,66] With an increase in the applied stress level, both intergranular and intragranular deformation processes occur in nanocrystalline SiC.[66] In this event, intragranular deformation is realized through emission of lattice dislocations from the interfacial phase, their slip in nanoscale grains, and absorption at interfaces.[66] (This process is very similar to plastic flow involving emission of lattice dislocations from either low- or high-angle GBs in nanocrystalline metals.[57,67,68] The active role of the specific (alternative) deformation modes in plastic flow in nanoceramic materials is capable of causing specific toughening mechanisms in these materials.

Let us consider the effects of plastic deformation in nanoceramics on their fracture toughness. In general, one expects conventional and new toughening mechanisms to operate in nanocrystalline single-phase and composite ceramics. The conventional toughening mechanisms—ductile phase toughening, fiber toughening, and transformation toughening—have their analogs operating in microcrystalline ceramics.[1] At the same time, the action of these conventional mechanisms in nanoceramics has its specific features due to the specific structural features (first of all, the presence of nanoscale crystallites and a large amount of the interfacial phase) of the nanoceramics.

Besides the conventional mechanisms, special (new) toughening mechanisms are expected to effectively operate in nanoceramics due to the nanoscale and interface effects. In particular, strong candidates for such special mechanisms are the toughening by interface-mediated plastic deformation, the toughening by local lattice rotation in nanoscale crystallites, the toughening by stress-driven GB migration, and the toughening by enhanced interfacial diffusion. The special toughening mechanisms in nanoceramics should be controlled by the processes occurring on the scales of elementary defects carrying plastic flow and fracture. These processes result in nanoscale structural inhomogeneities releasing, in part, the stresses near the propagating crack tip. In particular, intense interfacial sliding and Coble creep processes carried by interfacial defects, as well as transformations of such defects (e.g., nucleation and climb of interfacial dislocations), create nanoscale defect configurations whose stress fields, in part, accommodate the stresses near the propagating crack tip. Also, local lattice rotation in nanoscale crystallites and stress-driven GB migration create nanoscale configurations of disclinations (defects of the rotational type), causing a similar stress relaxation near the propagating crack. All these relaxation mechanisms enhance the toughening behavior of ceramic nanocomposites and are not typical in microcrystalline ceramics.

Recently, Pozdnyakov and Glezer[69] have theoretically examined the fracture toughness of a quasi-brittle nanomaterial, where GB deformation occurs and a large crack propagates through the consecutive generation of nanocracks near its tip. The GB deformation has been accounted for by the introduction of a strengthening coefficient, and the nanocrack near the tip of a large propagating crack has been assumed to form if the local stress exceeds a critical value. As a result, Pozdnyakov and Glezer[69] concluded that plastic deformation carried by GBs can significantly increase the fracture toughness of nanocrystalline ceramics. However, the exact value of fracture toughness depends on the strengthening coefficient (which is, in turn, determined by plastic deformation modes operating near the crack tip) and cannot be found within this approach.

Along with the examination of the step-by-step crack growth along GBs, Pozdnyakov and Glezer[69] theoretically analyzed the propagation of cracks through the regions occupied by grain interiors, GBs, and triple junctions of GBs in nanocrystalline metals and ceramics. They indicated that the fracture toughness depends on the volume fractions of grain interiors, GBs, and triple junctions. Also, they assumed that the critical energy release rate can be averaged over the volume of the three phases (grain interiors, GBs, and triple junctions) to obtain the fracture toughness. The validity of the suggested mixture rule for fracture toughness is, however, not evident.

Now let us consider the effect of stress-driven migration of GBs on fracture toughness of nanoceramics.[70,71] The stress-driven GB migration is treated as a special deformation mode operating in nanocrystalline materials.[72,73] The stress-driven migration creates GB disclinations (line defects dividing GB fragments with different misorientation parameters[74,75]), and the disclination stress fields hamper crack growth.[72,73] Following Ref. 72, let us discuss the geometry of the stress-induced GB migration and its accommodation in nanocrystalline ceramics. In doing so, in order to simplify our analysis, we consider a two-dimensional model arrangement of nanoscale grains with pure-tilt GBs, including a vertical GB that migrates in a rectangular grain, as shown in Fig. 9.3. The tilt GBs are either low- or high-angle boundaries containing discrete or continuously distributed dislocations, as schematically shown in Fig. 9.3.

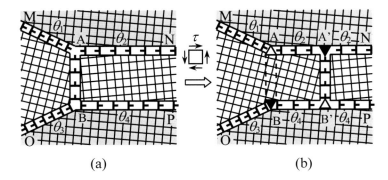

(a)   (b)

Figure 9.3 Stress-induced migration of a tilt boundary in a nanocrystalline specimen (two-dimensional model). The vertical tilt boundary migrates from its initial position *AB* (a) to the final position *A'B'* (b). The shear coupled to migration is accommodated by formation of disclinations (triangles) at GB junctions *A*, *A'*, *B*, and *B'*.

In particular, the vertical GB contains a finite wall-like ensemble of dislocations that provide its tilt misorientation, ω. In its initial state, the vertical GB (Fig. 9.3a) terminates at the GB junctions A and B, which are supposed to be geometrically compensated. There are no angle gaps at these triple junctions, or in other words, the sum of tilt misorientation angles of all GBs joining at each of these junctions is equal to zero, where summation of the angles is performed clockwise along a circuit surrounding a triple junction.[76,77]

In the case under consideration, the following balance equations are valid:

$$\theta_1 + \theta_2 = -\omega \quad \text{(for GB junction } A\text{),} \tag{9.1}$$

$$\theta_3 + \theta_4 = \omega \quad \text{(for GB junction } B\text{),} \tag{9.2}$$

Following the geometric theory of triple junctions,[76,77] the geometrically compensated GB junctions A and B (Fig. 9.3a) do not create long-range stresses.

Let us consider the situation where the vertical GB migrates from the position AB to the position $A'B'$ (Fig. 9.3b) and no other accommodating structural transformations occur. As a result of the migration, the angle gaps $-\omega$ and ω appear at the GB junctions A and B, respectively[72] (Fig. 9.3b). That is, the sum of the misorientation angles at the junction A after GB migration is $\theta_1 + \theta_2$, which is equal to $-\omega$ according to Eq. 9.1. The sum of the misorientation angles at the junction B after GB migration is $\theta_3 + \theta_4$, which is equal to ω according to Eq. 9.2. Also, GB migration results in the formation of two new triple junctions A' and B' characterized by the angle gaps ω and $-\omega$, respectively. That is, the sum of the misorientation angles at the junction A' is equal to ω, as it is directly seen in Fig. 9.3b. The sum of the misorientation angles at the junction B' is equal to $-\omega$ (Fig. 9.3b).

Within the theory of defects in solids, the GB junctions A, B, A', and B' characterized by the angle gaps $\pm\omega$ are defined as partial wedge disclinations that serve as powerful stress sources characterized by the disclination strengths $\pm\omega$.[59,75–79] The disclinations at the points A, B, A', and B' form a quadrupole configuration whose formation accommodates GB migration coupled to shear in a nanocrystalline specimen. The disclination quadrupole creates internal stresses screened at distances being around the largest size (largest interspacing between the disclinations) of the quadrupole.

Morozov *et al.*[71] have calculated the effect of GB migration near the tip of a preexistent crack on the fracture toughness of nanocrystalline ceramics. They considered migration of a GB near the tip of a mode *I* crack in a nanocrystalline ceramic solid (Fig. 9.4). Following Morozov *et al.*,[71] the crack has a length *l* and occupies the region $(-l < x < 0, y = 0)$, while the crack tip

$x = y = 0$ lies in a junction of two GBs. One of these GBs of length d, denoted as u (Fig. 9.4), is supposed to make an angle $\alpha + \pi/2$ with the crack surface and a right angle with the neighboring GB. Due to the action of tensile applied load and its concentration near the crack tip, GB u migrates by a distance s to a new position u' (Fig. 9.4). GB migration results in the formation of a disclination quadrupole, which shields the crack tip and thereby increases fracture toughness.

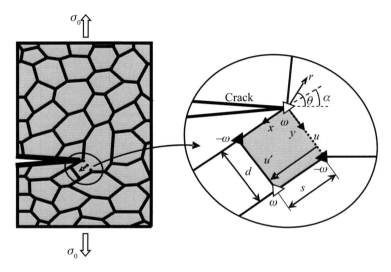

Figure 9.4 GB migration and formation of a disclination quadrupole in a deformed nanoceramic solid.

Let K_{IC}^{σ} denote the fracture toughness for brittle fracture (in the absence of GB migration), s_0 designate the equilibrium migration length, G and v specify the shear modulus and Poisson's ratio, respectively, and γ_e be the effective specific surface energy. The dependences of the normalized fracture toughness K_{IC}/K_{IC}^{σ} and normalized equilibrium migration length s_0/d on the parameter $\omega\sqrt{Gd/[4\gamma_e(1 - v)]}$ are shown in Fig. 9.5 for the case $\alpha = \pi/6$. As follows from Fig. 9.5, for a GB of specified length d, the equilibrium GB length s_0 decreases with an increase in ω, while the fracture toughness K_{IC} increases with ω.

Using Fig. 9.5, let us calculate the fracture toughness for the case of nanocrystalline ceramic silicon carbide (3C-SiC) with the following parameter values: G = 217 GPa, v = 0.23, γ = 1.84,[80] and d = 15 nm. Then for the case of an intragrain crack and ω = 30°, one obtains K_{IC}/K_{IC}^{σ} = 1.32, while for ω = 5°, we have K_{IC}/K_{IC}^{σ} = 1.20. Therefore, in 3C-SiC, GB migration can increase fracture toughness by 20% – 30%.

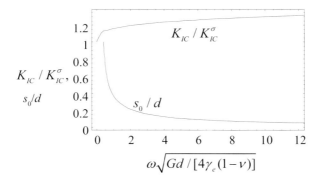

Figure 9.5 Dependences of the normalized fracture toughness K_{IC}/K_{IC}^{σ} and normalized migration length s_0/d on the parameter $\omega\sqrt{Gd/[4\gamma_e(1-v)]}$ that characterizes the disclination strength, ω, for $\alpha = \pi/6$.

Recently, nanoscale twin deformation carried by partials emitted from GBs[81] has been theoretically shown to be a specific toughening mechanism in nanocrystalline ceramics and metals. The model suggested in Ref. 81 describes the generation of deformation twins near cracks of mixed I and II modes in nanocrystalline metals and ceramics. In the framework of the model, a deformation twin nucleates through stress-driven emission of twinning dislocations from a GB distant from the crack tip. The deformation of twin nucleation releases, in part, high local stresses near the crack tip and thereby enhances fracture toughness.

Also, creep deformation conducted by GB processes (Ashby–Verall creep carried by intergrain sliding accommodated by GB diffusion and grain rotations[82,83]) has been suggested as a toughening mechanism inherent to nanocrystalline metals.[84,85] Although the focus of Refs. 84 and 85 is on nanocrystalline metals, their approach can be extended to the situation with nanoceramics. At the same time, the creep deformation considered in these references is a slow diffusion-controlled process which can effectively contribute to toughening of nanocrystalline materials only in the limiting partial cases where diffusion is very fast (at high temperatures) and/or crack growth is extremely slow.

Nevertheless, there are expectations that GB sliding can effectively contribute to toughening of nanocrystalline materials in a more general situation with comparatively wide ranges of crack tip velocity and temperature. For instance, let us consider the specific toughening mechanism associated with GB sliding in nanoceramics with amorphous GBs.[86] It is known that the mechanical behavior of nanocrystalline ceramics is strongly influenced by GB deformation processes in amorphous GBs often existing in such materials; see, for example, experimental data[1,20–22] and computer simulations.[66,87] For instance, following experimental data on Si$_3$N$_4$-based ceramics and ceramic composites, compressive creep in these materials takes place through GB sliding.[88–93] Also, following computer simulations[66,87] of the evolution of the nanocrystalline cubic phase of 3C-SiC under a mechanical load, plastic deformation intensively occurs within amorphous GBs in these nanocrystalline ceramics.

Bobylev *et al.*[86] have studied the effect of GB sliding on crack growth in nanocrystalline ceramics. They assumed that cracks initiate GB sliding which leads to the generation of immobile dislocations at triple junctions of GBs. The GB-sliding-produced dislocations cause partial stress relaxation in the vicinities of crack tips and thereby hamper crack growth. A dislocation (Fig. 9.6) is generated due to GB sliding along an amorphous GB in a nanocrystalline ceramic specimen. The Burgers vector magnitude of the dislocation at the boundary gradually grows[86] (because of GB sliding arrested at the triple junction) in parallel with local plastic strain carried by GB sliding in this boundary. The direction of the dislocation Burgers vector is parallel with the boundary line (plane in the three-dimensional case). Such dislocations are treated in terms of the Volterra theory of dislocations in continuum media[94,95] as dislocation-like sources of internal stresses (but not conventional lattice dislocations). In doing so, such dislocations are commonly immobile at triple junctions of amorphous GBs (Fig. 9.6).

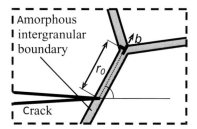

Figure 9.6 Edge dislocation at the amorphous intergranular boundary near the tip of a long crack that intersects the boundary.

Following Bobylev *et al.*,[86] consider a nanocrystalline ceramic solid consisting of nanoscale grains divided by GBs and containing a long and flat mode *I* crack of length *l*. Let the crack intersect the boundary at the point distant by r_0 ($r_0 < l$) from the nearest triple junction (Fig. 9.6). Also, let θ denote the angle between the crack plane and the amorphous intergranular boundary (see Fig. 9.6). The local stresses near the crack tip in the solid under a tensile load initiate intergrain sliding that results in the formation of an immobile dislocation at the nearest triple junction. The dislocation shields the crack tip and thus increases fracture toughness.

Bobylev *et al.*[86] have calculated the effect of the intergrain-sliding-produced dislocation at a GB or amorphous intergranular boundary (Fig. 9.6) on the fracture toughness K_{IC} of a nanocrystalline solid. The dependences of the ratio K_{IC}/K_{IC}^{σ} (where K_{IC}^{σ} is the fracture toughness for brittle fracture, as earlier) on the angle θ, for $r_0 = 5$ nm, are presented in Fig. 9.7a, for nanocrystalline 3C-SiC. From this figure it follows that the local fracture toughness K_{IC} reaches its maximum value when the boundary plane (and thereby the Burgers vector of the dislocation) makes the angle of around 70° with the direction of crack growth. This is related to the fact that the shear stress acting in the boundary plane is maximum when the plane makes the angle $\approx$70° with the direction of crack growth. The dependence of K_{IC} on the distance r_0 between the dislocation and the crack tip, for $\theta = 70°$, is presented in Fig. 9.7b. From this figure it follows that at $r_0 = 1$ nm, dislocation formation leads to the increase of K_{IC} by $\approx$30%. This effect is significant, but the increase in K_{IC} rapidly falls with increasing r_0.

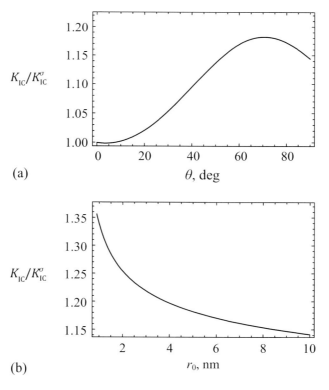

Figure 9.7 Dependence of the ratio K_{IC}/K_{IC}^{σ} in the case of edge dislocation located near crack tip in 3C-SiC on (a) angle θ made by the Burgers vector (or boundary plane) with crack growth direction, for $r_0 = 5$ nm; and (b) distance r_0 between the crack tip and the dislocation, for $\theta = 70°$.

To summarize, several mechanisms associated with plastic deformation in nanoceramics can contribute to enhancement of their fracture toughness characteristics due to the shielding of large preexistent cracks. In doing so, GB deformation processes play the dominant role as toughening mechanisms. In general, along with enhancement of the fracture toughness of nanoceramics, plastic deformation carried by GBs in nanoceramics can also initiate the generation of nanocracks, the nuclei of brittle fracture. This subject will be discussed in the next section.

9.5 GENERATION OF NANOCRACKS AND NANOVOIDS AT GRAIN BOUNDARIES AND THEIR TRIPLE JUNCTIONS IN DEFORMED NANOCRYSTALLINE CERAMICS

In most cases, nanoceramics show low ductility/machinabilty, which significantly limits their practical utility.[1,2,5] The key factors responsible for low ductility/machinabilty at ambient temperatures are the suppression of lattice dislocation slip, high level of stresses operating during the deformation process, and especially the presence of fabrication-produced flaws.

Despite the fact that most nanoceramics have low ductility, there are several intriguing examples of nanoceramics showing good ductility or even superplasticity at lower temperatures and/or higher strain rates, compared with their conventional microcrystalline counterparts.[9,11,18–22,39,96–105] The enhanced ductility in these examples is concerned, in particular, with a low amount or even absence of fabrication-produced flaws. In this context, there is large interest in understanding the micromechanisms for generation of cracks and voids in initially flaw-free nanoceramics.

In Ref. 106, a theoretical model has been suggested describing the generation and growth of nanocracks at triple junctions in nanocrystalline materials in which GB sliding significantly contributes to plastic flow. GB sliding in nanocrystalline metals and ceramics commonly occurs by either local shear events[107–109] or movement of GB dislocations[107,110,111] and leads to accumulation of sessile dislocations at triple junctions.[109,112] In the case of the dislocation mode of GB sliding, mobile GB dislocations (with the Burgers vectors parallel to GB planes) move, causing the sliding in a mechanically loaded specimen (Fig. 9.8a). They are stopped at the triple junctions of GBs, where GB planes are curved and thereby dislocation movement is hampered (Fig. 9.8a). Further movement of GB dislocations needs an increase of the applied stress. When the applied stress increases, GB dislocations reach a triple junction and come into a dislocation reaction resulting in the formation of a sessile dislocation at the junction[109,111] (Fig. 9.8b). This process is an elementary act of plastic deformation involving GB sliding in nanocrystalline materials, where the number of triple junctions is large. The process repeatedly occurs in a deformed nanocrystalline specimen and gives rise to an increase of the Burgers vector of the sessile

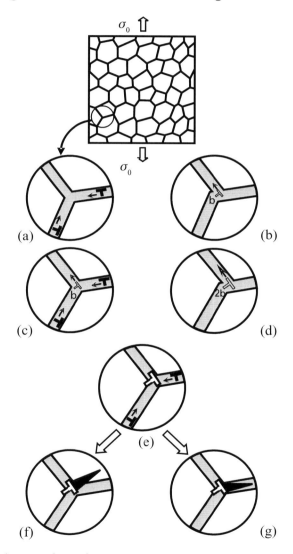

Figure 9.8 Nucleation of a triple-junction nanocrack in a nanocrystalline specimen. GB sliding causes (a–e) formation of triple-junction dislocation whose stress field induces the generation of a nanocrack either (f) in the grain interior or (g) along a GB.

dislocation (Fig. 9.8c–e). (Also, accumulation of the sessile dislocations at triple junctions occurs in a similar way in the situation where GB sliding is carried by local shear events.[109]) Following Ref. 106, a nanocrack at the triple junction is generated to release the strain energy of the sessile dislocation when its Burgers vector magnitude reaches a critical value (Fig. 9.8f,g). In doing so, the nanocrack may nucleate either in the grain interior (Fig. 9.8f) or along a GB adjacent to the triple junction (Fig. 9.8g).

In general, the energetically favorable generation of a crack/nanocrack in the stress field of a dislocation with a large Burgers vector is characterized by its equilibrium length L_e corresponding to the minimum energy of the system.[106,113,114] In this case, the generation and growth of a nanocrack with the length $L < L_e$ are energetically favorable until its length L reaches L_e. That is, the equilibrium state of the nanocrack with the length $L = L_e$ is stable. This case includes, in particular, the generation and growth of a triple-junction nanocrack (Fig. 9.8f,g).

The authors of Ref. 106 calculated the dependences of the equilibrium length L_e of a nanocrack at a dislocated triple junction on the parameter n, the number of passes of mobile GB dislocations through the triple junction. These dependences are presented in Fig. 9.9, for the shear stress $\tau = 0.01G$, Poisson's ratio $\nu = 0.3$, and different values of the angle α between the nanocrack plane and one of GB planes.

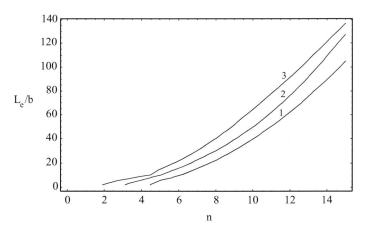

Figure 9.9 Dependencies of nondimensional equilibrium nanocrack length L_e/b on parameter n, for the applied shear stress $\tau = 0.01G$, $\nu = 0.3$, and $\alpha = -\pi/3$, 0, and $2\pi/3$ (see curves 1, 2 and 3, respectively).

As follows from Fig. 9.9, the equilibrium length L_e rapidly grows with an increase of the parameter n, for the nanocrack growing both in the grain interior and along a GB. The equilibrium length L_e rapidly falls[106] with increasing the specific surface energy γ_s. In general, the equilibrium length L_e^{GB} of a nanocrack nucleating along a GB plane can be either larger or lower than the equilibrium length L_e^{vol} of a nanocrack nucleating in the grain interior. The former case ($L_e^{GB} > L_e^{vol}$) is realized at large values of the specific GB energy γ_{GB}. In doing so, the nucleation of the triple-junction nanocrack growing along a GB is energetically preferred compared with the nanocrack growing in the grain interior. The second case ($L_e^{GB} < L_e^{vol}$) is realized at low values of γ_{GB}. In this case, the nucleation of the triple-junction nanocrack growing in the grain interior is preferred.

Besides the formation of nanocracks in the stress field of triple-junction dislocation with large Burgers vectors, nanocracks in nanocrystalline ceramics can also form in the stress field of disclination dipoles or quadrupoles formed due to GB sliding and GB migration, respectively. The formation of nanocracks in the stress field of disclination dipoles has been theoretically analyzed.[70,115,116] In Ref. 117, the formation of a nanocrack in the stress field of both a dislocation with a large Burgers vector and a disclination dipole formed due to GB sliding has been theoretically studied.

The model geometry of a nanocrack generated at a disclination dipole (formed due to GB sliding) is shown in Fig. 9.10.

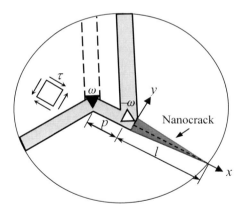

Figure 9.10 Geometry of a nanocrack formed at a disclination dipole.

The calculations[70,115,116] have demonstrated that the nanocrack at a disclination dipole is energetically favored to grow in a certain interval $l_{e1} < l < l_{e2}$ of nanocrack lengths l. The dependences of the critical nanocrack lengths l_{e1} and l_{e2} in the nanocrystalline ceramic materials α-Al$_2$O$_3$ and 3C-SiC on the disclination strength ω and the dipole arm p are plotted in Fig. 9.11a and Fig. 9.11b, respectively. As can be seen from Fig. 9.11, the critical lengths l_{e1} and l_{e2} are not determined at small values of ω and/or p. In this case, nanocrack growth is energetically unfavorable for any nanocrack length; that is, nanocracks are not formed. It can also be seen from Fig. 9.11 that an increase in the disclination strength ω and/or the dipole arm p leads to an expansion in the length range $l_{e1} < l < l_{e2}$ in which nanocrack growth is energetically favorable. It should be noted that at high disclination strengths, ω, the critical length l_{e1} is very small, whereas the equilibrium length l_{e2} of the nanocrack is comparable to the grain size.

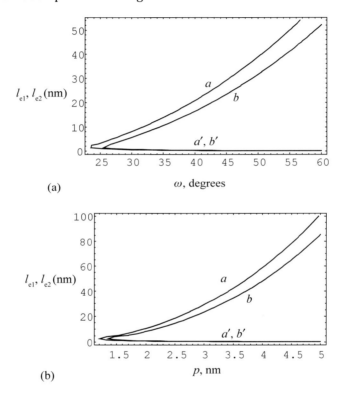

(a)

(b)

Figure 9.11 Dependences of the critical lengths (a', b') l_{e1} and (a, b) l_{e2} of the nanocrack on (a) the disclination strength ω and (b) the dipole arm p for the cases of the nanocrystalline materials (a, a') 3C-SiC and (b, b') α-Al$_2$O$_3$. Dependences are constructed for the parameters $\tau = 1$ GPa, (a) $p = 3$ nm, and (b) $\omega = \pi/4$.

In particular, for nanocrystalline α-Al$_2$O$_3$ at $\omega = \pi/6$, $p = 5$ nm, and the grain size $d = 50$ nm, we have $l_{e1} \approx 0.3p = 0.15$ nm (i.e., in fact, $l_{e1} = 0$) and $l_{e2} \approx 5p = d/2$. For the nanocrystalline ceramics 3C-SiC at $\omega = \pi/6$, $p = 5$ nm, and $d = 50$ nm, we obtain $l_{e1} \approx 0$ and $l_{e2} \approx 6.5p = 0.65d$. In other words, the equilibrium nanocrack length l_{e2} in these cases is equal to half the grain size or even more. Thus, disclination dipoles can be powerful enough stress sources to induce the formation of nanocracks with the length of the order of grain size. The formation and coalescence of these nanocracks can result in the intergranular fracture of nanocrystalline ceramics.

We have considered the nucleation of nanocracks at triple junctions of GBs in nanocrystalline ceramic materials in the situation where GB sliding effectively operates. Recently, theoretical models[118,119] have been suggested, describing the generation and evolution of nanovoids at GBs in such nanocrystalline materials. In particular, in the framework of the model suggested in Ref. 118, nanovoids are generated in the stress fields of dipoles of dislocations characterized by large Burgers vectors and formed at both GB steps and junctions due to intensive GB sliding (Fig. 9.12).

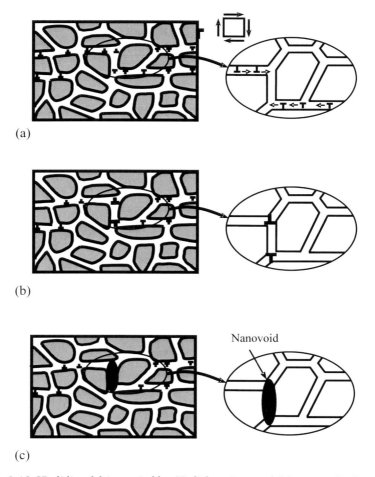

(a)

(b)

(c)

Figure 9.12 GB sliding (a) is carried by GB dislocations and (b) causes the formation of dipole of dislocations with large Burgers vectors at a GB step. (c) Stress fields of dislocations at a GB step induce the generation of a nanovoid.

Generally speaking, nanocrack generation can be suppressed in nanoceramics at high enough temperatures. The suppression is naturally attributed to the effects of diffusion on the generation of triple-junction nanocracks treated as typical elemental carriers of fracture in nanomaterials.[120] Diffusivity of nanocrystalline materials is highly accelerated, because of accelerated diffusivity along GBs (compared with the bulk diffusivity) whose number is large in such materials. The accelerated GB diffusion gives rise to the three following effects responsible for the suppression of nucleation of triple-junction nanocracks:

(i) The enhanced diffusion provides both intensive flow of vacancies from the local regions where high tensile stresses of the sessile triple-junction dislocations exist and intensive flow of interstitial atoms in the opposite direction (Fig. 9.13). In these circumstances, the tensile stresses are, in part, relaxed, and the nucleation of nanocracks is hampered.[120] The discussed effect of diffusion gives rise to an increase of the critical plastic strain at which triple-junction nanocracks (Fig. 9.8f,g) are generated. In certain ranges of parameters of a nanocrystalline ceramic solid, diffusion can even suppress nanocrack generation near triple junctions during the extensive stage of plastic deformation.[120]

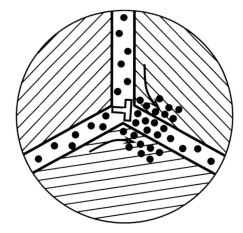

Figure 9.13 Diffusional flow of interstitial atoms causes a partial relaxation of tensile stresses created by triple-junction dislocation.

(ii) Sessile dislocations formed at triple junctions due to GB sliding come into reactions with dislocations that intensively climb along GBs. These dislocation reactions diminish the Burgers vector magnitudes of the sessile dislocations and thereby decrease the stress concentration at triple junctions. Consequently, GB dislocation climb, whose rate is controlled by GB diffusion, hampers nanocrack generation at triple junctions of GBs.[121]

(iii) The enhanced GB diffusion provides the effective action of the Coble creep[122–124] and triple-junction diffusional creep,[125] which compete with GB sliding. Consequently, the contribution of GB sliding to plastic flow decreases, and the growth of the Burgers vectors of sessile dislocations— nuclei of triple junction nanocracks (Fig. 9.8f,g)—slows down or stops.

It is natural to think that the three effects, (i)–(iii), of GB diffusion give rise to good ductility exhibited by some nanocrystalline ceramics.[9,11,18–20,39, 105,126] The enhanced diffusion suppresses nucleation of GB-sliding-induced nanocracks in nanocrystalline ceramic materials, which can thereby exhibit good ductility.[120, 127]

Recent molecular dynamics simulations[87] have demonstrated a nondiffusional mechanism for suppression of void formation in nanocrystalline SiC under tensile deformation. This mechanism is based on unconstrained plastic flow in amorphous GBs, leading to atomic-level necking at such GBs (Fig. 9.14). In doing so, reduction of grain size in nanocrystalline SiC specimens promotes simultaneous enhancement of ductility, toughness, and strength.[87]

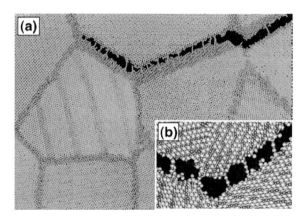

Figure 9.14 (a) Intercrystalline fracture of the d = 16 nm sample along GBs normal to the direction of applied strain. (b) Atomic-level necking leads to formation of one-atom-thick chains of alternating Si (blue) and C (green) atoms. Reprinted with permission from Mo and Szlufarska.[87] Copyright 2007, American Institute of Physics.

To summarize, nanocracks at triple junctions of GBs serve as typical elemental carriers of fracture in nanocrystalline ceramic materials. They are generated due to GB sliding, which effectively operates in nanoceramics at ambient and high temperatures. The triple-junction nanocracks are stable against their rapid growth and thereby can conduct ductile fracture. The nucleation of nanocracks can be effectively suppressed by diffusion in nanoceramics showing good ductility. A nondiffusional mechanism (based on unconstrained plastic flow in amorphous GBs) can contribute to suppression of void formation in nanocrystalline SiC under tensile deformation (Fig. 9.14).

9.6 SUMMARY

Thus, we have reviewed experimental data and theoretical models of fracture processes in nanocrystalline ceramics and ceramic nanocomposites. The typical elemental carriers of fracture in deformed nanocrystalline ceramic materials are nanocracks nucleated at GBs and their triple junctions due to GB sliding. Besides such deformation-produced nanocracks, fabrication-produced nanocracks and pores often exist in these materials and decrease their ductility. Evolution of typical nanocracks and nanovoids in a mechanically loaded nanocrystalline specimen causes fracture mode operating in the specimen and depends on its structural characteristics (first of all, grain size), diffusivity, and material parameters.

In nanocrystalline ceramic materials, plastic flow (conducted mostly by GB processes) is not intensive. Deformation mechanisms mediated by GBs in such materials do not provide effective relaxation of stress buildup at GB steps, triple junctions, and elemental flat nanocracks. In these circumstances, new flat nanocracks are easily generated at triple junctions due to GB sliding and stress concentration at neighboring flat nanocracks. Consequently, intergranular brittle fracture tends to occur in nanocrystalline ceramics with narrow grain size distributions.

The structural features and phase content cause significant effects on fracture processes in single-phase and composite nanoceramics, in particular, on their fracture toughness. Conventional and new toughening mechanisms are expected to operate in nanoceramics. The conventional toughening mechanisms—ductile phase toughening, fiber toughening, and transformation toughening—have their analogs operating in microcrystalline ceramics.[10] The most effective approaches to create ceramic nanocomposites with high fracture toughness appear to be the implantation of SWNTs in ceramics or of the use of nanocrystalline zirconia as one of the components of ceramic nanocomposites.

Besides conventional toughening mechanisms, new toughening mechanisms are expected to effectively operate in single-phase and composite nanocrystalline ceramics due to the nanoscale and interface effects. In particular, strong candidates for such special mechanisms are the toughening by interface-mediated plastic deformation, the toughening by local lattice rotation in nanoscale crystallites, the toughening by stress-driven migration of GBs, and the toughening by enhanced interfacial diffusion. The calculations of the effects of GB migration, lattice rotation, and GB sliding on fracture toughness demonstrated that these plastic deformation mechanisms can increase the fracture toughness of nanoceramics by 20%–30%.

Acknowledgments

The work was supported, in part, by the Russian Ministry of Education and Science (contract 14.740.11.0353), the National Science Foundation (grant CMMI #0700272), the Office of Naval Research (grant N000140–81--0405), the Russian Academy of Sciences Program "Fundamental Studies in Nanotechnologies and Nanomaterials," and the Russian Federal Agency of Science and Innovations (grant MK-1702.2008.1).

References

1. J.D. Kuntz, G.-D. Zhan and A.K. Mukherjee, *MRS Bull.*, **29**, 22 (2004).

2. A. Mukhopadhyay and B. Basu, *Int. Mater. Rev.*, **52**, 257 (2007).

3. R.A. Andrievski and A.M. Glezer, *Physics–Uspekhi*, **52**, 315 (2009).

4. A.V. Sergueeva, D.M. Hulbert, N.A. Mara and A.K. Mukherjee, in *Nanostructured Materials*, Ed. G. Wilde (Elsevier, 2008), p. 127.

5. S.C. Tjong and H. Chen, *Mater. Sci. Eng. R*, **45**, 1 (2004).

6. C.C. Koch, I.A. Ovid'ko, S. Seal and S. Veprek, *Structural Nanocrystalline Materials: Fundamentals and Applications* (Cambridge University Press, Cambridge, 2007).

7. M.A. Morris and M. Leboeuf, *Mater. Sci. Eng. A*, **224**, 1 (1997).

8. C.C. Koch, D.G. Morris, K. Lu and A. Inoue, *MRS Bullet.*, **24**, 54 (1999).

9. G.-D. Zhan, J.D. Kuntz, J. Wan and A.K. Mukherjee, *Nature Mater.*, **2**, 38 (2003).

10. G.-D. Zhan and A.K. Mukherjee, *Int. J. Appl. Ceram. Technol.*, **1**, 161 (2004).

11. Z. Xia, L. Riester, W.A. Curtin, H. Li, B.W. Sheldon, J. Liang, B. Chang and J.M. Xu, *Acta Mater.*, **52**, 931 (2004).

12. S. Bhaduri and S.B. Bhaduri, *Nanostruct. Mater.*, **8**, 755 (1997).

13. A.A. Kaminskii, M.Sh. Akchurin, R.V. Gainutdinov, K. Takaichi, A. Shirakava, H. Yagi, T. Yanagitani and K. Ueda, *Crystallography Reports*, **50**, 869 (2005).

14. Y. Zhao, J. Qian, L.L. Daemen, C. Pantea, J. Zhang, G.A. Voronin and T.W. Zerda, *Appl. Phys. Lett.*, **84**, 1356 (2004).

15. Y.T. Pei, D. Galvan D and J.Th.M. De Hosson, *Acta Mater.*, **53**, 4505 (2005).

16. R.W. Siegel, S.K. Chang, B.J. Ash, J. Stone, P.M. Ajayan, R.W. Doremus and L.S. Schadler, *Scr. Mater.*, **44**, 2061 (2001).

17. H. Liu, C. Huang, X. Teng and H. Wang, *Mater. Sci. Eng. A*, **487**, 258 (2008).

18. X. Zhou, D.M. Hulbert, J.-D. Kuntz, R.K. Sadangi, V. Shukla, B.H. Kear and A.K. Mukherjee, *Mater. Sci. Eng. A*, **394**, 353 (2005).

19. G.-D. Zhan, J.E. Garay and A.K. Mukherjee, *Nano Lett.*, **5**, 2593 (2005).

20. X. Xu, T. Nishimura, N. Hirosaki, R.-J. Xie, Y. Yamamoto and H. Tanaka, *Acta Mater.*, **54**, 255 (2006).

21. D.M. Hulbert, D. Jiang, J.D. Kuntz, Y. Kodera and A.K. Mukherjee, *Scr. Mater.*, **56**, 1103 (2007).

22. A. Dominguez-Rodriguez, D. Gomez-Garcia, E. Zapata-Solvez, J.Z. Chen and R. Chaim, *Scr. Mater.*, **56**, 89 (2007).

23. J.C.M. Li, *J. Appl. Phys.*, **33**, 2958 (1962).

24. M.Yu Gutkin, K.N. Mikaelyan and I.A. Ovid'ko, *Scr. Mater.*, **58**, 850 (2008).

25. K. Niihara, A. Nakahira and T. Sekino, in *Nanophase and Nanocomposite Materials*, MRS Symp. Proc., Eds. S. Kormaneni, J.C. Parker and G.J. Thomas (MRS, Pittsburg, 1993), **286**, p. 405.

26. K. Niihara, *J. Ceram. Soc. Jpn.*, **99**, 974 (1991).

27. K. Niihara and A. Nakahira, in *Advanced Structural Inorganic Composites*, Ed. P. Vincenzini (Elsevier, Amsterdam, 1991), p. 637.

28. R.A. Andrievski, in *Functional Gradient Materials and Surface Layers Prepared by Fine Particles Technology*, Eds. M.-I. Baraton and I. Uvarova (Kluwer, Dordrecht, 2002), p. 17.

29. S. Jiao, M.L. Jenkins and R.W. Davidge, *Acta Mater.*, **5**, 149 (1997).

30. M.Yu. Gutkin and I.A. Ovid'ko, *Scripta Mater.*, **59**, 414 (2008).

31. P.J.F. Harris, *Int. Mater. Rev.*, **49**, 31 (2004).

32. N.P. Padture, *Adv. Mater.*, **21**, 1 (2009).

33. A. Peigney, C. Laurent, F. Dobigeon and A. Rousset, *J. Mater. Res.*, **12**, 613 (1997).

34. R. Z. Ma, J. Wu, B. Q. Wei, J. Liang and D. H. Wu, *J. Mater. Sci.*, **33**, 5243 (1998).

35. T. Seeger, T. Köhler, T. Frauenheim, N. Grobert, M. Rühle, M. Terrones and G. Seifert, *Chem. Commun.*, 34 (2002).

36. P. Vincent, A. Brioude, C. Journet, S. Rabaste, S. T. Purcell, J. Le Brusq and J. C. Plenet, *J. Non-Cryst. Sol.*, **311**, 130 (2002).

37. X. Wang, N.P. Padture and H. Tanaka, *Nat. Mater.*, **3**, 539 (2004).

38. J. Sun, L. Gao, M. Iwasa, T. Nakayama and K. Niihara, *Ceram. Int.*, **31**, 1131 (2005).

39. G.-D. Zhan and A.K. Mukherjee, *Rev. Adv. Mater. Sci.*, **10**, 185 (2005).

40. Z. Burghard, D. Schon, J. Bill and F. Aldinger, *Int. J. Mater Res.*, **97**, 1667 (2006).

41. J. P. Fan, D.M. Zhuang, D.Q. Zhao, G. Zhang, M.S. Wu, F. Wei and Z.J. Fan, *Appl. Phys. Lett.*, **89**, 121910 (2006).

42. T. Wang, Z. Iqbal and S. Mitra, *Carbon*, **44**, 2804 (2006).

43. A.L. Vasiliev, R. Poyato and N.P. Padture, *Scr. Mater.*, **56**, 461 (2007).

44. D. Jiang, K. Thomson, J.D. Kuntz, J.W. Ager and A.K. Mukherjee, *Scr. Mater.*, **56**, 959 (2007).

45. E.L. Corral, J. Cesarano, A. Shyam, E. Lara-Curzio, N. Bell, J. Stuecker, N. Perry, M. Di Prima, Z. Munir, J. Garay and E.V. Barrera, *J. Am. Ceram. Soc.*, **91**, 3129 (2008).

46. P.A. Calvert, *Nature*, **399**, 210 (1999).

47. B.A. Cottom and M.J. Mayo, *Scr. Mater.*, **34**, 809 (1996).

48. A. Bravo-Leon, Y. Morikawa, M. Kawahara and M.J. Mayo, *Acta Mat.*, **50**, 4555 (2002).

49. O. Vasylkiv, Y. Sakka and V. Skorokhod, *Key Eng. Mater.*, **317–318**, 615 (2006).

50. L. Shu-quan, T. Xiao-ping, L. Shao-qiang and T. Yan, *J. Cen. South Univ. Technol.*, **14**, 1 (2007).

51. B. Zhao, B. Y. Du and T. L. Duan, in *Advanced Design and Manufacture to Gain a Competitive Edge*, Eds. X.-T. Yan, C. Jiang and B. Eynard (Springer, London, 2008), p. 215.

52. S.-M. Choi and H. Awaji, *Sci. Technol. Adv. Mater.*, **6**, 2 (2005).

53. R.W. Williams, *Mechanical Behaviour of Ceramics* (Cambridge University Press, London, 1979).

54. H. Gleiter, *Acta Mater.*, **48**, 1 (2000).

55. S. Veprek and A.S. Argon, *J. Vac. Sci. Technol.*, **20**, 650 (2002).

56. K.S. Kumar, S. Suresh and H. Van Swygenhoven, *Acta Mater.*, **51**, 5743 (2003).

57. W.W. Milligan, in *Mechanical Behavior of Bulk Nanocrystalline and Ultrafine-Grain Metals*, *in Comprehensive Structural Integrity*, Eds. I. Milne, R.O. Ritchie and B. Karihaloo (Elsevier, Amsterdam, 2003), p. 529.

58. R.Z. Valiev, *Nature Mater.*, **3**, 511 (2004).

59. M.Yu. Gutkin and I.A. Ovid'ko, *Plastic Deformation in Nanocrystalline Materials* (Springer, Berlin-Heidelberg-NY, 2004).

60. B.Q. Han, E. Lavernia and F.A. Mohamed, *Rev. Adv. Mater. Sci.*, **9**, 1 (2005).

61. I.A. Ovid'ko, *Int. Mater. Rev.*, **50**, 65 (2005).

62. I.A. Ovid'ko, *Rev. Adv. Mater. Sci.*, **10**, 89 (2005).

63. D. Wolf, V. Yamakov, S.R. Phillpot, A.K. Mukherjee and H. Gleiter, *Acta Mater.*, **53**, 1 (2005).

64. M.A. Meyers, A. Mishra and D.J. Benson, *Progr. Mater. Sci.*, **51**, 427 (2006).

65. C.S. Pande and K.P. Cooper, *Progr. Mater. Sci.*, **54**, 689 (2009).

66. I. Szlufarska, A. Nakano and P. Vashishta, *Science*, **309**, 911 (2005).

67. J.C.M. Li, *Phys. Rev. Lett.*, **96**, 215506 (2006).

68. S.V. Bobylev, M.Yu. Gutkin and I.A. Ovid'ko, *Acta Mater.*, **52**, 3793 (2004).

69. V.A. Pozdnyakov and A.M. Glezer, *Phys. Sol. State*, **47**, 817 (2005).

70. I.A. Ovid'ko, A.G. Sheinerman and E.C. Aifantis, *Acta Mater.*, **56**, 2718 (2008).

71. N.F. Morozov, I.A. Ovid'ko, A.G. Sheinerman and E.C. Aifantis, *Mater. Phys. Mech.*, **8**, 155 (2009) (in Russian).

72. M.Yu. Gutkin and I.A. Ovid'ko, *Appl. Phys. Lett.*, **87**, 251916 (2005).

73. M. Dao, L. Lu, R.J. Asaro, J.Th.M. De Hosson and E. Ma, *Acta Mater.*, **55**, 4041 (2007).

74. J.C.M. Li, *Surf. Sci.*, **31**, 12 (1972).

75. A.E. Romanov and V.I. Vladimirov, in *Dislocations in Solids*, Ed. F.R.N. Nabarro (North Holland, Amsterdam, 1992), **9**, p. 191.

76. W. Bollmann, *Philos. Mag. A*, **49**, 73 (1984).

77. W. Bollmann, *Philos. Mag. A*, **57**, 637 (1988).

78. G. Dimitrakopulos, P. Komminou, T. Karakostas and R.C. Pond, *Interface Sci.*, **7**, 217 (1999).

79. P. Klimanek, V. Klemm, A.E. Romanov and M. Seefeldt, *Adv. Eng. Mater.*, **3**, 877 (2001).

80. Z. Ding, S. Zhou and Y. Zhao, *Phys. Rev. B*, **70**, 184117 (2004).

81. M.Yu. Gutkin, I.A. Ovid'ko and N.V. Skiba, *Philos. Mag.*, **88**, 1137 (2008).

82. M.F. Ashby and R.A. Verall, *Acta Metall.*, **21**, 149 (1973).

83. W. Yang and H.T. Wang, *J. Mech. Phys. Sol.*, **54**, 875 (2004).

84. W. Yang and H.T. Wang, *Int. J. Sol. Struct.*, **45**, 3897 (2008).

85. F. Yang and W. Yang, *J. Mech. Phys. Sol.*, **57**, 305 (2009).

86. S.V. Bobylev, A.K. Mukherjee, I.A. Ovid'ko and A.G. Sheinerman, *Rev. Adv. Mater. Sci.*, **21**, 99 (2009).

87. Y. Mo and I. Szlufarska, *Appl. Phys. Lett.*, **90**, 181926 (2007).

88. J. Crampon, R. Duclos and N. Rakotoharisoa, *J. Mater. Sci.*, **25**, 1203 (1990).

89. R.D. Nixon, D.A. Koester, S. Chevacharoenkul and R.F. Davis, *Sci. Technol.*, **37**, 313 (1990).

90. J. Crampon, R. Duclos, and N. Rakotoharisoa, *J. Mater. Sci.*, **28**, 909 (1993).

91. S.Y. Yoon, T. Akatsu and E. Yasuda, *J. Mater. Sci.*, **32**, 3813 (1997).

92. J.-L. Besson, M. Mayne, D. Bahloul-Hourlier and P. Goursat, *J. Eur. Ceram. Soc.*, **18**, 1893 (1998).

93. K.J. Yoon, S.M. Wiederhorn and W.E. Luecke, *J. Am. Ceram. Soc.*, **83**, 2017 (2000).

94. V. Volterra, *Ann. Sci. Ecole Norm. Sup.*, **24**, 401 (1907).

95. J.P. Hirth and J. Lothe, *Theory of Dislocations* (Wiley, NY, 1982).

96. F. Wakai, S. Sakaguchi and Y. Matsuno, *Adv. Ceram. Mater.*, **1**, 259 (1986).

97. F. Wakai and H. Kato, *Adv. Ceram. Mater.*, **3**, 71 (1988).

98. T.-G. Nieh and J. Wadsworth, *Mater. Sci. Eng. A*, **147**, 229 (1991).

99. N. Kondo, F. Wakai, T.Nishioka and A. Yamakawa, *J. Mater. Sci. Lett.*, **14**, 1369 (1995).

100. O.D. Sherby, T.G. Nieh and J. Wadsworth, *Mater. Sci. Forum*, **243–245**, 11 (1997).

101. R.S. Mishra, T.R. Bieler and A.K. Mukherjee, *Acta Mater.*, **45**, 561 (1997).

102. T.G. Langdon, *Mater. Sci. Forum.*, **3043–06**, 13 (1999).

103. B.-N. Kim, K. Hiraga, K. Morita and Y. Sakka, *Nature*, **413**, 288 (2001).

104. A.K. Mukherjee, *Mater. Sci. Eng. A*, **322**, 1 (2002).

105. L. Gao, X. Jin, J. Li, Y. Li and J. Sun, *Mater. Sci. Eng. A*, **415**, 145 (2006).

106. I.A. Ovid'ko and A.G. Sheinerman, *Acta Mater.*, **52**, 1201 (2004).

107. A.P. Sutton and R.W. Balluffi, *Grain Boundaries in Crystalline Materials* (Oxford Sci., Oxford, 1996).

108. H. Conrad and J. Narayan, *Scr. Mater.*, **42**, 1025 (2000).

109. S.V. Bobylev, M.Yu. Gutkin and I.A. Ovid'ko, *Phys. Rev. B*, **73**, 064102 (2006).

110. A.A. Fedorov, M.Yu. Gutkin and I.A. Ovid'ko, *Acta Mater.*, **51**, 887 (2003).

111. M.Yu. Gutkin, I.A. Ovid'ko and N.V. Skiba, *Acta Mater.*, **52**, 1711 (2004).

112. M.Yu. Gutkin and I.A. Ovid'ko, *Philos. Mag. Lett.*, **84**, 655 (2004).

113. V.I. Indenbom, *Sov. Phys. Sol. State*, **3**, 1506 (1961).

114. M.Yu. Gutkin and I.A. Ovid'ko, *Philos. Mag. A*, **70**, 561 (1994).

115. I.A. Ovid'ko and A.G. Sheinerman, *Appl. Phys. Lett.*, **90**, 171927 (2007).

116. I.A. Ovid'ko and A.G. Sheinerman, *Phys. Sol. State*, **50**, 1044 (2008).

117. I.A. Ovid'ko and A.G. Sheinerman, *Phys. Rev. B*, **77**, 054109 (2008).

118. I.A. Ovid'ko and A.G. Sheinerman, *Philos. Mag.*, **86**, 3487 (2006).

119. I.A. Ovid'ko and A.G. Sheinerman, *Philos. Mag.*, **86**, 1415 (2006).

120. I.A. Ovid'ko and A.G. Sheinerman, *Acta Mater.*, **53**, 1347 (2005).

121. I.A. Ovid'ko and A.G. Sheinerman, *Rev. Adv. Mater. Sci.*, **6**, 21 (2004).

122. R.A. Masumura, P.M. Hazzledine and C.S. Pande, *Acta Mater.*, **46**, 4527 (1998).

123. H.S. Kim, Y. Estrin and M.B. Bush, *Acta Mater.*, **48**, 493 (2000).

124. V. Yamakov, D. Wolf, S.R. Phillpot and H. Gleiter, *Acta Mater.*, **50**, 61 (2002).

125. A.A. Fedorov, M.Yu. Gutkin and I.A. Ovid'ko, *Scr. Mater.*, **47**, 51 (2002).

126. J. Wan, R.-G. Duan, M.J. Gasch and A.K. Mukherjee, *J. Am. Ceram. Soc.*, **89**, 274 (2006).

127. I.A. Ovid'ko and A.G. Sheinerman, *Acta Mater.*, **57**, 2217 (2009).

Chapter 10

ON CHALLENGING PROBLEMS IN BRIDGING THE SCALES FROM THE ATOMISTIC TO THE CONTINUUM

Mohammed Cherkaoui

School of Mechanical Engineering, Georgia Institute of Technology,
Atlanta, GA 30332, USA
E-mail: mcherkaoui@me.gatech.edu

10.1 INTRODUCTION

Although some understanding seems to be emerging on the influence of grain size on the strength of nanocrystalline (NC) materials, it is not presently possible to accurately model or predict their deformation, fracture, and fatigue behavior, as well as the relative trade-offs of these responses with changes in the microstructure. Even empirical models predicting deformation behavior do not exist due to lack of reliable data. Also, atomistic modeling has been of limited utility in understanding behavior over a wide range of grain sizes ranging from a few nanometers (~5 nm) to hundreds of nanometers due to inherent limitations on the computation time step, leading to unrealistic applied stresses or strain-rates, and scale of calculations. Moreover, the sole modeling of the microstructures is hindered by the need to characterize defect densities and understand their impact on strength and ductility. For example, NC materials processed by ball milling of powders or extensive shear deformation (e.g., equal channel angular extrusion [ECAE]) can have high defect densities, such as voids, and considerable lattice curvature. Accordingly, NC materials are often highly metastable and are subject to coarsening. Recently, processing techniques such as electrodeposition have

Mechanical Properties of Nanocrystalline Materials
Edited by James C. M. Li
Copyright © 2011 Pan Stanford Publishing Pte. Ltd.
www.panstanford.com

advanced to the point to allow the production of fully dense, homogeneous, and low-defect material that can be used to measure properties reliably and reduce uncertainty in modeling associated with initial defect densities (Nieh and Wang 2005).

Identification of the fundamental phenomena that result in the "abnormal" mechanical behavior of NC materials is a challenging problem that requires the use of multiple approaches (e.g., molecular dynamics [MD] and micromechanics). The abnormal behavior in NC materials is characterized by a breakdown of the Hall–Petch relation (Hall 1951; Petch 1953), that is, the yield stress decreases for decreasing grain size below a critical grain diameter. Also, recent experiments (Wang, Hamza, *et al.* 2005) revealed that, in the case of face-centered cubic (FCC) NC materials, a decrease in the grain size engenders an increase in the strain-rate sensitivity. Recent work by Asaro and Suresh (2005) successfully modeled the size effect in the strain-rate sensitivity, or alternatively in the activation volume, by considering the effect of dislocation nucleation from stress concentrations at grain boundaries. Although experimental observations and MD simulations suggest the activity of local mechanisms (e.g., Coble creep, twinning, grain boundary dislocation emission, grain boundary sliding), it is rarely possible to directly relate their individual contributions to the macroscopic response of the material. This is primarily due to the fact that the scale and boundary conditions involved in molecular simulations are several orders of magnitude different from that in real experiment or of typical polycrystalline domains of interest. In addition, prior to predicting the global effect of local phenomena, a scale transition from the atomic scale to the mesoscopic scale must first be performed, followed by a second scale transition from the mesoscopic scale to the macroscopic scale. Micromechanical schemes have been used in previous models and have proven to be an effective way to perform the second scale transition (Jiang and Weng 2004; Capolungo, Cherkaoui, *et al.* 2005; Capolungo, Jochum, *et al.* 2005). However, the scale transition from the atomistic scale to the mesoscopic scale is a more critical and complex issue. The present chapter will raise the difficulties in performing systematic scale transitions between different scales, especially from an atomistic to a mesoscopic one. The chapter will also highlight succinctly the promising methodologies that can ensure such a challenging issue of bridging the scales. A few of these methodologies are developed and discussed in detail in the next chapters of the book.

10.2 ON VISCOPLASTIC BEHAVIOR OF NC MATERIALS

The viscoplastic behavior of NC materials has been subject to numerous investigations, most of which are focused on the role of interfaces (grain boundaries and triple junctions) and aimed at identifying the mechanisms responsible for the breakdown of the Hall–Petch relation. Within this context, the viscoplastic behavior of NC materials relies on a generic idea in which grain boundaries serve as softening structural elements providing the effective action of the deformation mechanisms in NC materials. Therefore any modeling attempts toward the viscoplastic behavior of NC materials face the problem of identification of the softening deformation mechanisms inherent in grain boundaries, as well as the description of their competition with conventional lattice dislocation motion.

The nature of the softening mechanism active in grain boundaries is still subject to debate (Konstantinidis and Aifantis 1998; Cai, Kong, *et al.* 2000; Cai, Kong, *et al.* 2001; Yin, Whang, *et al.* 2001; Kumar, Suresh, *et al.* 2003). Konstantinidis and Aifantis (1998) assumed that the grain boundary phase is prone to dislocation glide activities where triple junctions act as obstacles and has the properties of disclination dipoles. Tensile creep of nanograined pure Cu with an average grain of 30 nm prepared by the electrodeposition technique has been investigated at low temperatures by Cai *et al.* (2000). The obtained creep curves include both primary and steady state stages. The steady-state creep rate was found to be proportional to the effective stress. The activation energy for the creep was measured to be 0.72 eV, which is close to that of grain boundary diffusion in NC Cu. The experimental creep rates are of the same order of magnitude as those calculated from the equation for Coble creep. The existence of threshold stress implies that the grain boundaries of the nanograined Cu samples do not act as perfect sources and sinks of atoms (or vacancies). Hence the rate of grain boundary diffusion is limited by the emission and absorption of atoms (or vacancies). The results obtained suggest that the low-temperature creep of nanograined pure Cu in this study can be attributed to the interface-controlled diffusional creep of the Coble creep type. The creep of cold-rolled NC pure copper has been investigated in the temperature range of 20°C–50°C and different stresses by Cai *et al.* (2001). The average grain size of as-rolled samples was 30 nm. The author concluded that the creep behavior is attributed to grain boundary sliding accommodated by grain boundary diffusion. Coble creep–type behavior operating at room temperature was also revealed by the experimental studies of Yin *et al.* (2001) performed on porosity-free NC nickel with 30 nm grains produced by electrodeposition processing. Kumar *et al.* (2003) studied the mechanisms of deformation and damage evolution in electrodeposited, fully dense NC Ni with an average grain size of ~30 nm. Their experimental studies consist of (i) tensile tests performed *in situ* in the transmission electron microscope (TEM) and (ii) microscopic observations made at high resolution following *ex situ* deformation induced by compression, rolling, and nanoindentation. The obtained results revealed that deformation is instigated by the emission of dislocations at grain boundaries whereupon voids and/or wedge cracks form along grain boundaries and triple junctions as a consequence of transgranular slip and unaccommodated grain boundary sliding. The growth of voids at separate grain boundaries results in partial relaxation of constraint, and continued deformation causes the monocrystalline ligaments separating these voids to undergo significant plastic flow that culminates in chisel-point failure.

In overall, for NC materials with grain sizes ranging from ~100 nm down to ~15 nm, theoretical models, molecular simulations, and experiments suggest three possible mechanisms governing their viscoplastic responses.

First, the softening behavior of NC materials may be attributed to the contribution of creep phenomena, such as Coble creep (Coble 1963), accounting for the steady-state vacancy diffusion along grain boundaries (Sanders, Rittner, *et al.* 1997; Kim, Estrin, *et al.* 2000; Kim, Hiraga, *et al.* 2005; Kim and Estrin 2005). This hypothesis is motivated by several experimental observations and models which revealed that creep mechanisms could operate at room temperature in the quasistatic regime (Cai, Kong, *et al.* 2000; Cai, Kong, *et al.* 2001; Yin, Whang, *et al.* 2001). However, more recent work has suggested that the observation of creep phenomena could be due to the presence of flaws in the initial structure of the samples, leading to non-fully dense specimens (Li, Blum, *et al.* 2004).

Second, both MD simulations (Warner, Sansoz, *et al.* 2006) and experimental studies (Ke, Hackney, *et al.* 1995) have shown that solid motion of grains (e.g., grain boundary sliding or grain rotation) is one of the primary plastic deformation mechanisms in NC materials. For example, MD simulations on shear of bicrystal interfaces (Warner, Sansoz, *et al.* 2006) showed that grain boundary sliding could be appropriately characterized as a stick-slip mechanism. Moreover, grain boundary sliding could operate simultaneously with interface dislocation emission (Kumar, Suresh, *et al.* 2003; Wang, Hamza, *et al.* 2005). Discussion in the literature has focused on the possible accommodation of these mechanisms by vacancy diffusion (Van Swygenhoven and Caro 1997; Kumar, Suresh, *et al.* 2003). However, the most recent studies tend to show that the grain boundary sliding and grain rotation mechanisms are not accommodated by vacancy diffusion. For example, *ex situ* TEM observations of electrodeposited nickel (Kumar, Suresh, *et al.* 2003) clearly show the creation of cracks localized at grain boundaries. Recently, an interface separation criterion was introduced to predict the observed low ductility of NC materials with small grain sizes (<~50nm) (Wei and Anand 2004). The authors indicated that a detailed description of the dislocation emission mechanism could improve their model predictions.

Third, MD simulations on 2D columnar structures (Yamakov, Wolf, *et al.* 2001), 3D NC samples (Derlet and Van Swygenhoven 2002), and planar bicrystal interfaces (Sansoz and Molinari 2005; Spearot, Jacob, *et al.* 2005; Spearot, Jacob, *et al.* 2006) suggest that interfacial dislocation emission can play a prominent role in NC material deformation (Yamakov, Wolf, *et al.* 2001; Van Swygenhoven, Derlet, *et al.* 2004). The grain boundary dislocation emission mechanism was first suggested by Li in order to describe the Hall–Petch relation (Li 1963). In this model, dislocations are emitted by grain boundary ledges, which act as simple dislocation donors in the sense that a ledge can emit a limited number of dislocations equal to the number of extra atomic planes associated with the height of the ledge. Once the dislocation source is exhausted, the ledge is annihilated and the interface becomes defect free. Recent work has indicated that planar interfaces (without ledges or steps) can also emit dislocations, as exhibited by models based on energy considerations (Gutkin, Ovid'Ko, *et al.* 2003) and atomistic simulations on bicrystal interfaces (Sansoz and Molinari 2005; Spearot, Jacob, *et al.* 2005; Spearot, Jacob, *et al.* 2006). Moreover, MD simulations of two-dimensional (2D) columnar and three-dimensional (3D) NC geometries lead to similar conclusions regarding the role of the interface on dislocation emission (Yamakov, Wolf, *et al.* 2001; Van Swygenhoven, Derlet, *et al.* 2004). The latter have also shown that grain boundary dislocation emission is a thermally activated mechanism, although there are differences in the definition of the criterion for emission of the trailing partial dislocation. A mesoscopic model accounting for the effect of thermally activated grain boundary dislocation emission and absorption has recently been developed and shows that the breakdown of the Hall–Petch relation could be a consequence of the absorption of dislocations emitted by grain boundaries (Capolungo, Cherkaoui, *et al.* 2006). The model also raises the question of the identification of the primary interface dislocation emission sources (e.g., perfect planar boundary, ledge).

Clearly, atomistics are most useful to characterize the structure of grain boundaries and unit processes of dislocation emission, ledge formation, absorption, and transmission. The large length and time scales of polycrystalline responses preclude application of atomistics and necessitate a strategy for bridging scales based on continuum models. However, conventional continuum crystal plasticity is inadequate for this purpose for a number of reasons, most notably in its inability to distinguish the effects of grain boundary character on interfacial sliding and dislocation nucleation/absorption processes. Grain boundaries are treated as geometric boundaries for purposes of compatibility in conventional theory. Moreover, although continuum micromechanics approaches have been developed that incorporate grain boundary surface effects that play a role in the inverse Hall–Petch behavior in NC metals there are problems with conventional models such as inability to factor in dislocation sources in nucleation-dominated regimes and inability to predict appropriate concentrations of stress at grain boundary ledges and triple junctions.

Moving toward an appropriate theory of a cooperative response of NC materials requires a combination of three modeling elements: molecular statics/dynamics, continuum crystal plasticity theory, and self-consistent micromechanics (Fig. 10.1). Such a theory should be able to model kinetics of dislocation nucleation and motion properly, as well as coarsening and shear-banding phenomena. The latter is a challenge that requires the notion of cooperative slip localization to be introduced over many grains.

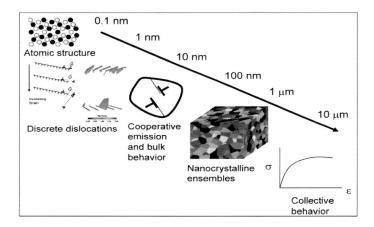

Figure 10.1 Multiple-length scales to be considered in mechanism-based self-consistent multisale modeling of NC materials.

Therefore, developing a framework that can link scales of atomic-level grain boundary structure with emission of dislocations, grain boundary–dislocation interactions, and grain boundary–sliding processes, informing the structure of a self-consistent modeling methodology of anisotropic elastic–plastic crystals that can handle both bulk dislocation activity and grain boundary sliding induced by atomic shuffling/rearrangement or grain boundary dislocation motion, is still a challenging problem to overcome. The resulting theory should be founded on consideration of the surface area to volume ratio in polycrystals, along with accurate accounting for surface energies and activation energy estimates for various nucleation sources, which affect the change to grain boundary–mediated deformation processes at grain sizes below several hundred nanometers. Also, the effect of grain size distribution has to be considered (Zhu and Asaro 2006). Figure 10.1 shows the scales involved in the multiscale modeling of such kind of frameworks, ranging from the interatomic scale that characterizes the grain boundary structure, region of excess energy, and ledges or triple junctions to individual grains that limit transit of dislocations to large sets of NC grains, producing collective strength, work hardening, and ductility properties of interest.

10.3 ON CHALLENGING PROBLEMS OF BRIDGING THE SCALES FROM THE ATOMISTIC TO THE CONTINUUM IN NC MATERIALS

The link between atomic-level and grain boundary structures in NC materials can be considered under the so-called field of mesomechanics, which focuses on the behavior of defects rather than that of atoms. Mesomechanics approaches are needed to complement atomistic methods and to provide information about defect interaction and the kinetics of plastic deformation. Such fundamental information can then be transferred to the continuum level to underpin the formulation of flow and evolutionary behavior of continuum-based constitutive equations. This type of multiscale material design capability will require a few challenges to be overcome. One of the most powerful mesomechanics methods is dislocation dynamics, where considerable progress has been made during the past two decades owing to a variety of conceptual and computational developments. It has moved from a curious proposal to a full and powerful computational method. In its present stage of development, dislocation dynamics has already addressed complex problems and quantitative predictions have been validated experimentally. Progress in 3D dislocation dynamics has contributed to a better understanding of the physical origins of plastic flow and has provided tools capable of quantitatively describing experimental observations at the nanoscale and microscale, such as the properties of thin films, nanolayered structures, microelectronic components, and micromechanical elements (Ghoniem, *et al.* 2003).

New and efficient computational techniques for processing and visualizing the enormous amount of data generated in mesomechanical and continuum multiscale simulations must be developed. Then, the issue of computational efficiency must be addressed so that truly large-scale simulations on thousands of processors can be effectively performed.

It should be noticed that the behavior of NC can be undertaken within the framework of nonlocal formulations that had originally been developed to predict size effects in conventional polycrystalline materials (e.g., Aifantis 1987; Fleck, *et al.* 1994; Fleck and Hutchinson 1997; Gao, *et al.* 1999; Huang, *et al.* 2000; Shu and Fleck 1999; Dai and Parks 1997; Dai 1997; Shi, *et al.* 2000). These approaches will require improved and more robust numerical schemes to deal with a more physical description of dislocation interaction with themselves and with grain boundaries or other obstacles in NC materials.

The issues discussed earlier, in addition to the ever-increasingly powerful and sophisticated computer hardware and software available, are driving the development of multiscale modeling approaches in NC. It is expected that, within the next decade, new concepts, theories, and computational tools will be developed to make truly seamless multiscale modeling a reality. In this chapter, we briefly outline the status of research in each component that enters in building a multiscale modeling tool to describe the viscoplastic behavior of NC materials. Two major components will be addressed in the present chapter and individually discussed in the coming chapters.

First, the chapter will discuss methodologies that rely on the ability of atomistic studies in computing structures and interfacial energies for boundaries to provide a link between the atomistic level and defects that govern the deformation mechanisms of NC materials.

Second, the chapter will discuss the possible ways of incorporating the mesocopic information generated by the above studies in classical continuum micromechanics frameworks to account for grain boundary structures.

10.3.1 Mesoscopic Studies

10.3.1.1 Computing structure and interfacial energies of boundaries

The computing structure and interfacial energies of boundaries is a required preliminary step to model kinetics of dislocation nucleation and motion properly. In view of the focus on building multiscale models for NC materials, avoiding for this purpose complexities associated with impurities, substitutional atoms, or second phases, simple FCC pure metals such as Cu and Al are mainly taken as model materials to perform the atomistic studies. For both materials, embedded atom potentials (EAM) have been developed previously (Mishin, Farkas, *et al.* 1999) that are appropriate for modeling dislocation nucleation and dissociation into Shockley partial dislocations associated with stacking faults. Accordingly, an algorithmic platform can be established that can serve as a useful basis for later extension to more complex alloy systems. The EAM describes the nondirectional character of bonding in Cu quite well and hence provides more realistic consideration of grain boundary and dislocation core structures. Hence, consideration of Cu facilitates thorough and rigorous characterization of the multiscale model from the atomistic scale up. Two critical properties that must be well characterized by the interatomic potential to model dislocation nucleation and defect structures are the intrinsic and unstable stacking fault energies. For example, Rittner and Seidman (1996) showed that the predicted interface structure can vary depending on the magnitude of the intrinsic stacking fault energy. Mishin *et al.* report an intrinsic stacking fault energy of 44.4 mJ/m^2 and an unstable stacking fault energy of 158 mJ/m^2 for Cu, both of which compare favorably with experimental evidence and quantum calculations presented in their work.

The computing structure and interfacial energies of boundaries and modeling kinetics of dislocation nucleation was recently the original work of Spearot *et al.* (2005). Their contribution relies on a methodology that builds grain boundaries in Cu and Al bicrystals through a two-step process: (1) nonlinear conjugate gradient energy minimizing using a range of initial starting positions and (2) equilibrating (annealing) to a finite temperature using Nosé–Hoover dynamics (Nosé 1984; Hoover 1985). The grain boundary energy is calculated over a defined region around the bicrystal interface after the energy minimization procedure. Figure 10.2a,b shows the grain boundary energy at 0 K as a function of misorientation angle for interfaces created by symmetric rotations around the [001] and [110] tilt axes, respectively. Grain boundary structures predicted from energy minimization calculations for several low-order coincident site lattice (CSL) interfaces in copper are shown in Fig. 10.3. Atoms shaded white are in the [001] plane, while atoms shaded black are in the [002] plane. The structural units for each grain boundary are outlined for clarity. The calculated structures for the S3 {111}/[110] and S11 {113}/[110] interface structures are mirror symmetric across the interface plane, while the S5 {210}/[001] structure shows a slight asymmetric character. The structural unit for the S5 {210}/[001] interface in Fig. 10.2b is commonly defined as B (cf. Bachurin, *et al.* 2003).

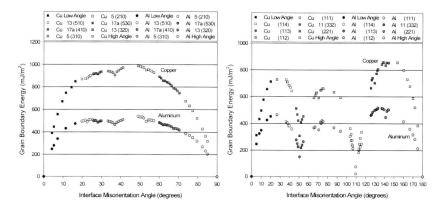

Figure 10.2 Interface energy as a function of the misorientation angle for symmertic tilt (a) [001] and (b) [110] copper and aluminum grain boundaries (Spearot *et al.* 2005).

Figure 10.3a,b shows detailed views of the planar and stepped S5 {210} interface structures. The viewing direction is along the [001] crystallographic axis (Z-direction), and atom positions are projected into the X–Y plane for clarity. Snapshots of the atomic configurations at the interface are taken after the isobaric–isothermal equilibration procedure at 0 bar and 10 K. The structure of each interface can be readily identified by shading atoms according to their respective {001} atomic plane, as indicated in the legend of Fig. 10.3. The planar 53.1° interface in Fig. 10.3a is composed entirely of B structural units, in agreement with previous atomistic simulations that employ embedded-atom method interatomic potentials (Bachurin, Murzaev, *et al.* 2003). It is noted that two configurations are commonly observed for this structural unit, the other being termed the B structural unit (Sutton and Vitek 1983). The B structural unit is identical to that shown in Fig. 10.3a; however, an additional atom is located in the center of the "arrowhead-shaped" feature. Supplementary energy minimization calculations are performed to verify that the copper S5 {210} boundary composed entirely of B structural units is accurate. Energy minimization calculations report an interfacial energy of 950 mJ/m² for the boundary composed entirely of B structural units, which is lower than all other potential configurations for this particular misorientation. Thus, the interface configuration shown in Fig. 10.3a is appropriate.

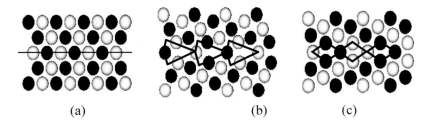

(a)   (b)   (c)

Figure 10.3 Grain boundary interface structures for low-order CSL interfaces in copper: (a) Σ3 111}/[110], (b) Σ5 210}/[001], and (c) Σ11 113}/[110] (Spearot *et al.* 2005).

Figure 10.4 shows the grain boundary structure for a copper 41.1° [001] interface. The interface is comprised of structural units from both S5 {210}/[001] and S5 {310}/[001] interfaces. The S5 {310}/[001] structural unit is commonly defined as C; thus the 41.1° **grain** boundary interface has a |CCB.CCB| structure (Sutton and Vitek 1983).

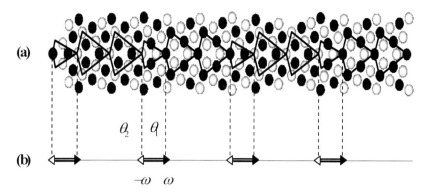

Figure 10.4 (a) Grain boundary structure for a 41.1° [001] interface in copper. The interface is comprised of structural units from Σ5 {210} and Σ5 {310} boundaries. (b) Disclination/dislocation representation of the interface (Spearot *et al.* 2005).

10.3.1.2 Kinetics of dislocation nucleation and motion

MD simulations are also adopted to (1) observe dislocation nucleation from the planar and stepped bicrystal interfaces, (2) compute the stress required for dislocation nucleation, and (3) estimate the change in interfacial energy associated with the nucleation of the first partial dislocation during the deformation process. The aim is to have MD simulations provide an appropriate set of values for use in the proposed continuum model for NC deformation. Dislocation nucleation from ledges or steps along the interface plane is considered a primary cause of the initiation of plastic deformation in the model of Spearot *et al.* (2005).

Figure 10.5 shows emission of dislocations computed from MD within a periodic unit cell for a S11 symmetric tilt boundary in Cu. Clearly we are interested in stresses and activation energies necessary for dislocation nucleation/emission from both planar and stepped boundaries. To compute the stress required for dislocation nucleation, both the planar and stepped interface models are subjected to a sequence of steps of increasing applied uniaxial tension perpendicular to the interface plane. A similar procedure has been used in the literature to determine the stress required for dislocation nucleation in NC samples (cf. Yamakov, Wolf, *et al.* 2001). For example, Fig. 10.6 shows the essentially athermal (10 K) nucleation of a dislocation at a ledge in a S5 {210} 53.1° boundary symmetric tilt boundary [001]. Clearly, MD simulations are capable of capturing the first partial dislocation as it is nucleated from the interface. This partial dislocation is nucleated on one of the primary {111}/<112> slip systems, in agreement with that predicted using a

Schmid factor analysis of the lattice orientation (cf. Hosford 1993). The core of the nucleated partial dislocation (which is shown in blue) has both edge and screw character, while the leading partial dislocation is connected back to the interface by an intrinsic stacking fault (shown in green). Nucleation of the trailing partial dislocation from the interface is not observed during the simulation time.

Figure 10.5 Snapshots of dislocation emission during uniaxial tension of Cu for the Σ11 (113) 50.5° grain boundary model for a depth of 32.52 lattice units (Spearot *et al.* 2005).

This is characteristic of MD simulations of dislocation nucleation in copper and has been discussed at length by Van Swygenhoven *et al.* (Derlet and Van Swygenhoven 2002; Froseth, Derlet, *et al.* 2004). To determine the magnitude of the resolved shear stress that acts on the slip plane in the direction of the partial dislocation nucleation, the unixial state of stress is resolved onto the activated {111} plane in the <112> slip direction. This stress is calculated as 2.58 GPa. If additional tensile deformation is applied to the interface model, it is noted that dislocation nucleation will occur at other sites along the interface plane. In addition, the nucleated dislocation shown in Fig. 10.3c will propagate through the periodic boundary.

Images of partial dislocation nucleation from the stepped interface with 53.1° misorientation are shown in Fig. 10.6. MD simulations reveal that dislocation nucleation originates from the interface ledge and occurs on one of the primary {111}<112> slip systems. The leading partial dislocation, which has both edge and screw components, is connected back to the interface via an intrinsic stacking fault. Even though the dislocation is nucleated at the interface step, the dislocation moves along the activated slip plane, eventually incorporating regions of the interface away from the ledge (as shown in Fig. 10.6 [bottom right]). To determine the magnitude of the stress that acts on the slip plane in the direction of the partial dislocation nucleation, the unixial state of stress is resolved onto the activated {111} plane in the <112> slip direction. The stress required for partial dislocation nucleation was calculated as 2.45 GPa.

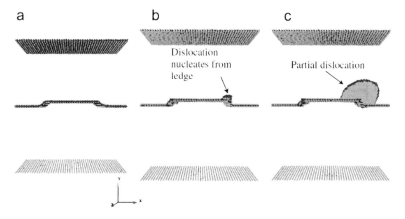

Figure 10.6 Nucleation of a partial dislocation loop during uniaxial tension of the planar S5 (210) 53.1° interface model at 10 K. Atoms are colored by the centrosymmetry parameter (Capolungo, *et al.* 2007).

10.3.1.3 Mesoscopic simulations of nanocrystals

There are two objectives to be met by atomistic modeling of ensembles of nanocrystals under mesoscopic simulations. The first objective is to build representative polycrystalline structures by energy minimization to determine the distribution of grain boundary character. Since deformation of nanocrystals tends toward control of interfaces, it is necessary to understand whether there is an expectation in the change of grain boundary character with mean grain size in polycrystals. We may speculate that the fraction of special boundaries will increase as average grain size decreases, because the system energy becomes increasingly dependent upon minimization of the boundary energy. For example, certain CSL boundaries have been shown to have a substantially lower energy than those boundaries with non-CSL orientations (Wolf 1990; Rittner and Seidman 1996; Spearot *et al.* 2004a). Commensurate with a higher fraction of special boundaries would be a more faceted nature of boundaries. This will provide direct input into a continuum model in terms of the statistical distribution of dislocation sources, since each grain boundary source and mediation effect will have different activation energy barrier strength. Moreover, the activation volume depends on grain size or feature spacing (cf. Asaro and Suresh 2005) and can be estimated with atomistics. The second objective is to validate the continuum micromechanics model over a relatively limited range of NC grain size lying in the range of the transition from bulk to boundary-mediated deformation.

Mesoscopic simulations of nanocrystals can be carried out by MD methods that rely on building Voronoi tesselated 3D grain structures with appropriate grain distribution functions. A conjugate gradient energy minimization procedure is then required, followed by finite temperature equilibration. Within tessellations of microstructure and assignment of misorientation distribution, a misorientation-dependent interfacial energy penalty function may be introduced to build the initial structure prior to energy minimization, with the goal of enhancing existing algorithms that consider only facet size and no differential energies among facets in the tesselation. The use of a columnar NC structures can be adopted for better visualization and interpretation of the mechanisms that contribute to grain growth, diffusion, and deformation processes at high temperatures with respect to the [110] CSL boundaries. After obtaining the minimum energy configuration for the NC grains, further simulations are necessary to highlight the effect of application of and reaction to mechanical deformation on the atomic structure. This portion is of interest in answering more questions concerning the effect of length scales in nanomaterials: (1) how does dislocation nucleation occur as a function of grain size? (2) What are the specific deformation mechanisms, and how do these compare with the findings of Yamakov et al. with respect to their columnar NC structures? (3) How do these compare with the recent findings of Van Swygenhoven *et al.* (cf. Froseth *et al.* 2004) with respect to their 3D Voronoi NC structures? (4) How do the CSL boundaries affect the nucleation of full and partial dislocations from grain boundaries as a function of grain size?

10.3.2 Continuum micromechanics modeling

Significant advances in multiscale modeling are essential to understand and model larger-scale cooperative deformation phenomena among grains, including strengthening mechanisms and localization of shear deformation. Methods that combine molecular and continuum calculation are still a challenging problem to model relevant deformation phenomena across length scales ranging from tens of nanometers to hundreds of nanometers. However, it must be emphasized that this must be done in the context of a rigorous continuum defect field theory capable of accepting quantitative information from atomistic calculations and high-resolution experiments. New, specialized modeling tools must be developed since existing bulk plasticity models, including conventional crystal plasticity, are of limited use in modeling the behavior of sets of NC grains (say 10–100 grains) since they are too phenomenological in character to accept detailed information regarding grain boundary structure. Moreover, the use of dislocation dynamics to bridge the atomistic and continuum descriptions has its own fundamental limitations of time and length scales, not to mention the difficulty to incorporate the complex variety of dislocation nucleation mechanisms and interactions with grain boundaries that characterize NC materials. The present section will discuss how molecular statics and dynamics calculations performed in the mesoscopic studies can support development of continuum models for dislocation nucleation, motion, and interaction of statistical character, which can then serve in the context of a micromechanics scheme as a viable alternative to explicit simulations in NC materials. Recent contributions that rely on the concept of combining atomistics and continuum micromechanics are developed in a few chapters of the present book.

10.3.2.1 Thermodynamic construct for activation energy of nucleation and competition of bulk and interface dislocation structures

As mentioned earlier, plastic deformation in NC materials results from the competitive activity of grain boundary sliding (Ke, Hackney, *et al.* 1995) and grain boundary dislocation emission (Van Swygenhoven, Spaczer, *et al.* 1999; Van Swygenhoven, Derlet, *et al.* 2004). Recent experimental studies on physical-vapor-deposited NC materials also suggest the possible accommodation of grain boundary sliding by the penetration of a dislocation, nucleated and emitted from a grain boundary, into the grain boundary opposite the dislocation source (Markmann, Bunzel, *et al.* 2003).

Let us recall here that NC materials with small grain sizes in the order of ∼30nm, in which grain boundary dislocation emission is expected to be active, have been experimentally reported to be initially dislocation free (except for the dislocation structural units, or structural dislocation units, constructing the grain boundaries) (Zhu and Langdon 2005). The dislocation emission process is fairly complex (Van Swygenhoven, Caro, *et al.* 2001; Yamakov, Wolf, *et al.* 2001; Spearot, Jacob, *et al.* 2005).

First, a leading partial dislocation is nucleated and propagates by growth within the grain cores on favorable slip systems. As shown in the work by Warner *et al.* (2006), on the basis of a quasicontinuum study coupling both finite elements and molecular statics, prior to the emission of the leading partial dislocation, the grain boundary can sustain significant atomic shuffling. This is the case of grain boundaries containing E structural units.

Second, the emitted dislocation will propagate into the grain core. In the case of NC materials produced by physical vapor deposition, electrodeposition and ball milling followed by compaction, which are to date the three only fabrication processes enabling the fabrication of fine-grained NC materials, the grain cores are defect free. Let us note that depending on the fabrication process, twins can observed within the initial structure of grain cores (Kumar, Suresh, *et al.* 2003). However, these twins can be treated as mobile grain boundaries, and their presence will consequently lead to lower mean free paths of mobile dislocations. Let us acknowledge recent molecular simulations of the interaction of screw dislocation with twin boundaries, which revealed that a screw dislocation can either be absorbed in a twin boundary or cut through the twin boundary (Jin, Gumbsch, *et al.* 2006). Moreover a criterion function of the faults difference was introduced to predict the interaction of twin boundaries and dislocations. This study can be considered a first approach in order to understand the details of the dislocation/grain boundary collision process. In all cases an emitted dislocation will propagate until it reaches either a grain boundary or a twin boundary. Since postmortem observation of NC Ni samples produced by electrodeposition have revealed solely the occasional presence of dislocation within the grain cores, the emitted dislocation must penetrate into the grain boundary (Kumar, Suresh, *et al.* 2003).

Following the penetration of the leading partial dislocation the grain boundary dislocation source can nucleate a trailing partial dislocation which will annihilate the stacking fault upon propagating within the grain core. However, in most cases and even in high-stacking-fault-energy materials such as Al, molecular simulations do not predict the emission of the trailing partial dislocation (Wolf, Yamakov, *et al.* 2005). Experimentally, an increase in the number of stacking faults has been measured during plastic deformation (Markmann, Bunzel, *et al.* 2003). However, this increase is not pronounced enough to confirm predictions from molecular simulations. Hence, to date there is no accepted theory or model able to rigorously define a criterion for the emission of the trailing partial dislocation. The molecular simulations on a bicrystal interface will clearly help bringing a new element to the debate, as discussed in the previous section on mesoscopic studies.

Simultaneously, several other issues related to the grain boundary dislocation emission process deserve special attention:

1. What are the most prominent grain boundary dislocation sources (e.g., perfect planar grain boundaries, grain boundary ledges, triple junctions)?

2. Does grain boundary sliding affect the emission of dislocations?

3. What is the macroscopic effect of dislocation emission from grain boundaries?

Clearly, the dislocation emission mechanism is much localized, and a continuum model of the mechanism must take into account the local nature of the phenomenon (e.g., dependence on grain boundary misorientation angle). Since molecular simulations are the only tool able to provide the required details on the dislocation emission process, it is capital to develop a methodology capable of receiving information from molecular simulations.

MD simulations on 2D columnar structures (Yamakov, Wolf, *et al.* 2001; Yamakov, Wolf, *et al.* 2002), fully 3D structures (Yamakov, Wolf, *et al.* 2001; Yamakov, Wolf, *et al.* 2002; Froseth, Van Swygenhoven, *et al.* 2004; Froseth, Derlet, *et al.* 2004), and bicrystal interfaces (Spearot 2005; Spearot, Jacob, *et al.* 2005; Spearot, Jacob, *et al.* 2006) have revealed the thermally activated nature of the dislocation emission process. Consequently the dislocation emission mechanism can be described at the continuum level with well-accepted theories based on statistical mechanics.

Locally, the emission of a dislocation by a grain boundary source, which could be either a typical disclination unit or a grain boundary ledge (Murr and Venkatesh 1978; Venkatesh and Murr 1978; Murr 1981), will have two effects: (1) from the conservation of the burger vector, it should lead to a net strain (significant or not) on the structure of the grain boundary and (2) it will create a dislocation flux from the grain boundary to the grain core. Therefore, appropriate tools are required for modeling the effect of dislocation emission on the strain within the grain boundary, as well as for kinetics of boundary–bulk interactions, emission, and absorption.

From statistical mechanics (Ashby 1975), the effect of a given process is typically written as the product of an activation rate term, accounting for the probability of success of the process and for the frequency at which the phenomenon occurs, and of second term describing the average effect of the phenomenon. Hence, in a general case the strain-rate engendered by the activity of a thermally activated mechanism, noted as γ, can be written as follows (Capolungo *et al.* 2007):

$$\gamma = \gamma^\circ \nu P \qquad (10.1)$$

Here γ° is the average strain engendered by the event considered, ν is the frequency of attempt, and P denotes the probability of successful emission.

Adopting the thermodynamic description proposed in early work by Gibbs, the probability of success given by a Boltzmann distribution and noted as P is described in a phenomenological manner as follows (Capolungo *et al.* 2007):

$$P = \left(\exp - \frac{\Delta G_0}{kT} \left(1 - \left(\frac{\sigma}{\sigma_c} \right)^p \right)^q \right) \tag{10.2}$$

Here ΔG_0, σ_c, p, and q represent the free enthalpy of activation, the critical emission stress, and two parameters describing the shape of the grain boundary dislocation emission resistance diagram. Physically, the free enthalpy of activation represents the energy that must be brought to the system at a given temperature for an event, in our case a dislocation emission, to be successful. The event is said to be successful if a dislocation initially in a stable configuration reaches an unstable configuration with positive driving force.

As discussed here next, the statistical description of thermally activated mechanism appears to be well suited for receiving information directly from molecular simulations. This provides an opportunity to perform the scale transition from the atomistic scale to the scale at which continuum micromechanics can be adopted.

In recent MD simulations on a perfect planar (210) Σ5 bicrystal interface and on a bicrystal interface with the same misorientation but containing a ledge, it was shown that the difference in the excess energy of the grain boundary at the initial undeformed state and at the state in which the emitted dislocation has reached an unstable configuration with a positive driving force can provide a good estimate of the free enthalpy of activation (Capolungo, Spearot, *et al.* 2007).

The details of the calculation of the excess energy are given here next. Figure 10.7 presents a schematic of the bicrystal constructed in an MD simulation and a schematic of the energy profile. The excess energy is given by (Nozieres and Wolf 1988; Wolf and Nozieres 1988; Muller and Saul 2004)

$$E^{\text{int}} = \sum_{i=1}^{N_A} \left[e_i - e_A \right] + \sum_{i=1}^{N_B} \left[e_i - e_B \right] \tag{10.3}$$

Here, N_A and N_B are the number of atoms in regions A and B, respectively. The bulk energies, e_A and e_B, are determined by averaging the individual atomic energies of a "slice" of atoms positioned sufficiently far away from the interface such that the presence of the boundary is not detected (beyond y_A or y_B in Fig. 10.7). Also, as mentioned in the previous section, it was shown that the critical emission stress at 0 K, denoted by σ_c, can be calculated from a simple tensile simulation on the *NPT* ensemble.

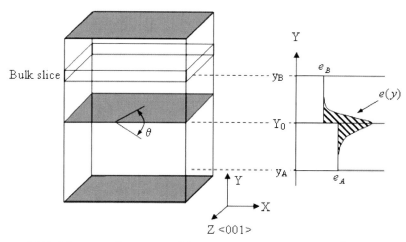

Figure 10.7 Schematic illustration of the calculation of interface "excess" energy.

The evaluation of the free enthalpy of activation and of the critical emission stress at 0 K enables the estimation of the probability of successful emission presented in Fig. 10.8. The parameters $p = 1$ and $q = 1.5$ are chosen so that the dislocation emission resistance diagram has a rather abrupt profile. It is shown that for this particular geometry, the dislocation emission process is activated at very high values of the local stress in the grain boundaries, ranging from ~2,450 MPa in case of a stepped interface to ~2,580 MPa in case of a perfect planar interface.

Hence, these simulations reveal that grain boundary ledges are more prone to emit dislocation than perfect planar grain boundaries. However, let us note here that these simulations are limited to the case of a single misorientation angle, and consequently only one of the five degrees of freedom of the grain boundary is not null. Obviously, these simulations will be extended to a wider range of grain boundary misorientations in order to draw conclusions. Also, as mentioned earlier, to date the parameters p and q have not yet been calculated from molecular simulations.

The frequency of attempt of dislocation emission could be calculated from molecular statics simulations. However, as mentioned in the discussion by Van Swygenhoven *et al.* (2004), molecular simulations often predict the emission of a single leading partial dislocation within the grain cores of NC materials, leaving behind a stacking fault in the materials. Also, an increase in the total stacking faults of NC materials was measured experimentally (Markmann, Bunzel, *et al.* 2003). However, if as predicted by molecular simulations, dislocation activity is incomplete, the amount of stacking faults measured will be much higher. Let us recall here that the little experimental

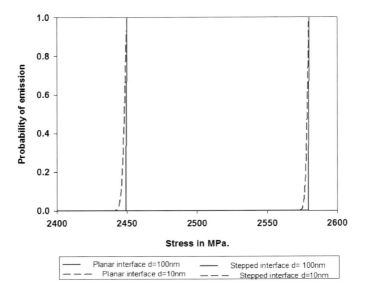

Figure 10.8 Predicted probability of successful dislocation emission with respect to the Von Mises stress.

data available revealed that NC materials with small grain sizes in the order of ~30 nm (which are either produced by ball milling, electrodeposition, or physical vapor deposition) have an initial microstructure, which is virtually dislocation free. Moreover no conclusive experimental data has shown that grain boundary sliding, whether or not accommodated by diffusion mechanisms, is active in the size range. Hence, the earlier discussion suggests that molecular simulation cannot yet quantitatively capture the complete activity of dislocation emission.

Hence, it is proposed that a continuum model be developed in order to approximate as reasonably as possible the frequency of attempt of dislocation emission. As discussed by Ashby (1975), the emission frequency is bound by two extreme values, ν_0, representing the dislocation bound frequency in the case of discrete obstacles, and ω_A, representing the atomic frequency. Several models were already developed to approximate the frequency of activation in case of discrete obstacles (Granato, Lucke, *et al.* 1964; Friedel 1969). For example, Granato *et al.* (1964) predicts a frequency in the order of $10^{11}/s$.

It is proposed here to evaluate the average strain engendered by a single dislocation emission event from continuum-based reasoning. Typically grain boundaries are described with dislocation of disclination structural unit models. Let us recall here that disclinations, first introduced by Volterra in

1907 (Volterra 1907), are linear rotational defects (see Fig. 10.10) in which strength is given by Frank's vector, denoted in Fig. 10.10 (Romanov 2003).

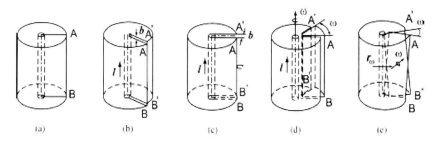

(a)   (b)   (c)   (d)   (e)

Figure 10.9 (a) Perfect cylindrical volume element, (b) twist dislocation, (c) edge dislocation, (d) wedge disclination, and (e) twist disclination (Romanov 2003).

Similarly to dislocation, which can have either a twist or an edge character, a disclination can have either a twist or a wedge character (see Fig. 10.10). Tilt boundaries are composed of a series of wedge disclinations.

The emission of a dislocation will lead to a change in the strength of the disclinations localized in the vicinity of the source, which engenders a net plastic strain. Plastic deformation in the grain boundary would accordingly occur via local rotation of the two adjacent grains composing the grain boundary. Theoretical work based on the disclination dipole construction of grain boundaries has already been developed and was able to discuss qualitatively the activity of grain boundary dislocation emission (Gutkin, Ovid'Ko, *et al.* 2003). However, these first studies need further extensions to estimate the net strain engendered by an emission process. Also, molecular simulations on bicrystal interfaces have revealed that upon emitting a dislocation a perfect planar interface can generate a ledge (Spearot, Jacob, *et al.* 2005).

An alternative approach consists of considering grain boundaries as regions of high concentrations of geometrically necessary dislocations. Indeed, grain boundaries are regions in the material presenting curvatures in the crystalline network. As described first by Nye (1953) and later in Ashby's work (1970), these curvatures can directly be related to the presence of dislocations, referred to as geometrically necessary. Hence, the net strain within the grain boundary engendered by the emission of a single dislocation could also be evaluated by investigating the effect of a decrease in the GND density on the curvature of the crystalline network. However let us note that this approach would be more suited for the description of low-angle grain boundaries in which dislocation cores can be identified.

10.3.2.2 Kinetics of boundary–bulk interactions, emission, absorption

As mentioned earlier, the dislocation emission process leads to a dislocation flux from the grain boundary region into the grain core. Also the converse, which corresponds to the penetration of a dislocation present within the grain core into the grain boundary, is strongly expected to occur. Moreover, it is of primary importance to characterize at the continuum level the effect of the presence of stacking faults on the emission and propagation of the trailing partial dislocation. Let us recall that these stacking faults are induced by the propagation of the leading partial dislocation within grain cores.

Fortunately, the initial dislocation density within the grain cores of NC materials is extremely low. Hence, dislocation networks interactions do not appear as being of primary importance. Consequently, typical strain hardening theories (Kocks 1976; Nes 1997; Kocks and Mecking 2003) based on the simultaneous activity of athermal dislocation storage, engendering a decrease in the mean free path of dislocations, and on the thermally activated dislocation annihilation mechanism are not suited in case of NC materials.

The kinetics of deformation cannot be appropriately described without rigorous models describing the coupling of dislocation emission from the grain boundaries, dislocation penetration within the boundaries, and dislocation glide within the grain cores and grain boundary. Three key aspects shall be considered:

1. Dislocation stability within grain cores
2. The effect of stacking faults on the emission of the trailing partial dislocation
3. The effect of dislocation penetration on the deformation of grain boundaries

Supposing an initial microstructure with grain cores devoid of dislocations, which is particularly the case in NC materials produced via electrodeposition, once a dislocation is nucleated and propagates within the grain cores, it can either be absorbed within the grain boundary opposite to the dislocation source or be stored within the grains. The latter is less likely to happen. Establishing a stability criterion for an emitted dislocation will directly let us evaluate the probability of dislocation absorption. Clearly, from the conservation of Burgers vector, the penetration of an emitted dislocation will lead to plastic deformation within the grain boundary. Simultaneously, the propagation of the leading partial dislocation leaves a stacking fault within the grain core, which could have two effects: (1) increase the resistance to dislocation glide within the grain cores and (2) impede or facilitate the nucleation and emission of the trailing partial dislocation.

Qin, Chen, *et al.* (1999) proposed a model for the stability of dislocation within grain cores. The proposed reasoning is fairly simple and based on the stress fields localized in the grain boundary area and engendered by the local lattice expansion present at grain boundaries. The lattice expansion was measured experimentally on samples produced with various processes (Van Petegem, Dalla Torre, *et al.* 2003). At equilibrium the stresses applied by the grain boundaries are equal to Peierls stresses. It is then shown (see Fig. 10.11) that a decrease in the grain size leads to a decrease in the surface area in which the dislocation can be stable (Qin, Du, *et al.* 2002). Note here that in Qin *et al.*'s work the elastic modulus of the grain boundaries is dependent on the excess volume within the grain boundaries and decreases with the grain size (Qin, Chen, *et al.* 1999).

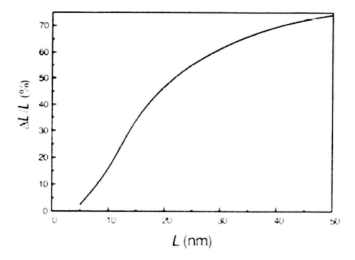

Figure 10.10 Ratio of the length of stability of a dislocation over the grain size, denoted by *L*, with respect to the grain size (Qin, Du, *et al.* 2002).

Recently Asaro and Suresh (2005) developed a model to predict the transition from typical dislocation glide dominated plasticity to grain boundary dislocation emission plasticity occurring in NC materials. By minimizing the energy of an extended dislocation, accounting for the energies of the two partial dislocations, of their interaction, and of the stacking fault, the authors derive the equilibrium distance between the two partial dislocations and the yield stress within the grain cores. While this study is the first of the kind to establish a criterion for the emission of full dislocation from grain boundaries, it does not account for the thermally activated nature of the grain boundary dislocation emission process.

The previously described model, coupled with the MD simulations, will facilitate the modeling of emission criterion for the trailing partial dislocation. Alternatively, Van Swygenhoven *et al.* (2004) have shown via MD simulations that the ratio of the stable stacking fault energy to the unstable stacking fault energy has an influence on the emission of the trailing partial dislocation.

Finally, following MD simulations focusing on the dislocation penetration process, a model will be developed at the continuum level to quantify the net strain resulting from dislocation penetration events. The model will be based on the disclination structural unit description of grain boundaries. Indeed, from the conservation of Burgers vector, the dislocation penetration mechanism will directly lead to an increase in the strength of the wedge disclination. However, *a priori* and without MD simulations, it is impossible to assess the details of the penetration process.

10.3.2.3 Incorporation of the grain boundary network into a self-consistent scheme

From the characterization of the grain boundary dislocation emission mechanism, of the stability of dislocation within grain cores (driving the penetration of an emitted dislocations), and of the effect of stacking faults on the emission of trailing dislocations, constitutive laws describing the behavior of both grain cores and grain boundaries can be established. In order to develop a model capable of predicting the behavior of NC materials and to be able to receive information on the microstructure, three issues must be addressed:

1. How to perform the scale transition from the mesoscopic scale to the macroscopic scale
2. How to introduce the grain boundary geometry within the continuum model
3. How to account for the effect of grain boundary sliding

Finite elements and micromechanics are the two possible ways to perform the scale transition. Although finite elements can reveal higher level of details than traditional micromechanics (e.g., nonhomogeneous stresses and strain fields within the grain cores and grain boundaries), its use is rather costly in terms of computation time. Moreover, it is fairly complex, if not impossible, to recreate an exact microstructure with the same statistical distribution of the grain boundaries as observed experimentally. Hence, recent micromechanical models, which ineluctably account for the statistical description of the microstructure, that have proven to be effective in the case of modeling of NC materials (Jiang and Weng 2004; Capolungo, Cherkaoui, *et al.* 2005; Capolungo, Cherkaoui, *et al.* 2006) can be of great interest. As will be discussed in the following section, the selected micromechanical scheme can be extended to account for the peculiarity of the geometry of grains and grain boundaries.

The micromechanical approach is based on a composite description of the material, which is typically represented as a two-phase material composed of (1) an inclusion phase representing grain cores and (2) a matrix phase representing grain boundaries and triple junctions. Also three-phase models have recently been used to predict the quasi-static purely viscoplastic response of NC materials (Benkassem, Capolungo, *et al.* 2006), where a coated inclusion is embedded in an effective homogeneous material and the coating represents both grain cores and triple junctions, while the inclusion represents grain cores. Three-phase models are well suited to describe materials in which diffusion mechanisms, such as Coble creep, and sliding of phases, such as grain boundary sliding, are activated.

The extension of Kroner's method to the case of inhomogeneous elastic–viscoplastic materials was used in past studies to predict the effect of the activity of Coble creep on the breakdown of the Hall–Petch law (Capolungo, Cherkaoui, *et al.* 2005). In this approach, the elastic response of the material is decomposed as the sum of the contribution of a spatially independent term and a fluctuation term. Similarly, the same decomposition is performed for the viscoplastic response. This scheme has the benefit of being fairly simple in its implementation but does lead to stiffer responses than the secant elastic–viscoplastic scheme used by Berbenni *et al.* (2005), which accounts for the spatial and time coupling of the solution fields.

The micromechanical scheme developed by Berbenni *et al.* (2005) was also used to estimate the effect of the combined effect of grain boundary dislocation emission and penetration (Capolungo, Cherkaoui, *et al.* 2006; Capolungo, Spearot, *et al.* 2006). The following method is used to homogenize the behavior of the NC material.

First, the elastic moduli are decomposed into a uniform part and a fluctuating part. In order to ensure the compatibility and equilibrium in the representative volume element, Kunin's projection operators (Kunin 1983) are used to transform the fields on the space of possible solutions. The self-consistent approximation is applied to the projected equations. In self-consistent schemes, the properties of the homogeneous equivalent medium are obtained by imposing that the spatial average of the nonlocal contributions is equal to zero. At this stage, the system cannot be solved because the viscoplastic strain field is still to be determined. The problem is solved by translating the local viscoplastic strain-rate about a nonnecessarily uniform but compatible strain-rate, which is chosen to be the self-consistent solution for a polycrystalline material displaying a purely viscoplastic behavior.

Second, the global behavior is obtained by performing the homogenization step, which consists of averaging the local fields over the volume and setting the averaged fields equal to the macroscopic fields. However, let us note that this scheme does not account for possible strain or stress jumps at the interface that occur during sliding of grains. Also, the previously mentioned models are based on Eshelby's (1957) solution to the inclusion problem in which inclusions are supposed ellipsoidal, which leads to homogeneous stress and strain states.

Micromechanical schemes are based on Eshelby's solution of the inclusion problem, which is obtained via the use of Green's functions (Eshelby 1957; Eshelby 1961; Mura 1993), and inclusions are assumed, for simplicity, to be ellipsoidal. This assumption leads to a homogeneous solution of the inclusion problem, which induces the homogeneity of the localization tensors. Hence with traditional micromechanical approaches the predicted stress and strain fields in all phases are homogeneous. Typically a higher level of refinement is not required to obtain acceptable predictions of the global behavior of the material. However, previous work has shown that (1) dislocation emission necessitates high values of stresses, which cannot be predicted with Eshelbian schemes (Capolungo, Cherkaoui, *et al.* 2006), and (2) triple junctions are regions of high stress concentrations (Benkassem, Capolungo, *et al.* 2006). Hence, to circumvent the limitations of Eshelbian's micromechanical schemes, new solutions to the inclusion problems are necessary to consider other grain shapes and to account for the effect of grain boundary ledges.

References

Agnew, S. R., B. R. Elliott, *et al.* (2000). "Microstructure and mechanical behavior of nanocrystalline metals." *Materials Science and Engineering A* 285(1–2): 391–396.

Aifantis, E. C. (1987). "The physics of plastic deformation." *International Journal of Plasticity* 3: 211–247.

Asaro, R. J., and S. Suresh (2005). "Mechanistic models for the activation volume and rate sensitivity in metals with nanocrystalline grains and nano-scale twins." *Acta Materialia* 53(12): 3369–3382.

Ashby, M. F. (1970). "The deformation of plastically non homogeneous materials." *Philosophical Magazine* 21: 399–424.

Bachurin, D. V., R. T. Murzaev, *et al.* (2003). "Atomistic computer and disclination simulation of [001] tilt boundaries in nickel and copper." *Fizika Metallov i Metallovedenie* 96(6): 11–17.

Benkassem, S., L. Capolungo, *et al.* (2006). "Mechanical properties and multi-scale modeling of nanocrystalline materials." Submitted.

Berbenni, S., V. Favier, *et al.* (2004). " Micromechanical modeling of the elastic viscoplastic behavior of polycrystalline steels having different microstructures." *Materials Science and Engineering A:* 128–136.

Cai, B., Q. P. Kong, *et al.* (2001). "Creep behavior of cold-rolled nanocrystalline pure copper." *Scripta Materialia* 45(12): 1407–1413.

Cai, B., Q. P. Kong, *et al.* (2000). "Low temperature creep of nanocrystalline pure copper." *Materials Science and Engineering A* 286(1): 188–192.

Capolungo, L., M. Cherkaoui, *et al.* (2005). "A self consistent model for the inelastic deformation of nanocrystalline materials." *Journal of Engineering Materials and Technology* 127: 400–407.

Capolungo, L., M. Cherkaoui, *et al.* (2007). "On the elastic-viscoplastic behavior of nanocrystalline materials." *International Journal of Plasticity* 23(4): 561-591.

Capolungo, L., C. Jochum, *et al.* (2005). "Homogenization method for strength and inelastic behavior of nanocrystalline materials." *International Journal of Plasticity* 21: 67–82.

Capolungo, L., D. E. Spearot, *et al.* (2007). "Dislocation nucleation from bicrystal interfaces and grain boundary ledges: relationship to nanocrystalline deformation." *Journal of the Mechanics and Physics of Solids* 55(11): 2300–2327.

Coble, R. L. (1963). "A model for boundary diffusion controlled creep in polycrystalline materials." *Journal of Applied Physics* 34(6): 1679–1682.

Dai, H., and D. M. Parks (1997). "Geometrically-necessary dislocation density and scale-dependent crystal plasticity." In A. S. Khan (ed.), *Proceedings of Plasticity '97: The Fifth International Symposium on Plasticity and its Current Applications.* Juneau, Alaska, Neat Press, 17–18.

Dai, H. (1997). "Geometrically-necessary dislocation density in continuum plasticity theory, FEM implementation and applications." Ph.D. thesis. Massachusetts Institute of Technology, Department of Mechanical Engineering.

Daymond, M. R., and H. G. Priesmeyer (2002). "Elastoplastic deformation of ferritic steel and cementite studied by neutron diffraction and self-consistent modelling." *Acta Materialia* 50(6): 1613–1626.

Derlet, P. M., and H. Van Swygenhoven (2002). "Length scale effects in the simulation of deformation properties of nanocrystalline metals." *Scripta Materialia* 47(11): 719–724.

Eshelby, J. D. (1957). "The determination of an ellispoidal inclusion and related problems." *Proceedings of the Royal Society of London* A241: 376–396.

Eshelby, J. D. (1961). *Elastic Inclusions and Inhomogeneities.* North Holland.

Esquivel, E. V., and L. E. Murr (2005). "Grain boundary contributions to deformation and solid-state flow in severe plastic deformation." *Materials Science and Engineering: A* 409(1–2): 13–23.

Fleck, N. A., G. M. Muller, M. F. Ashby, et al. (1994). "Strain gradient plasticity: theory and experiment." *Acta Metallurgica et Materialia* 42: 475–487.

Fleck, N. A., and J. W. Hutchinson (1997). "Strain gradient plasticity." *Advances in Applied Mechanics* 33: 295–361.

Friedel, J. (1969). *Physics of Strength and Plasticity*. M.I.T. Press.

Froseth, A., H. Van Swygenhoven, *et al.* (2004). "The influence of twins on the mechanical properties of nc-Al." *Acta Materialia* 52: 2259–2268.

Froseth, A. G., P. M. Derlet, *et al.* (2004). "Dislocations emitted from nanocrystalline grain boundaries: nucleation and splitting distance." *Acta Materialia* 52(20): 5863–5870.

Gao, H., Y. Huang, W. D. Nix, *et al.* (1999). "Mechanism-based strain gradient plasticity–I. Theory." *Journal of the Mechanics and Physics of Solids* 47: 1239–1263.

Ghoniem, N. M., E. P. Busso, *et al.* (2003) "Multiscale modelling of nanomechanics and micromechanics: an overview." *Philosophical Magazine* 83: 3475–3528.

Granato, A. V., K. Lucke, *et al.* (1964). *Journal of Applied Physics* 35: 2732.

Gutkin, M. Y., I. A. Ovid'Ko, *et al.* (2003). "Transformation of grain boundaries due to disclination motion and emission of dislocations pairs." *Materials Science and Engineering* A339: 73–80.

Hall, E. O. (1951). "The deformation and aging of mild steel." *Proceedings of the Physical Society of London* B64: 747.

Hosford, W. F. (1993). *The Mechanics of Crystals and Textured Polycrystals*. New York, Oxford University Press.

Hull, D., and D. J. Bacon (2005). *Introduction to Dislocations*. Butterworth Heinemann.

Jiang, B., and G. J. Weng (2004). "A generalized self consistent polycrystal model for the yield strength of nanocrystalline materials." *Journal of the Mechanics and Physics of Solids* 52: 1125–1149.

Jin, Z. H., P. Gumbsch, *et al.* (2006). "The interaction mechanism of screw dislocations with coherent twin boundaries in different face-centred cubic metals." *Scripta Materialia* 54(6): 1163–1168.

Ke, M., S. A. Hackney, *et al.* (1995). "Observations and measurement of grain rotation and plastic strain in nanostructured metal thin films." *Nanostructured Materials* 5: 689–697.

Kim, B.-N., K. Hiraga, *et al.* (2005). "Viscous grain-boundary sliding and grain rotation accommodated by grain-boundary diffusion." *Acta Materialia* 53(6): 1791–1798.

Kim, H. S., and Y. Estrin (2005). "Phase mixture modeling of the strain-rate dependent mechanical behavior of nanostructured materials." *Acta Materialia* 53: 765–772.

Kim, H. S., Y. Estrin, *et al.* (2000). "Plastic deformation behaviour of fine grained materials." *Acta Materialia* 48: 493–504.

Kocks, U. F. (1976). "Laws for work hardening and low temerature creep." *Transactions of the ASME*: 76–85.

Kocks, U. F., and H. Mecking (2003). "Physics and phenomenology of strain hardening." *Progress in Materials Science* 48: 171–273.

Konstantinidis, D. A., and E. C. Aifantis (1998). "On the 'anomalous' hardness of nanocrystalline materials." *Nanostructured Materials* 10: 1111–1118.

Kumar, K. S., S. Suresh, *et al.* (2003). "Deformation of electrodeposited nanocrystalline nickel." *Acta Materialia* 51: 387–405.

Kunin, I. A. (1983). *Elastic Media with Microstructure II: Three Dimensional Models.* Berlin, Heidelberg, New York, Tokyo.

Li, J. C. M. (1963). "Petch relation and grain boundary sources." *Transactions of the Metallurgical Society of AIME* 227: 239.

Li, Y. J., W. Blum, *et al.* (2004). "Does nanocrystalline Cu deform by Coble creep near room temperature?" *Materials Science and Engineering A* 387–389: 585–589.

Ma, Q., W. Mao, *et al.* (2006). "Rapid texture measurement of cold-rolled aluminum sheet by X-ray diffraction." *Scripta Materialia* 54(11): 1901–1905.

Markmann, J., P. Bunzel, *et al.* (2003). "Microstructure evolution during rolling of inert-gas condensed palladium." *Scripta Materialia* 49(7): 637–644.

Mishin, Y., D. Farkas, *et al.* (1999). "Interatomic potentials for monoatomic metals from experimental data and ab initio calculations." *Physical Review B* 59: 3393–3407.

Muller, P., and A. Saul (2004). "Elastic effects on surface physics." *Surface Science Reports* 54(5–8): 157.

Mura, T. (1993). *Micromechanics of Defects in Solids.* Dordrecht/Boston/London, Kluwer Academic Publisher.

Murr, L. E. (1981). "Strain induced dislocation emission from grain boundaries in stainless steel." *Materials Science and Engineering* 51: 71–79.

Murr, L. E., and E. Venkatesh (1978). "Contrast phenomena and indentification of grain boundary ledges." *Metallography* 11: 61–79.

Nes, E. (1997). "Modelling of work hardening and stress saturation in FCC metals." *Progress in Materials Science* 41: 129–193.

Nieh, T. G., and J. G. Wang (2005). "Hall Petch relationship in nanocrystalline Ni and Be-B alloys." *Intermetallics* 13: 377–385.

Nozieres, P., and D. E. Wolf (1988). "Interfacial properties of elastically strained materials. I. Thermodynamics of a planar interface." *Zeitschrift fur Physik B (Condensed Matter)* 70(3): 399.

Nye, J. F. (1953). "Some geometric relations in dislocated crystals." *Acta Metallurgica* 1: 153–162.

Petch, N. J. (1953). "The cleavage strength of polycrystals." *Journal of Iron and Steel Institute* 174: 25–28.

Qin, W., Z. Chen, *et al.* (1999). "Dislocation pileups in nanocrusyalline materials." *Journal of Alloys and Compounds* 289: 285–288.

Qin, W., Z. H. Chen, *et al.* (1999). "Crystal lattice expansion of nanocrystalline materials." *Journal of Alloys and Compounds* 292: 230–232.

Qin, W., Y. W. Du, *et al.* (2002). "Dislocation stability and configuration in the crystallites of nanocrystalline materials." *Journal of Alloys and Compounds* 337: 168–171.

Qu, J. (1993). "The effect of slightly weakened interfaces on the overall elastic properties of composite materials." *Mechanics of Materials* 14(4): 269–281.

Rittner, J. D., and D. N. Seidman (1996). "<110> symmetric tilt grain-boundary structures in FCC metals with low stacking-fault energies." *Physical Review B: Condensed Matter* 54(10): 6999.

Romanov, A. E. (2003). "Mechanics and physics of disclinations in solids." *European Journal of Mechanics A/Solids* 22(5): 727–741.

Sanchez-Bajo, F., A. L. Ortiz, *et al.* (2006). "Novel analytical model for the determination of grain size distributions in nanocrystalline materials with low lattice microstrains by X-ray diffractometry." *Acta Materialia* 54(1): 1–10.

Sanders, P. G., M. Rittner, *et al.* (1997). "Creep of nanocrystalline Cu, Pd, and Al-Zr." *Nanostructured Materials* 9(1–8): 433–440.

Sansoz, F., and J. F. Molinari (2005). "Mechanical behavior of Sigma tilt grain boundaries in nanoscale Cu and Al: a quasicontinuum study." *Acta Materialia* 53: 1931–1944.

Shi, M. X., Y. Huang, *et al.* (2000). Plastic flow localization in mechanism-based strain gradient plasticity. *International Journal of Mechanical Sciences* 42: 2115–2131.

Shu, J. Y., and N. A. Fleck (1999). "Strain gradient crystal plasticity: size-dependent deformation of bicrystals." *Journal of the Mechanics and Physics of Solids* 47: 297–324.

Spearot, D. E. (2005). "Atomistic calculations of nanoscale interface behavior in FCC metals." *Woodruff School of Mechanical Engineering*. Atlanta, Georgia Institute of Technology, 276.

Spearot, D. E., K. I. Jacob, *et al.* (2005). "Nucleation of dislocations from [001] bicrystal interfaces in aluminum." *Acta Materialia* 53: 3579–3589.

Spearot, D. E., K. I. Jacob, *et al.* (2006). "Dislocation nucleation from bicrystal interfaces with dissociated structure." *International Journal of Plasticity* 23(1): 143–160.

Spearot, D. E., K., L. Capolungo, *et al.* (2008) "On the elastic tensile deformation of [100] bicrystal interfaces in copper", *Computational Materials Science* 42(1): 57–67.

Sutton, A. P., and V. Vitek (1983). "On the structure of tilt grain boundaries in cubic metals. I. Symmetrical tilt boundaries." *Philosophical Transactions of the Royal Society of London A* 309(1506): 1–36.

Van Petegem, F. Dalla Torre, *et al.* (2003). "Free volume in nanostructured Ni." *Scripta Materialia* 48: 17–22.

Van Swygenhoven, H., and A. Caro (1997). "Plastic behavior of nanophase Ni: a molecular dynamics computer simulation." *Applied Physics Letters* 71(12): 1652.

Van Swygenhoven, H., A. Caro, *et al.* (2001). "Grain boundary structure and its influence on plastic deformation of polycrystalline FCC metals at the nanoscale: a molecular dynamics study." *Scripta Materialia* 44: 1513–1516.

Van Swygenhoven, H., P. M. Derlet, *et al.* (2004). "Stacking fault energies and slip in nanocrystalline metals." *Nature Materials* 3: 399–403.

Van Swygenhoven, H., M. Spaczer, *et al.* (1999). "Microscopic description of plasticity in computer generated metallic nanophase samples: a comparison between Cu and Ni." *Acta Metallurgica* 47: 3117–3126.

Venkatesh, E. S., and L. E. Murr (1978). "The influence of grain boundary ledge density on the flow stress in nickel." *Materials Science and Engineering* 33: 69–80.

Volterra, V. (1907). *Ann. Ecole Normale Supérieure de Paris* 24: 401.

Wang, Y. M., A. V. Hamza, *et al.* (2005). "Activation volume and density of mobile dislocations in plastically deforming nanocrystalline Ni." *Applied Physics Letters* 86(24): 241917.

Warner, D. H., F. Sansoz, *et al.* (2006). "Atomistic based continuum investigation of plastic deformation in nanocrystalline copper." *International Journal of Plasticity* 22(4): 754.

Wei, Y. J., and L. Anand (2004). "Grain-boundary sliding and separation in polycrystalline metals: application to nanocrystalline fcc metals." *Journal of the Mechanics and Physics of Solids* 52(11): 2587.

Wolf, D., V. Yamakov, *et al.* (2005). "Deformation of nanocrystalline materials by molecular dynamics simulation: relationship to experiments?" *Acta Materialia* 53: 1–40.

Wolf, D. E., and P. Nozieres (1988). "Interfacial properties of elastically strained materials. II. Mechanical and melting equilibrium of a curved interface." *Zeitschrift fur Physik B (Condensed Matter)* 70(4): 507.

Wu, X., Y. T. Zhu, *et al.* (2006). "Twinning and stacking fault formation during tensile deformation of nanocrystalline Ni." *Scripta Materialia* 54(9): 1685–1690.

Yagi, N., A. Rikukawa, *et al.* (2006) "Deformation by grain rotations in nanocrystalline fcc-metals." *Materials Science and Engineering: A* 442(1–2): 323–327.

Yamakov, V., D. Wolf, *et al.* (2002). "Deformation twinning in nanocrystalline Al by molecular dynamics simulation." *Acta Materialia* 50: 5005–5020.

Yamakov, V., D. Wolf, *et al.* (2001). "Length-scale effects in the nucleation of extended dislocations in nanocrystalline Al by molecular-dynamics simulation." *Acta Materialia* 49(14): 2713–2722.

Yin, W. M., S. H. Whang, *et al.* (2001). "Creep behavior of nanocrystalline nickel at 290 and 373 K." *Materials Science and Engineering A* 301(1): 18–22.

Zhilyaev, A. P., B. K. Kim, *et al.* (2002). "Orientation imaging microscopy of ultrafine-grained nickel." *Scripta Materialia* 46(8): 575–580.

Zhu, B., R. J. Asaro, *et al.* (2006). "Effects of grain size distribution on the mechanical response of nanocrystalline metals: part II." *Acta Materialia* 54(12): 3307–3320.

Zhu, Y. T., and T. G. Langdon (2005). "Influence of grain size on deformation mechanisms: An extension to nanocrystalline materials." *Materials Science and Engineering: A* 409(1–2): 234–242.

Subject Index

Name Index

Abramowitz, M. 195–196, 206
Ager, J. W. 252, 273
Agnew, S. R. 49, 50, 51, 52, 53, 57–58,
     156, 161, 302
Aifantis, E. C. 98, 128, 257, 265, 274,
     279, 283, 302, 304
Aifantis, K. 204–206
Ajayan, P. M. 246, 252, 272
Akatsu, T. 260, 274
Akben, M. G. 91
Akchurin, M. Sh. 246, 271
Albe, K. 172, 184, 185, 186, 191
Alber, I. 104, 130
Aldinger, F. 252, 272
Alers, G. A. 64, 88
Alexandor, L. E. 217, 242
Alexandrov, I. V. 32, 33, 41, 42–45, 48,
     51, 52, 54, 58, 134, 159, 201, 210,
     214, 216, 218, 230, 240
Altshuler, T. L. 63, 87
Amura, T. 146, 160
Anand, L. 100, 129, 280, 307
Ando, T. 34, 54
Andrade, U. R. 68, 69, 89
Andrievski, R. A. 245, 246, 250, 251,
     271, 272
Ankem, S. 89
Aoki, K. 34, 55
Apps, P. J. 142, 144, 160
Argon, A. S. 95, 100, 120, 128, 129, 171,
     172, 190, 236, 242, 254, 273
Arima, H. 34, 54
Armstrong, R. W. 15, 28, 61, 62, 63, 64,
     65, 66, 67, 68, 70, 71, 72, 73, 74,
     75, 76, 77, 78, 79, 80, 81, 82, 83,
     84, 85, 86, 87, 88, 89, 90, 91, 93,
     95, 100, 104, 116, 117, 122, 123,
     124, 125, 126, 127, 128, 129, 131,
     180, 201, 204, 206, 209
Arnold, W. 122, 124, 126, 131
Arsenault, R. J. 27
Asaro, R. J. 3, 75, 76, 89, 99, 128, 191,
     209, 257, 274, 278, 283, 290, 299,
     302, 307
Ash, B. J. 246, 252, 272

Ashby, M. F. 5, 27, 72, 89, 91, 94, 100,
     127, 129, 259, 274, 293, 296, 297,
     302, 303
Atzmon, M. 204, 208
Aust, K. T. 91, 95, 127–128, 152, 161,
     201, 207
Awaji, H. 253, 273
Azushima, A. 34, 55

Babyak, J. W. 165, 170, 189
Bachurin, D. V. 285–286, 302
Bacon, D. J. 304
Bahloul-Hourlier, D. 260, 274
Bai, Y. 233, 234, 238, 242–243
Bailey, J. E. 4, 27, 29, 210
Bailon, J.-P. 91
Baily, J. E. 4
Baily, R. 24, 29
Baker, S. P. 157, 162
Baker, T. N. 87, 162
Balluffi, R. W. 262, 275
Balogh, L. 4, 27
Barai, P. 112, 119, 120, 121, 127, 131
Baraton, M.-I. 272
Barnes, A. J. 53, 59
Baro, G. 15, 166, 189
Barrera, E. V. 252, 253, 273
Barta-Schreiber, N. 51, 68
Baskes, M. I. 159, 162
Bassani, J. L. 104, 130
Basu, B. 245, 246, 250, 262, 271
Baudelet, B. 39, 55, 95, 128
Bechtold, J. H. 64, 88
Bell, J. F. 68, 88
Bell, N. 252, 253, 273
Bement, A. L. 87–89, 129
Benkassem, S. 301–302
Benson, D. J. 81, 89, 99, 128, 164, 204,
     205, 209, 214, 215, 235, 240, 254,
     273
Benveniste, Y. 105, 109, 130
Berbenni, S. 301, 303